Principles and Techniques of Electron Microscopy

Principles and Techniques of Electron Microscopy

BIOLOGICAL APPLICATIONS

Volume 8

M. A. HAYAT

Professor of Biology
Kean College of New Jersey
Union, New Jersey

 VAN NOSTRAND REINHOLD COMPANY

NEW YORK   CINCINNATI   ATLANTA   DALLAS   SAN FRANCISCO
LONDON   TORONTO   MELBOURNE

Van Nostrand Reinhold Company Regional Offices:
New York   Cincinnati   Atlanta   Dallas   San Francisco

Van Nostrand Reinhold Company International Offices:
London   Toronto   Melbourne

Library of Congress Catalog Card Number: 70-129544
ISBN 0-442-25693-0

Manufactured in the United States of America

Published by Van Nostrand Reinhold Company
135 West 50th Street, New York, N. Y. 10020

Published simultaneously in Canada by Van Nostrand Reinhold Ltd.

15 14 13 12 11 10 9 8 7 6 5 4 3 2 1

Library of Congress Cataloging in Publication Data

Hayat, M   A
  Principles and techniques of electron microscopy.

  Includes bibliographies.
  Vols. 8   edited by M.A. Hayat.
  1. Electron microscope—Collected works.   I. Title.
[DNLM: 1. Microscopy, Electron. OH205 H413p H2VO]
QH212.E4H38   578'.4'5   70-129544
ISBN 0-442-25693-0 (v.8)

PREFACE

This is the eighth volume of a multi-volume series on the principles and techniques employed for studying biological specimens with the aid of an electron microscope. Since its inception in 1970, the series has successfully reflected the growth of electron microscopy in instrumentation as well as in methodology. There was a pressing need to keep readers abreast of the remarkable expansion of the field in recent years and the ever growing importance of its contributions to the understanding of many problems in biological and medical sciences. This treatise serves as an international authoritative source in the field, and is designed to cover important new developments systematically. The treatise departs from the tradition that books on methodology present only the contemporary consensus of knowledge. It is written by scholars, and when they have anticipated the potential usefulness of a new method, they have so stated. The authors have not hesitated to include ideas in progress. The treatise should serve as a guide and survey, which can save a newcomer the tedious search for information scattered in biological journals.

This volume has developed over the years through the joint effort of ten distinguished author-scientists. As a result, a most comprehensive compilation of methods developed and used by a large number of competent scientists has been achieved. The book contains new viewpoints with particular regard to current problems. Areas of disagreement and potential research problems have been pointed out. It is hoped that the readers will become aware that correct interpretation of the information retrieved from electron micrographs is dependent upon an understanding of the principles underlying the methodology and instrumentation.

The basic approach in this volume is similar to that in the previous seven volumes, in that the methods presented have been tested for their reliability, and are the best of those currently available. The instructions for the preparation and use of various solutions, media, stains, and apparatus are straightforward and complete, and should enable the worker to prepare his or her specimens without outside help. Before undertaking the processing, one should read the entire procedure and prepare necessary solutions and other media. Each chapter is provided with an exhaustive list of references with complete titles. Full author and subject indexes are included at the end of the book.

It is encouraging to know that the previous volumes have been received favorably. It is my impression that this volume will also fulfill its purpose: to provide an understanding of the usefulness, limitations, and potential applications of special methods employed for studying the structure, composition, size, number, and location of cellular components, and to provide details of current improvements in the instrumentation.

M. A. Hayat
Berkeley Heights, N.J.

CONTENTS

2   PREPARATION AND ANALYSIS OF SERIAL SECTIONS IN ELECTRON MICROSCOPY
Robert L. Knobler, Jerome G. Stempack, and Mary Laurencin

5   INTERFERENCE PHENOMENON ON OSMIUM TETROXIDE-FIXED SPECIMENS FOR SYSTEMATIC ELECTRON MICROSCOPY
Karl Hermann Andres and Monika von Düring

6   COMPUTER PROCESSING OF ELECTRON MICROGRAPHS
P. W. Hawkes

Contents to

Contributors to This Volume

Karl H. Andres

Wolfgang Baumeister

Robert F. Dunn

Max Hahn

P. W. Hawkes

Robert L. Knobler

Mary Laurencin

D. L. Misell

Jerome G. Stempak

Monika von Düring

1. SPECIMEN SUPPORTS

W. Baumeister and M. Hahn

Institut für Biophysik und Elektronenmikroskopie der Universität Düsseldorf,
Düsseldorf, Germany

INTRODUCTION

Despite the achievement of great insights into the structural organization of matter and the outstanding contribution of electron microscopy to the development of modern biology, there remains an obvious gap between the actual performance and the capability of modern instruments. Although the best commercially available instruments routinely provide resolutions of 0.2 to 0.3 nm with biological specimens, significant structural information is usually restricted to 2-3 nm. Bridging this gap is exactly what biologists should seek to achieve in order to investigate the finest details of macromolecular and molecular organization.

There are two main reasons why a breakthrough in the field of molecular microscopy has hitherto not been achieved. First, progress in preparatory techniques has fallen far behind progress in instrumental development, for reasons which seem too complex to be discussed here in greater detail. No doubt, the problems of specimen preparation have not been attacked with the same converging intensity afforded the improvement of the resolving power. This problem is, at least partially, a consequence of the multitude and diversity of experimental requirements involved. Second, a more fundamental reason for this gap is the radiation sensitivity of biological specimens which makes it difficult to obtain electron micrographs bearing a sufficiently close resemblance to the original structure. Significant progress can only be expected if the damaging effects of the electron irradiation can be circumvented or drastically reduced.

This improvement, in turn, needs to be preceded by a thorough reconsideration of conventional preparatory procedures with regard to achieving optimal preservation. The problem of radiation damage has attracted wide attention over the past years, but we are still far from completely understanding the interactions involved among the specimen, stain, and support film during electron bombardment.

Since the early days of electron microscopy, the problem of obtaining adequate specimen supports has been one of the major obstacles to achieving a maximum utilization of instrumental capabilities. Ideally, specimen supports should meet three basic criteria: (1) high electron transparency, (2) minimum intrinsic structure, and (3) high mechanical stability under the electron beam. Of only slightly less importance are the ease and reliability needed in preparing a given specimen support. The first electron micrographs taken at higher than light-microscopical magnifications were obtained from "self-supporting" specimens, such as metal foils or cotton fibers spanning over the apertures of specimen diaphragms (Ruska, 1934). However, as early as 1935, Marton used "thin" aluminum foils as supports for microtome sections of biological specimens. Shortly thereafter, plastic (nitrocellulose) films, previously developed for experiments with ultrasoft X-rays (Trenktrog, 1923) and cathode rays (Kirchner, 1930), were introduced for electron microscopy (Marton, 1936, 1937).

In 1939, H. Ruska published the first review on the preparatory techniques. In the same paper he described an improved method and an apparatus (Fig. 1.1) for the routine production of thin nitrocellulose films. This progress in support

Fig. 1.1 Apparatus for spreading plastic films. (*From H. Ruska, 1939.*)

film techniques was immediately followed by a great success in biological electron microscopy, the visualization of virus particles (Kausche *et al.*, 1939).

In 1942, Schaefer and Harker introduced a new polymer, polyvinylformal, which, because of its improved mechanical stability in the electron beam, became the most commonly used material for support films. At moderate resolution levels criteria (1) and (2) are satisfactorily met by these conventional plastic films; nevertheless, the radiation sensitivity remains unsatisfactory. It seems astonishing that the potential of extremely radiation-resistant polymers developed in the last two decades has only scarcely been explored.

The progress in instrumental resolution, the growing insight into the mechanism of contrast and image formation, and the need for more radiation-resistant supports stimulated the development of various techniques for the production of low mass thickness metal films. Hass and Kehler (1941) described a method for the preparation of thin aluminum oxide films by anodizing an aluminum foil, a method that was later refined by several workers. Hast (1947), Cosslett (1948), and Kaye (1949) described techniques to produce thin aluminum, beryllium, or aluminum-beryllium alloy films by evaporation onto solid surfaces. Hast (1948) presented an elegant method to prepare extremely thin and smooth aluminum films by evaporation onto a glycerol surface. More than 20 years later, Müller and Koller (1972) revived this method of producing aluminum oxide films, which proved suitable for electron microscopy at atomic resolution.

Further progress was initiated by the development of a method to evaporate carbon to form thin, highly electron-transparent, and exceptionally stable amorphous films (Bradley, 1954). The deposition of thin evaporated carbon layers on top of plastic films became the most successful way of stabilizing these supports against drift and shrinkage under electron bombardment. Presently, thin, pure carbon films are by far the most common specimen supports used for high resolution electron microscopy. However, when the goal of visualizing single atoms was envisaged during the sixties, it became obvious that their expected faint contrast would be obscured by the pronounced phase contrast structure of carbon films.

As early as 1960, Fernández-Moran had put forward the attractive suggestion to use exfoliated graphite or mica single crystals as specimen supports, thus avoiding random variations of mass thickness which lead to the undesired granular appearance. Subsequently, Beer and coworkers explored various techniques for the production of suitable graphite crystals. The superiority of graphite over carbon films in the field of molecular microscopy has now convincingly been demonstrated (Wiggins and Beer, 1972; Hashimoto *et al.*, 1974; Johansen, 1975). Nevertheless, difficulties in achieving large scale production of sufficiently thin graphite films, the relatively small size of crystallites obtained, and entirely new adsorption properties necessitating new surface conditioning and specimen mounting procedures have hitherto prevented widespread application.

Some progress in the large scale production of very thin crystalline layers has been achieved recently with vermiculite, a native hydrated alumino-silicate with a layered structure similar to that of the micas. It has been shown that the specimen information can further be enhanced if the periodic background displayed by these crystalline supports is separated from the specimen structure by optical filtering (Baumeister and Hahn, 1974a). One might suspect now that electron irradiation randomizes atom positions in the crystal and thus creates new random pictorial noise, which, because of its "white" spatial spectrum, cannot be discriminated any more from the usually extended object spectrum. Loss of crystallinity can, in fact, be observed, but it is nevertheless of minor importance, because the doses tolerated by crystalline supports without significant changes of crystalline order are greater by several orders of magnitude than those leading to the "steady state" of destruction in biomolecules.

More serious at present are the problems of surface conditioning of "low noise" supports for proper adsorption of biomolecules and avoidance of surface contamination during mounting and observing biogenic materials. Contaminated areas show a noise pattern quite similar to that of amorphous carbon films, which, if superimposed onto the specimen structure under investigation, obliterate the finest details.

PREPARATION AND PROPERTIES OF SPECIMEN SUPPORTS

Specimen Grids and Specimen Apertures

A wide variety of grid types for mounting very thin specimen supporting films or, whenever possible, for directly supporting the specimen is commercially available. The majority of the grids are supplied in two standard sizes, 2.3 mm and 3.05 mm in diameter. Usually grids are made of copper; for special fields of application (e.g., where reactive reagents are involved, for high-temperature investigations, for X-ray and microprobe analysis) titanium, stainless steel, nickel, molybdenum, rhodium, palladium, silver, tungsten, platinum, gold, and carbon-coated nylon grids are available. For some applications it may be sufficient to use the cheaper gilded, silvered, or platinized copper grids, instead of those manufactured completely of the respective noble metal.

Figure 1.2 shows a selection of commonly used specimen grid patterns. The standard square mesh grids are offered in various mesh sizes, ranging from 50-mesh/inch to 500-mesh/inch. The open area decreases with the increasing number of mesh/inch. The 50-mesh grids have an open area of approximately 80%, while the open area of the 500-mesh grids is usually as low as 35%. The exact values depend upon the specific bar thickness, which may vary between 15 μm and 50 μm. It is clearly economical to keep the ratio between the open area and the total area as high as possible. On the other hand, one has to consider that very large open areas with thin bars unduly weaken the grid and enhance the effects of specimen heating and charging.

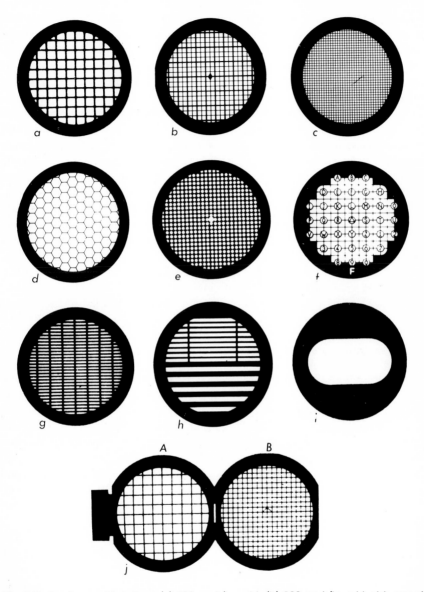

Fig. 1.2 Specimen grid designs: (a) 100 mesh/in. grid; (b) 200 mesh/in. grid with central mark; (c) 400 mesh/in. grid; (d) hexagonal type grid; (e) round hole grid with central mark; (f) finder grid; (g) rectangular mesh grid, 75 × 300 mesh/in.; (h) multiple slot grid; (i) single slot grid; (j) folding grid—side A 100 mesh/in., side B 200 mesh/in.

Although no general rule can be formed, the optimum choice should take into consideration the shape and dimension of the structure under investigation, the stability of the support film mounted on the grid, and the beam diameter used. Where it is desirable to examine large uninterrupted areas of the specimen, hexagonal type (Fig. 1.2d) and multiple slot type (Fig. 1.2h) grids are particularly suited. For the inspection of serial sections, single slot grids (Fig. 1.2i) are advantageous; however, they require extraordinarily stable support films. Grids with central marks (Fig. 1.2b, e) facilitate the orientation in the electron microscope, and finder grids where each opening is identified by letters or numbers (Fig. 1.2f) allow a rapid reidentification and location of specific points of interest. For poorly adhering materials, e.g., metallic microgrids, it is advisable to sandwich them between folding grids (Fig. 1.2j).

In high resolution electron microscopy, specimen apertures used as supports have certain advantages over specimen grids. Even a very careful handling of the delicate grids cannot avoid a slight bending, which, because of the limited depth of focus in high resolution electron microscopy, renders exact focusing difficult. If the plane of the grid within the area under observation has a slope against the plane perpendicular to the optical axis, conditions vary over a single micrograph. Moreover, defocus and consequently contrast transfer of the spatial frequencies of structural elements are raised by tilted projection. This may lead to a unidirectional loss of the finest structural details (highest spatial frequencies) by coincidence with transfer gaps surrounding zero-crossings of the transfer function of an electron microscope. Lack of perpendicularity especially prevents taking minimal-beam-exposure micrographs by focusing and correcting astigmatism on one spot and, after deactivating deflection prisms for beam and image, imaging an area a few micrometers away (Fig. 1.3), if the deflection cannot be positioned to run perpendicularly to the slope of the specimen.

Single-hole and multi-hole apertures (Fig. 1.4) (with hole diameters between 20 and 750 μm) manufactured from molybdenum, platinum-iridium, platinum-iridium-tantalum, and platinum-iridium-tantalum-molybdenum are commercially available. Since specimen apertures are relatively expensive, it is usually necessary to clean and repeatedly use them. This practice is not worthwhile for specimen grids. For the removal of thin carbon films used without an underlying plastic film, a short ultrasonic treatment is usually sufficient. Unirradiated plastic films can easily be removed from the apertures with suitable organic solvents.

After the polymers are crosslinked under the electron beam, the cleaning procedure becomes more troublesome. For thorough cleaning of platinum-iridium apertures from strongly adhering polymers and superimposed carbon or metal layers, it is necessary to decompose the layers by means of a potassium bisulfate melt in a platinum crucible. The platinum-iridium apertures are heated in the melt for ~5 min. After cooling, the melt is dissolved in hot diluted sulfuric acid. The released apertures are then repeatedly washed in distilled water,

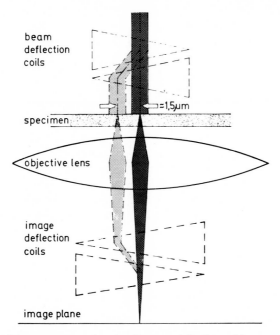

Fig. 1.3 Minimal beam exposure by simultaneous deflection of beam and image for all operations except the actual recording. (*From Baumeister and Hahn, 1975.*)

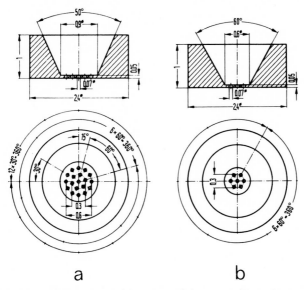

Fig. 1.4 Specimen diaphragms with (a) 19 holes; (b) 7 holes. (*From Siemens Bulletin*)

and, after a short ultrasonic treatment in this medium, they are carefully examined under a stereomicroscope for cleanliness.

Cleaning of molybdenum apertures is less tedious: after precleaning in hot dilute alkali the apertures are heated to "red heat" in a tantalum "shuttle" for $\sim\frac{1}{2}$ hr in a vacuum $\sim10^{-5}$ mbar or to "white heat" ($\sim1{,}200\text{--}1{,}300°C$) in two pulses of ~1 min duration.

It is advisable to store the clean apertures in glass tubes and not in plastic containers, in order to avoid noticeable contamination due to recondensation or migration of plasticizers. For coated grids and apertures, a wide variety of storage units is commercially available. The units should be made from inert material, and they should be designed in a way that keeps the specimens dust-free, minimizes the contact area between the grid and the box, and eases the handling of the grids without risk of damaging. Based on our experience, a storage in the frequently used gelatine capsules is not recommended, as it leads to contamination clearly recognizable on otherwise nearly "structureless" support films. A simple and safe way of grid storage is in petri dishes coated with filter paper, or even better and more convenient, on glass slides equipped with a set of parallel thin glass rods (Fig. 1.5).

grid

glass rod

glass slide

Fig. 1.5 Simple device for clean and safe storage of coated specimen grids.

Microgrids and Perforated Films

For many purposes the mesh number (50–400) of the photographically etched specimen grids and the hole diameters of drilled diaphragms (20–750 μm) are unsatisfactory. This applies to many types of self-supporting specimens as well as to the mounting of extremely thin and fragile support films and of relatively small flake-shaped crystalline supports. Silver microgrids with a mesh number up to 2,000 mesh/inch and with open areas of 5 X 5 μm are commercially available. For safe manipulation, these silver microgrids should be sandwiched between folding grids of large mesh size.

Boersch *et al.* (1959) developed an electron optical technique for the production of metallic microgrids with up to $\sim$8,000 mesh/inch (open areas 1.5 X 1.5 μm). An electron optically reduced image of a square mesh grid or a diaphragm is projected onto a copper target coated with a thin plastic film. The irradiated area of the plastic film is crosslinked by the incident electrons and is rendered insoluble in organic solvents. After removal of the nonirradiated parts of the plastic film, nickel is electrically deposited on the exposed copper surface.

Table 1.1   Relative Merits of the Major Methods for the Preparation of Perforated Films and Microgrids

Method	Authors	Approx. range of hole diameters obtained	Advantages–Disadvantages
Electron beam writing methods	Boersch et al. (1959) Martin and Speidel (1972)	.06 μm–2 μm	Well-defined geometries –Unsuitable for large scale production
Replica methods	DeSorbo and Cline (1970) Tesche (1973)	.1 μm–8 μm	Well-defined hole sizes –Unsuitable for large scale production
Successive water droplet methods	Sjöstrand (1956) Sakata (1958) Drahos and Delong (1960) Pease (1975)	.5 μm–20 μm	Simple preparation –Difficult to standardize all conditions
Dew freezing method	Fukami et al. (1972)	.5 μm–100 μm	Suitable for preparing very large holes –Relatively troublesome
Emulsion methods	Harris (1962) Johnston and Reid (1971) Baumeister and Seredynski (1976) Dowell (1970) Hoelke (1975) Moharir and Prakash (1975)	.5 μm–10 μm .5 μm–10 μm	Relatively simple preparation, fairly well controllable hole sizes –?
Etching methods	Möldner (1965) Lickfeld and Menge (1968)	.05 μm–10 μm .05 μm–10 μm	Simple preparation –Hole size variation and polymorphism

The resulting nickel microgrid is released by dissolving the copper and the remaining plastic material in chromic acid. Very small carbon microgrids were produced by Loeffler (1964), utilizing radiation-induced mass-loss of carbon in the presence of residual oxygen and water vapor. A demagnified image of an electron source is scanned over a carbon film of ~10 nm thickness following a suitable pattern until holes appear. This process is controlled by electron optical magnification. Up to mesh numbers of 10,000 per inch the same ratio of bar area to open area as for usual commercial grids can be achieved, while as a maximum circular holes of ~100 nm diameter ~200-nm-apart can be obtained.

Martin and Speidel (1972) described the production of ultrafine carbon microgrids utilizing the otherwise undesirable effect of specimen contamination due to cracking of hydrocarbon deposits during electron bombardment in a scanning electron microscope. This procedure is certainly not suited for large scale production of microgrids, although it might be useful for some special applications to "write" microgrids with a predictable structure using a small diameter electron beam. Mesh numbers of ~50,000 mesh/inch and bar widths of 65 nm can be achieved. The same technique was used to "write" half-tone pictures (Müller, 1970) as an example for high density information storage, as well as to generate phase zone plates (Thon and Willasch, 1970) for correction of spherical aberration.

As early as the late forties, several methods for the preparation of perforated films had been developed. Jaffe (1948) suggested perforating polyvinylformal films by condensing water vapor on a glass slide. Watts (1949) perforated collodion films on a water surface by sparks from a high-frequency Tesla transformer. The diameters of holes obtained by the sparking method varies between 0.1 μm and 3.5 μm. The basic preparation principles for perforated films had become apparent—insertion and development of local faults in plastic films and generation of pits by locally destructive physical action. Later these concepts were augmented by the replica method. The inflation of modifications of these basic techniques described in the literature is remarkable insofar as it obviously reflects the difficulties in obtaining reproducible results.

The first routine method for the preparation of perforated Formvar films was described by Sjöstrand (1956). Basically, minute water droplets are introduced into a thin layer of a polyvinylformal solution covering a glass slide. If the Formvar layer is sufficiently thin, the water droplets will produce holes in the layer. Too thick a polyvinylformal film favors the formation of blowholes or "pseudoholes" (Fig. 1.6); i.e., the film is not really perforated. The following procedure was recommended by Sjöstrand (1956):

A glass slide is dipped into a polyvinylformal solution (2% Formvar in ethylenedichloride as solvent) and transferred to a small glass vessel with a saturated atmosphere of ethylenedichloride. The saturation is achieved by covering the walls of the vessel with a filter paper wetted with the solvent. The slide is left

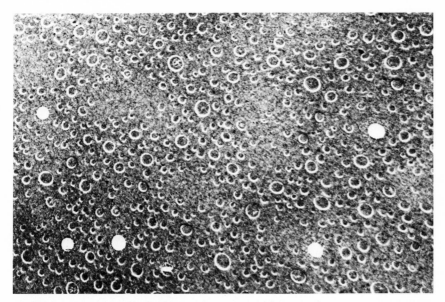

Fig. 1.6 Gold coated perforated film with few holes and many blowholes or "pseudoholes" due to incomplete perforation. ×8,000.

in a vertical position in this vessel for 20 min for draining off the Formvar solution and thinning of the layer. A stream of warm air saturated with water vapor is blown into the vessel. The slide is now removed from the glass vessel and after evaporation of the ethylenedichloride and the water, the perforated plastic film is floated off onto a clean water surface and transferred to specimen grids as usual.

The same principle, i.e., the introduction of minute water droplets into nascent plastic films, has been applied by Sakata (1958) to prepare perforated nitrocellulose films. The water droplets are inserted by gently breathing upon a freshly spread nitrocellulose film. Since the film is in most cases not yet really perforated after this treatment, the "pseudoholes" are converted into real holes by heating the glass plate on the flame of a Bunsen burner. Holes obtained by this procedure have a mean diameter of ~1 μm. Sakata (1958) reported that perforated films with larger holes (10-20 μm) can be obtained by spraying water droplets onto the surface by means of a nebulizer followed by an analogous "developing" procedure.

Drahos and Delong (1960) presented an improved version of the "breathing" method. They dipped a glass slide into a solution of 0.2-0.3% nitrocellulose in isoamyl acetate, and following the withdrawal they held it above the solution surface in a vertical position for ~30 sec for further thinning. Then, the slide was removed from the vessel and breathed upon after an interval of a few seconds.   The number and size of the holes depends upon the duration and intensity of breathing.   Perforated areas are easily recognizable because of their "foggy" appearance.   If the film is sufficiently thin, it is completely perforated by this treatment and can be used without any further treatment.

Fukami *et al.* (1972) devised a method to make films with very large (10–100 μm) perforations.   A clean glass surface is hydrophobized using a strong water repellent (Softex-KWO).   The glass plate is cooled below −5°C in a refrigerator.   Then the plate is taken out in the room atmosphere where one forms dew drops on the surface by breathing upon it.   The plate is immediately immersed in liquid nitrogen so that the drops are instantly frozen.   Formvar solution (0.3–0.5%) in ethylenedichloride is quickly poured and spread on the surface of the plate.   The plate is inclined to drain excess solution, and it is kept horizontal until the solvent and the water vaporize, leaving a plastic replica of the frozen dew drops on the glass slide.   To aid complete perforation of thin areas replicating the original dew drops the replica film is mounted on a supporting screen and exposed to ethylenedichloride vapor for 5–15 min at 60°C.

Recently, Pease (1975) described a similar procedure for preparing perforated films with large open areas (micronets).   Adapting a procedure well-known from monolayer work, he recommended that glass slides be hydrophobized by a protracted soaking in a saturated solution of ferric stearate in benzene.   After washing off excess ferric stearate and vigorously rubbing the slides with thin paper wipes with successive changes of pure benzene leaving little but a firmly adsorbed monolayer of the soap, one obtains a highly hydrophobic surface.

A nitrocellulose (Parlodion) film is prepared by dipping the pretreated slide in 0.4–0.6% amyl acetate solution.   Draining of the freshly cast film is done in the usual way by holding its face vertically and tipping its bottom edge so that excess solution is drained away rapidly.   When the upper edge and corner of the wet film start to exhibit interference colors indicative of drying, water is condensed on the surface by exposure to furiously boiling water.   Slides are exposed to the steam for 2–3 min; on the slides, thin but still wet areas of the film become milky.   Exposure to the steam is continued until the milkiness disappears, indicative of complete evaporation of the solvent.   The residual net now exhibits a frosted appearance.   Next, suitable regions of the film, which is nonhomogenous because of the drainage pattern, are selected (thin regions with large holes near the top, thicker regions with smaller holes near the bottom).   The chosen area is outlined by painting the rest of the slide with 0.4–0.6% nitrocellulose solution; this relatively thick collar facilitates flotation of the delicate micronets.

Grids with the dried nets on them are baked at 170–180°C in order to open up pseudoholes.   The concomitant substantial loss of nitrocellulose, due to decomposition and subsequent volatilization, markedly attenuates the micronets; coating the nets with a thick layer of carbon, gold, or—where a hydrophilic surface is needed—a final layer of silicon monoxide is, therefore, required to achieve sufficient stability. Lower temperatures (125-140°C) lead to incomplete perforation of pseudoholes.

Although excellent results have been obtained with the techniques described so far, it seems generally preferable to add the perforating liquid directly to the polymer solution to form an emulsion (simultaneous techniques); the main reason for this is that the successive techniques require condensation conditions which are difficult to control and, moreover, involve a rather delicate timing for the insertion of the water microdroplets into the drying film. Harris (1962) developed a simultaneous method for making perforated films, which in its various modifications proved to yield quite reproducible results:

A 0.25% solution of polyvinylformal in chloroform or ethylene dichloride is prepared, to which an appropriate quantity of glycerol (the glycerol content controls the number and size of the holes) is added. The mixture is shaken to form an emulsion. A microscope slide is dipped into the emulsion, and after it is slowly withdrawn and drained in a vertical position, it is allowed to dry for ~10 min. The slide is exposed to the steam for 1 min. The steam can simply be generated by immersing a heater in a water-filled beaker. The tendency to produce predominantly pseudoholes is substantially reduced by immersing the slide in acetone for a few seconds (Bradley, 1965). The film is now floated off on a water surface and can be mounted on specimen grids or specimen apertures by any suitable method.

Obviously, the diameter of the perforations depends, to a large extent, upon the degree of dispersion of the glycerol in the Formvar solution. Johnston and Reid (1971) increased the dispersion by ultrasonic treatment, which indeed yielded quite reproducible films with small-diameter perforations. Perforated films with fairly constant hole densities over the entire area and narrow hole size distributions can be obtained by varying the glycerol content and a controlled ultrasonic treatment. The hole diameter can be varied between 0.05 and 5 μm, which is suitable for nearly all applications (Baumeister and Seredynski, 1976).

A solution of 0.2 gm Formvar in chloroform is prepared. The quantity of glycerol added may be varied between 0.01 and 5 ml, according to the desired mean hole size. The solution is magnetically stirred until the polymer is completely dissolved. Then the Erlenmeyer flask containing the solution is sonicated either for 30 min in an ultrasonicator tank, filled with water, or intermittently (3 to 5 times per minute) by means of an ultrasonic disintegrator. Microscope slides cleaned in a detergent solution and thoroughly rinsed with distilled water are rubbed dry with a clean cloth. The slides are dipped into the emulsion for 5-10 sec, and then slowly withdrawn (speed ~0.5 in./sec) and allowed to dry under dust-free conditions. If a high reproducibility is required, motor-driven dipping devices with a variable speed as used for handling monolayers (see, e.g., Rothen, 1968) or an apparatus described by Revell *et al.* (1955, Fig. 14) can be used. After drying and draining in the usual way, the slides are placed for ~10 sec

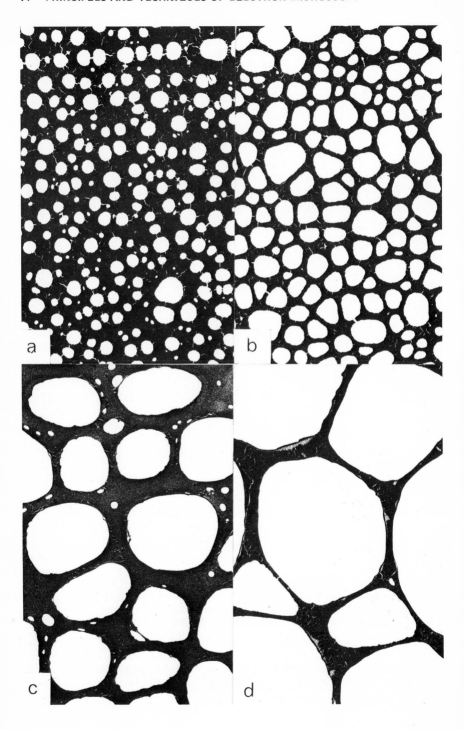

**Table 1.2   Relationship Between Glycerol Content and
Open Area**

Glycerol content (ml glycerol/90 ml Formvar solution)	Open area (% of total area)
0.01	4.2
0.03	13.5
0.1	34.1
0.3	35.9
1.0	52.0
3.0	63.2
5.0	78.4

in steam generated by boiling water and then allowed to dry again. To aid complete perforation, the slides are dipped into a beaker containing acetone for 10–20 sec. After the final drying, the films are floated off on a water surface and transferred onto specimen grids in the conventional way. If the rubbing, after the slides have been cleaned with the detergent, is performed strictly in one direction (e.g., parallel to the long axis of the slide) and if, in addition, the slide is kept in an inclined position (parallel to the direction of rubbing) during the first drying step, the formation of "hole avenues" is favored (Fig. 1.9).

Perforated films prepared by means of this relatively simple procedure are quite uniform in appearance and show a relatively narrow size range of perforations (Fig. 1.7) (Baumeister and Seredynski, 1976). Evidently the open area increases with increasing glycerol concentrations (Table 1.2). Plots of relative frequencies of hole size classes reveal distinct maxima, which are concomitantly shifted to larger hole sizes. It may, at a first glance, be surprising that in every case the smallest hole sizes dominate in the relative hole frequency distributions, although, apparently, the area occupied by this order of size is relatively small. This, however, simply means that the glycerol is completely dispersed to quite uniform microdroplets which at increasing concentrations show a higher probability of secondary fusion. The second and the third maxima (at higher concentrations) reflect the probability of fusion for a given concentration. Broadening of the maxima reflects the increasing variation of hole diameters with higher glycerol contents.

Figure 1.8 shows the percentage of total area occupied by distinct hole size classes for various glycerol contents. Optimum conditions fitting particular experimental requirements can be derived therefrom. At glycerol concentrations higher than 3 ml/90 ml organic solvent the micronets become rather fragile.

Fig. 1.7 Gold coated perforated films prepared with the glycerol method. The average hole size increases with the concentration of glycerol in the polyvinylformal-chloroform/glycerol emulsion. ×10,000. (a) Glycerol content 0.1 ml (per 90 ml polyvinylformal solution); (b) glycerol content 0.3 ml; (c) glycerol content 1 ml; (d) glycerol content 3 ml.

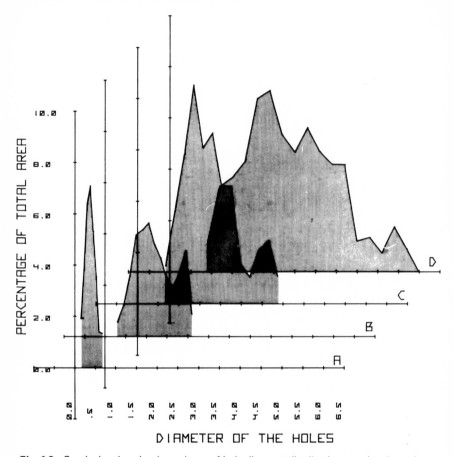

DIAMETER OF THE HOLES

Fig. 1.8 Graph showing the dependence of hole diameter distributions on the glycerol concentration of the emulsion. (a) Glycerol content 0.1 ml (per 90 ml polyvinylformal solution); (b) glycerol content 0.3 ml; (c) glycerol content 1 ml; (d) glycerol content 3 ml. *(From Baumeister and Seredynski, 1976.)*

With glycerol additions below 0.1 ml the hole size remains nearly constant, but the number of holes and consequently the open area decrease. Between 0.3 and 0.6 ml of glycerol, "hole-clusters" dominate. "Hole-avenues" (Fig. 1.9), obtained as described above, are of considerable advantage for minimal beam exposure techniques, where focusing and correction of astigmatism must be performed on a specimen area a few micrometers away from the area actually to be micrographed (see also Fig. 1.3), since they provide a reasonably high probability of blindly hitting a hole with similar dimensions and specimen areas as viewed for focusing.

Dowell (1970) and Hoelke (1975) described a method according to which

Fig. 1.9 Representative area of a gold-coated perforated film with "hole-avenues." Glycerol content 0.3 ml/90 ml polyvinylformal solution. ×8,000.

light alkanes (hexane or octane) are added to a solution of Formvar in chloroform (alkane concentration is 0.05–0.1 vol %). After the film is cast on a glass slide, the alkane microdroplets can be dissolved in petroleum ether or other organic solvents, which do not dissolve the polymer. Thus, perforations are formed with diameters below 1 μm.

Moharir and Prakash (1975) recommended preparing two Formvar solutions: (1) 0.18–0.25% in chloroform and (2) 0.18–0.25% in ethylene dichloride. After standing overnight, 1 ml of solution (2) is added to 99 ml of solution (1), and the mixture is shaken vigorously. Films are cast in the conventional way on freshly cleaved mica and floated off. While films obtained in this way have satisfactory concentrations of small isolated (0.05–1 μm) holes, films obtained with a mixture of 1 ml of solution (1) and 99 ml of solution (2) show a network structure. The mechanism of hole formation is discussed by the authors in terms of surface tension and viscosity differences of the solvents.

Another very simple method to perforate Formvar films was suggested by Möldner (1965). Ordinary nonperforated Formvar films, mounted on specimen grids, are placed on a sheet of filter paper, which should be in an inclined arrangement. By means of a standard nebulizer, methanol is sprayed onto the films. The minute droplets locally dissolve the plastic film to form holes. It is difficult, however, to control the number and size of microdroplets so that a reasonable number of holes is obtained and so that the methanol droplets do not spread to form a coherent film. Consequently, Lickfeld and Menge (1968) suggested a modified procedure. The grids coated with nonperforated ordinary Formvar films are immersed in a 2% solution of uranyl acetate in methanol. The contribution of the negative stain to the perforation process is not quite clear, and, to our experience, immersion in pure methanol (3–5 min) yields similar results. Holes obtained by this method appear to be quite polymorphous (Fig. 1.10), circular holes with smooth contours coexist with network-like perforations.

Another alternative procedure for the production of perforated films has been proposed by Weichan (1970). Local faults in a collodion film formed by bromication, are "developed" upon irradiation in an electron microscope. This procedure necessarily prevents routine production of these holey films.

When quite uniform hole sizes are required, replica techniques should be employed. Tesche (1973) used commercially available filters (nuclepore) with nearly cylindrical pores (0.1–8 μm in diameter) produced by etching fission tracks as templates. These filters are cleaned in concentrated hydrochloric acid and then repeatedly washed in distilled water. After drying, a layer of carbon or metal is evaporated onto the surface of the filter. Because of the relatively low thermal stability of the filters, thermal stress should be minimized. The filter material is dissolved using chloroform, and the remaining perforated carbon or metal films are scooped up on specimen grids. Alternatively, the films can be separated from the filter by evaporating a water-soluble interlayer onto the filter. The coarse structure of such interlayers does, however, influence the structure of the final carbon or metal layer.

Similar perforated films with holes of relatively uniform size and distribution, but with a larger number of holes per unit area, can be obtained by replication of porous metal wafers (Fig. 1.11). The porous wafers are made, according to DeSorbo and Cline (1970), by selectively etching the rod phase of the directionally solidified eutectics NiAl-Cr or NiAl-Mo. Replication can be performed in a one- or two-step process. In the one-step replication, a water-soluble interlayer is evaporated on the porous wafer, and then a carbon or metal layer is evaporated over it, coaxially to the whiskers. In the two-step replication, a plastic (Triafol, Bioden) replica is used to replicate the eutectic master. After stripping from the master, a gold layer is deposited on the plastic replica, which

Fig. 1.10 Gold coated perforated film prepared by "methanol etching." The holes are polymorphous with varying diameters. X 10,000.

can be dissolved. The success of the latter method, which has the advantage that the eutectic master can be used repeatedly, depends critically upon the depth of etching and the length of the protrusions in the plastic replica.

Tanaka *et al.* (1974) developed a method to prepare "ultrafine" (hole diameters 10–100 nm) perforated films from tropomyosin, a muscle protein. A drop of a solution containing 0.1–0.2 mg/ml of the protein, 0.2 M KCl and 10mM sodium acetate buffer (pH 5.6) is placed on a specimen grid precoated with a conventional perforated film. After ~30 sec, excess protein is washed away with the same protein-free buffer solution. This preparation is fixed in 0.5% osmium tetroxide and coated with an evaporated carbon layer.

Fig. 1.11 Perforated gold film obtained by replicating a master obtained from a selectively etched eutectic metal-alloy. ×13,000. (*From DeSorbo and Cline, 1970.*)

Because of the radiation sensitivity and the low electric and thermal conductivity of all the above-mentioned films, it is usually necessary to evaporate a thin carbon or metal layer onto them. Since most specimens must at least after irradiation be considered semiconductors, the electron tunneling properties of the supporting film play an important role. In this respect, especially gold coated perforated films seem to be a good choice. If, owing to recrystallization processes, the perforated gold films become too fragile, double (gold-carbon) or triple (gold-carbon-silicon monoxide) layers can be used (Johansen, 1974). Where it seems advantageous to use pure carbon or metal perforated films, the plastic substrate can be removed by dissolving it in suitable solvents. Nitrocellulose backings, which decompose at higher temperatures, can be removed by a prolonged (several hours) baking at 180°C (Reimer, 1967).

Plastic Films

For the suitability of plastic support films two basic criteria are decisive: (1) high electron transparency and (2) high mechanical stability (even during irradiation). Owing to the relatively low density of organic materials, criterion (1) depends merely upon the geometrical thickness of the films. Achieving a suitable thickness, in turn, is mainly a question of suitable preparation techniques. The stability of plastic support films in the electron beam is a very complex phenomenon, which is discussed in greater detail later. Evidently the radiation sensitivity is determined by the chemical structure (aromaticity, elementary composition, crosslinking or scission tendency) of a given compound, which should be considered carefully in selecting a material to prepare the film. Hitherto the potentials of thin organic layers as supporting films have not yet been explored systematically, and, most likely, considerable advantage could be derived from recent progress in polymer chemistry and thin layer engineering.

Materials for Plastic Films. The most generally used materials for the formation of organic supporting films are nitrocellulose and polyvinylformal. Nitrocellulose (trade names: Collodion, Parlodion, Zapon) is obtained by esterification of highly polymeric cellulose with nitric acid. It is heterogeneous with respect to the degree of nitration and polymerization. The nitrogen content of nitrocellulose cotton may vary between 10.7% and 12.6%, which means that 1.9-2.45 of the three hydroxyl functions of the D-glucopyranose units are converted into the ester. For a stylized structure, see Table 1.3.

Polyvinylformal (trade names: Formvar, Pioloform F, Movital F, Rhovinal F) is prepared from formaldehyde and polyvinyl alcohol. Polyvinyl alcohols are synthetic resins of high molecular weight containing various proportions of hydroxyl and acetate groups produced by hydrolysis of polyvinyl acetate. The conditions of the acetal reaction and the concentration of the formaldehyde and polyvinyl alcohol used are closely controlled to form polymers containing predetermined proportions of hydroxyl groups, acetate groups, and acetal groups, which are randomly distributed along the molecule. For a stylized structure, see Table 1.3.

The physico-chemical properties of polyvinylformals are largely determined by the ratio of PV-acetate: PV-acetal: PV-alcohol. The most generally used polyvinylformals (e.g., Formvar, Type 15/95E, Monsanto, Inc.) have approximately the following composition: PV-acetate, 9.5-13%; PV-acetal, 82-83%; and PV-alcohol, 5-6%. As a general rule, the substitution of formal groups for acetate groups results in a more hydrophobic polymer with a higher heat distortion temperature. Polyvinylformal is easily crosslinked by heating as well as ionizing

Table 1.3   Physical and Chemical Properties of Some Plastic Film Materials

	Nitrocellulose	Poly-(vinylformal)	Poly-(vinylcarbazole)	Polystyrene
Trade names	Collodion Parlodion	Formvar Mowital F Pioloform F Rhovinal	Luvican	Polystyrol
Structural unit				
Elementary composition, %	C 27.1 H 3.4 N 10.5 O 59.0	C 58.5 H 8.1 O 33.4	C 86 H 6.5 N 7.5	C 92.3 H 7.7
Density, g/cm^3	1.58	1.20	1.19	1.05
Elastic modulus $\times 10^{-5}$ N/mm^2	15–20	20–40	31–35	33
Distortion temperature, °C	50	50	150	75
Radiation resistance [Dose for $1/e$ decay of mechanical properties (vacuum), Mrad]	1–10	~100	>1000	>1000
Maximum mass loss upon irradiation (vacuum), %	70–85 (Reimer, 1965)	30–40 (Reimer, 1965)	? 	5 (Brockes, 1957)

radiation.   Crosslinking involves trans-acetalization and mechanisms such as reactions between acetate or hydroxyl groups on adjacent chains (Monsanto Technical Bull. No. 6070).

Kölbel (1966) has claimed that the addition of organo-polysiloxanes stabilizes nitrocellulose and polyvinylformal films upon irradiation. The siloxanes are either added to the resin solutions in concentrations of 0.2–0.5% (resin concentration 1%) or evaporated under normal atmospheric pressure at $170°C$ onto plastic films previously formed in the conventional way. Stabilization is ascribed to a mutual crosslinking of the polymers. It seems likely that the incorporation of thermosetting resins such as phenolics could additionally reduce the radiation sensitivity of the film because such additives may also act as acceptors in the intermolecular transfer of ionization or excitation energy released by primary events; this point is also discussed on page 77.

Further search for improved plastic support films should also take into consideration plastic materials of high inherent radiation resistance. Polystyrene (Table 1.3) and poly(N-vinyl carbazole) are, e.g., polymers which due to their inherent aromatic stabilization are extraordinarily radiation-resistant. Polystyrene especially, has been used with considerable success for mounting thin sections, unstained and negatively stained viruses, and macromolecules (Baumeister and Hahn, 1975; Hahn and Baumeister, 1976). Because of the high radiation resistivity no additional stabilization by evaporated carbon layers is required even for working at high beam intensities. Electron transparency is high relative to geometrical thickness because of the low density of polystyrene (1.05 gcm^{-3}) and spatial variations of mass thickness are low (Fig. 1.12). The smoothness of their surface yields quite homogenous distributions of negative stains, so that resolutions can be achieved which are more or less limited by the structure of the negative stain and the radiation sensitivity of the specimen (Fig. 1.13) and far less by properties of the supporting film.*

Solvents for Preparation of Plastic Films. Selection of suitable solvent systems involves a great number of factors, and although it is beyond the scope of this chapter to present a complete survey of them, a brief discussion is given here. One has to consider the complex phenomena associated with the solubility of polymers, which depend upon the type and number of functional groups present in a given resin as well as on their molecular weight and degree of branching. The otherwise useful rule of thumb to dissolve a given compound in an analogous solvent, is of very limited value for polymers, and complete miscibility between a given solvent and a monomer does not mean much for the solubility of the corresponding polymer in the same solvent.

Furthermore, one has to consider the solution viscosity, which depends upon the concentration of the solution as well as upon the conformation of the

*A certain drawback of polystyrene films is their stiffness and brittleness necessitating a very cautious manipulation. Sometimes adhesion to grids is poor, which can be overcome by using slightly bent (convex) grids.

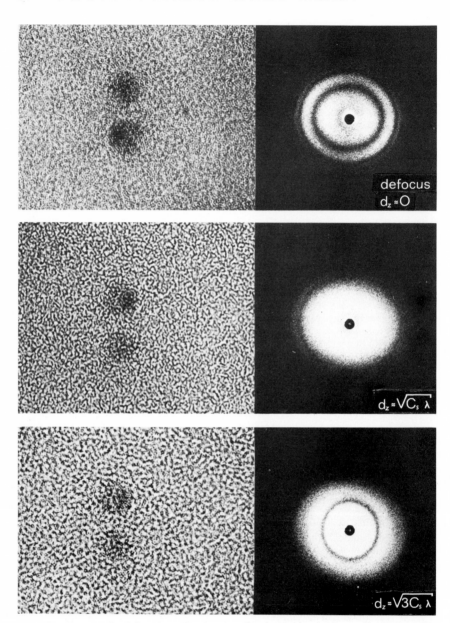

Fig. 1.12 Micrograph of ferritin molecules on a thin (~70 Å) polystyrene film with corresponding light optical diffractograms. Contrast equals that of a 35–40-Å-thick evaporated carbon film owing to the lower density of polystyrene; defocus dependent phase contrast granularity is similar to that of carbon films.

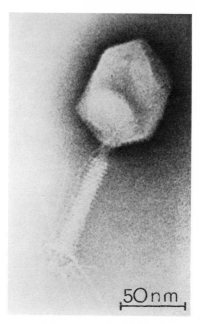

Fig. 1.13 T$_2$-bacteriophage on a polystyrene film, negatively stained by potassium phosphotungstate.

polymer. The latter in turn is a function of the solvation of the polymer. If the solvation in a given solvent system is poor, the polymers are expected to adopt a contracted random coil conformation (low viscosity of the solution). In contrast, a high degree of solvation of the polymer leads to a more extended and open conformational form (high viscosity of the solution), which, after spreading at an interface, favors a strongly interconnected structure of the plastic film.

As has been pointed out by Schuster and Fullam (1946), a selection of suitable solvent systems must be adapted to the particular requirements of each of the two basic modes of film formation: (1) spreading on a water surface and (2) casting onto a solid surface. The thickness of plastic films prepared by casting onto glass is controlled by the conformation of the polymer in the solution, the concentration of the solution, its viscosity, the volatility of the solvents, and the rate of withdrawal of the glass plate or the fall of the liquid meniscus.

For spreading of films on a water surface, a low solubility of the solvent in the subphase and the spreading coefficient of the solution also become important. The spreading coefficient (Adam, 1941) is a function of the surface tension of the subphase (e.g., water), the surface tension of the solution, and the interfacial tension of the two liquids. If the spreading coefficient is positive, spreading occurs. On the other hand, if the coefficient is negative, the solution dropped onto the subphase forms "lenses." For example, polyvinylformal in ethylene dichloride will not spread at a water-air interface, while polyvinylformal in benzene: ethyl alcohol (2 : 1) spreads fairly well. Polyvinylformal in ethylene dichloride will spread, however, only if a minute quantity of a fatty acid (e.g., oleic acid) is added to the solution. The fatty acid molecules diffuse to the interface, reducing the interfacial tension and causing a positive spreading coefficient.

It is pointed out in this context, that "spreading" of most polymers at water–air interfaces is of an entirely different nature as compared to spreading of heteropolar low molecular weight substances. The latter orient themselves at the water–air interface because of the strong interaction between the polar groups and the water molecules, and exhibit a well-defined configuration. The thickness of such layers never exceeds the length of the molecules. These principles are not valid for polymer films. Even if the concentration of the polymer is sufficiently low so that the macromolecules can exist as isolated "random coils," they do not necessarily flatten out on the water surface to a stretched or meandering linear configuration. The measured areas per monomeric unit do not correspond to those calculated on the basis of stereochemical considerations assuming an ideal monolayer (Müller, 1954). Apart from some monolayers forming semicrystalline films (Crisp, 1949), most of them show an amorphous structure consisting of conglomerates of "deformed coils."

Table 1.4 gives a survey of recommended solvents for polyvinylformal, nitrocellulose, and polystyrene and indicates their relative volatility as compared to diethyl ether. Only water-free solvents (stored over molecular sieves) should be used, especially for casting films; otherwise holey films will result. While all the solvents listed in Table 1.4 can be used for casting films on glass, only a few of them are suitable for spreading on water surfaces. Benzene, for example, can be used for both spreading and casting polyvinylformal films, whereas acetone, al-

Table 1.4 Film Materials and Recommended Solvents

Plastic material	Solvents for spreading	Relative* volatility	Solvents for casting	Relative* volatility
Nitrocellulose	Methyl isobutyl ketone	6.7	Acetone	2.1
	Butyl acetate	12.5	as well as all sol-	–
	Amyl acetate	13	vents suitable for spreading	
Polyvinylformal	Benzene/ethyl alcohol	–	Chloroform	2.5
	2 : 1 (by weight)	–	Ethylene dichloride	4.1
			Dioxane	7.3
			Tetrachloroethane	33
Polystyrene	Diethyl ether	1	Chloroform	2.5
	Benzene	3	Dioxane	7.3
	Methyl isobutyl ketone	6.7	as well as all sol- vents suitable for spreading	

*Relative to diethyl ether = 1.

though an excellent solvent for nitrocellulose, is unsuitable for spreading owing to its solubility in water.

An optimization of the manifold factors influencing the spreading process will usually necessitate the use of solvent mixtures. By oversimplifying, one can consider three cases: (*A*) the polymer is soluble in two solvents, (a) and (b), of the ternary system; (*B*) the polymer is soluble only in one of the solvents (a) and insoluble or not completely soluble in the other solvent (b); (*C*) the polymer is insoluble or not completely soluble in both solvents (a) and (b), but soluble in a mixture of (a) and (b). In case *B*, the diluent (b) should be more volatile than the solvent (a), and in case *C* the volatility of solvents (a) and (b) should be approximately equal; otherwise the polymer will precipitate during evaporation, giving rise to a nonuniform texture of the film. For polyvinylformal, mixtures of aromatic solvents provide a satisfactory starting point for the development of solvent systems (e.g., benzene: ethyl alcohol or toluene: ethyl alcohol (95%)).

Preparation of Films

SPREAD FILMS. This method, which originally was devised by Ruska (1939) for the formation of nitrocellulose support films, can in an analogous way be employed for most other plastic materials. Basically, a small amount of resin solution is dropped onto a liquid subphase, where it spreads spontaneously to form a solid film after evaporation of the solvent. Usually distilled water is used as a subphase.

To avoid any impurities, which will concentrate at the water–air interface and thus lower the surface tension, it is advisable to clean the surface immediately before spreading. This can easily be accomplished by filling the trough with water until the meniscus rises above the edge of the trough. Then the water surface is swept clean by means of a glass rod or a Teflon bar. If a solvent is used which is slightly soluble in water (e.g., amyl acetate), it is recommended that the water subphase be saturated with the respective solvent prior to spreading.

No general rule can be given concerning the concentration of the resins in spreading solutions. The concentration determines the film thickness, and an optimization must also consider the conformation of the polymer in solutions, its behavior at the interface, and the volatility of the solvents. Usually resin concentrations of 0.2–1% are employed.

Instead of simply dropping a small amount of the solution from a pipette or syringe onto the water surface, one can use an apparatus as described by Ruska (1939) (Fig. 1.1) for preparing the films. The latter approach has the advantage that film preparation and the subsequent deposition of the film on specimen grids and apertures can be accomplished under dust-free conditions. The apparatus (Fig. 1.1) essentially consists of an ordinary separatory funnel containing the resin solution and a covered dish filled with the subphase. The specimen grids are mounted on a wire mesh on the bottom of the dish. By opening the stopcock of the separatory funnel, one spreads a small quantity of the resin solution on the water surface. After complete evaporation of the solvent (as indicated by the disappearance of interference colors), the water surface is lowered by opening the out-flow of the dish, thus depositing the film onto the specimen grids.

CAST FILMS. Plastic support films can alternatively be prepared by casting onto a glass surface. The unpretentious requirements with regard to the solvent system and the relative ease with which thin films can be produced have meanwhile made the "casting" or "immersion" method as originally described by Schaefer and Harker (1942) the most common method of film preparation.

Basically, a glass plate (such as a microscope slide) is immersed in a solution of the polymer. After a few seconds the slide is withdrawn and the solvent is allowed to evaporate. Then the slide covered with the plastic film is slowly immersed at a shallow angle into a dish of distilled water to float off the film onto the water surface (see also below). New microscope slides free from scratches should be used, since the scratches would be replicated in the plastic film. The slides should be cleaned prior to use in a detergent solution, thoroughly rinsed in distilled water, and wiped dry with a clean soft cloth (see Hayat, 1970).

Nonuniform rates of withdrawal and varying conditions of draining and drying obviously give rise to a streaked appearance of the plastic films. Local variations in film thickness and differences in the average thickness of successive films from the same solution can be eliminated by using motor-driven dipping mechanisms as employed in monolayer techniques (see, for example, Rothen, 1968) or a simple film casting apparatus as devised by Revell et al. (1955).

This apparatus (Fig. 1.14a) is based on an evaporation funnel. The tubing below the ground-glass top is replaced by a suitable length of capillary tubing, which reduces the flow rate. A loose-fitting metal or glass plate glass on top of the funnel keeps the solution and the slide dust-free, minimizes atmospheric disturbances during drainage and drying of the film, and reduces the rate of evaporation of solvent. The cover plate is fitted with a spring clip, which holds the microscope slide, on which the plastic film is to be cast, vertically in the solution. The initial level of the resin solution in the funnel should be arranged to be clearly above the top of the microscope slide, so that steady flow conditions are achieved before the meniscus reaches the slide.

The films are allowed to dry in situ; then the slide is removed to float off the adhering plastic film. As already pointed out, the thickness of the film is partly dependent upon the rate of fall of the liquid meniscus, which, in turn, is governed for a given solution by the dimension of the apparatus. Once the apparatus has been calibrated, films of uniform and predictable thickness (over a range of 15–130 nm) can be prepared.

To ease the repeated use of the stock solution, Savdir (1963) suggested combining the apparatus as described by Revell et al. (1955) with an ordinary stock bottle equipped with a ground-glass joint, a pressure tube, and a raised tube (Fig. 1.14b). By blowing into the pressure tube, one forces the resin solution up the

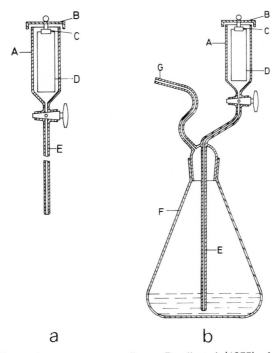

a   b

Fig. 1.14 (a) Film casting apparatus according to Revell *et al.* (1955). *A*, evaporation funnel; *B*, brass plate; *C*, clip for fixing the slide; *D*, slide; *E*, capillary tube. (b) Modified film casting apparatus according to Savdir (1963). *A–D*, same as in (a); *E*, rising tube; *F*, 100-ml flask containing the polymer solution; *G*, pressure inlet tube.

raised tube into the evaporation funnel, wetting the vertically mounted microscope slide. The stopcock is now closed for 10–30 sec then opened again so that the resin solution flows back to the stock bottle.

After casting and drying, the films are scored around the edge with a dissecting needle or with the edge of a razor blade. Then the slides are dipped at a shallow angle into a dish filled with distilled water, film side uppermost, and slowly moved in a lengthwise direction. The plastic film peels off the slide and floats on the water surface. Breathing on the film immediately prior to immersion and during the operation will, if necessary, facilitate the peeling.

Transfer of Plastic Films onto Specimen Grids and Specimen Apertures. "Dry" stripping and transfer of cast films by means of Scotch tape as described by Schuster and Fullam (1946) is applicable only for relatively thick films or necessitates a "framing" of the film. For films floating on a water surface, several transfer techniques have been developed, which are discussed below.

Lowering the water surface as described above for the Ruska apparatus is the most convenient method for spread films which cover most of the available water surface. However, a shortcoming of this method is that films spread in the usual way are nonuniform in thickness (increasing from the outer fringe to the center). Hence, some of the specimen grids or apertures are usually coated with unsuitable film areas. These grids must, therefore, be discarded in an additional assorting procedure.

Both the cast and spread films can be transferred to grids and apertures by means of a scoop, which essentially consists of a platform equal to the size of a microscope slide and an angle-shaped handle. Figure 1.15 shows a scoop specially designed for specimen apertures by Dowell (1964). Such a scoop with the grids or apertures on it is brought to a horizontal position underneath the film floating on the water surface. By lifting the scoop slowly upward to bring it against the underside of the film, one raises it through the water–air interface at a slight

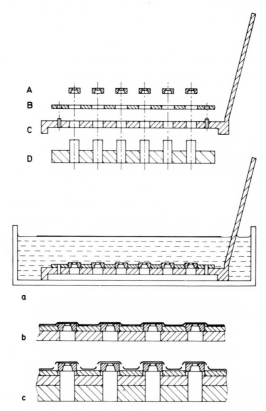

Fig. 1.15 Apparatus for depositing plastic films on specimen diaphragms according to Dowell (1964). *A*, specimen diaphragms; *B*, centering insert; *C*, platform with handle; *D*, ejector plate. The whole assembly is brought into a position underneath the floating film (a) and then lifted upward, depositing the film on the specimen diaphrams. After drying (b) the diaphragms with the film on top are ejected so that they can easily be picked up with a forceps.

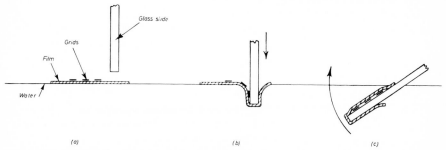

Fig. 1.16 Deposition of plastic films on specimen grids according to Hall (1953). The specimen grids are placed on top of the floating plastic film (a); a glass slide is pushed vertically through the interface so that the plastic film adheres to the slide, sandwiching the grids (b); then the slide is lifted upwards (c).

angle. Excess water is blotted from the underside with a filter paper, and after complete drying the film-coated specimen grids can be removed from the scoop with a sharp pair of forceps without prior scoring of the film around the edge of the grids. The scoop for specimen apertures (Fig. 1.15) is equipped with a special ejector plate.

Another procedure for transferring plastic films (Hall, 1953) is to place specimen grids on top of the film floating on the water surface. As shown in Fig. 1.16, a glass slide is brought to a vertical position over the film and pushed down past the surface, thus folding the film against the slide. Finally, the glass slide is turned and lifted through the interface. After complete drying, the grids may be lifted free. Oven drying at 40-60°C shortens the procedure.

Alternatively, a sheet of paper (filter paper or blank newspaper is convenient) is laid on the floating film with the grids on top of it. After the sheet of paper is completely wet, it is lifted along with the adhering grids and film from the water and allowed to dry completely. Finally, the grids are removed as described above (Fig. 1.17).

Storage of Polymer Solutions and Films. Plastic films may be stored in the dark under ordinary conditions for months or even years without significant quality loss. There are some indications that nitrocellulose films deteriorate when

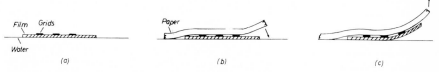

Fig. 1.17 Alternative procedure for the deposition of plastic films: The specimen grids are placed on top of the floating plastic film (a), a sheet of paper is laid upon it (b), and after complete wetting the paper with the adhering plastic film is lifted from the liquid surface (c). After drying, the grids are removed from the paper by means of a forceps.

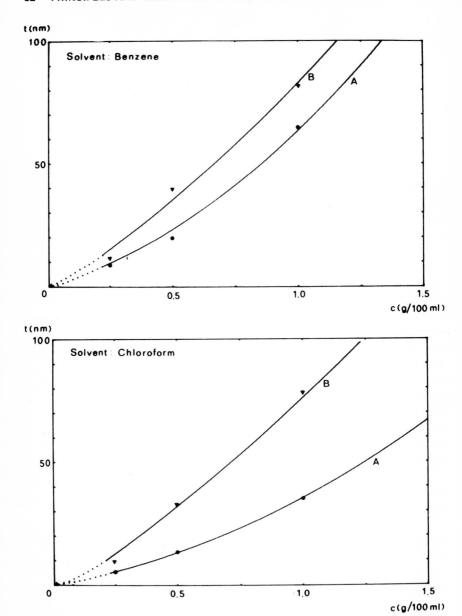

Fig. 1.18 Calibration curves for polystyrene films (thickness determined interferometrically) cast from benzene and chloroform. Speed of withdrawal: 5 mm/sec, 10 mm/sec. Note that film thickness is not strictly a linear function of polymer concentration c in the solvent, because of viscosity effects.

Table 1.5 Relationship Between Film Thickness (D) and
Concentration (c) of the Solution for Cast Films

Film material	Solvent	a
Formvar	Chloroform	2000
Formvar	Dioxan	1500
Collodion	Amyl acetate	1260
Collodion	Butyl acetate	1370

$D = a \cdot c$ (rate of withdrawal 5 cm/sec). Concentration c in % ($1\% = 1$ g polyvinylformal/nitrocellulose per 100 cm^3 solvent).
From Reimer (1967).

exposed to sunlight (Hall, 1957). The polymer solutions should be kept in bottles equipped with ground-in joints to avoid excess evaporation of the solvent.

Thickness Estimation for Plastic Films. As already pointed out, the thickness of films prepared by means of the Revell and Agar apparatus can be estimated with reasonable accuracy for a given resin, solvent system, and solution concentration, once the apparatus has been calibrated. Calibration is performed by inter-ferometric methods. Figure 1.18 shows a calibration curve for cast polystyrene films withdrawn from the polymer solution at constant speed by means of a motor-driven apparatus. Obviously film thickness is not a linear function of polymer concentration. The deviation from linearity is ascribed to viscosity effects. Reimer (1967) has presented a table which can be used to derive the thickness of polyvinylformal and nitrocellulose cast films and some frequently used solvents, assuming a nearly constant rate of withdrawal of ~5 cm/sec and a linear relationship between concentration and film thickness (Table 1.5).

The film thickness can roughly be estimated by examining the films in re-flected light when floating on a water surface. Over a preferably dark background, films showing any interference colors are too thick for most applications; a bluish coloration indicates a thickness of 75-100 nm. For conventional electron microscopy at moderate resolution levels, films exhibiting a pale, first-order gray color (30-50 nm) are suitable. "Colorless" films which can be distinguished from the free water surface only because of their different reflec-tivity are assumed to be 15-30 nm in thickness, but are usually too fragile to be used with wide-mesh specimen grids.

Supports for Unobstructed Mounting of Large Specimen Areas

The examination of serial sections and correlative light and electron microscopic investigations of large sections are greatly facilitated by mounting the specimen on strong films stretched over holes of ~1 mm in diameter (Dowell, 1959; Reale and Luciano, 1965). The application to correlative histochemistry is important because for numerous cellular components only light microscopical specific staining reactions are known, and electron microscopy of large semithin sections provides valuable complementary topographic information (Lange, 1972). The true picture of selective staining for electron microscopy has begun to emerge (Hayat, 1975).

Dowell (1957, 1959, 1964) found that composite films of evaporated carbon covered with plastic are suitable for mounting the above specimens without the frequently interfering grid bars. While conventional plastic films spanning large holes shrink and may finally break upon electron irradiation, pure carbon films are too brittle to be used with large-diameter holes. Composite films largely avoid these difficulties; they are mechanically stable even at high beam intensities. Dowell (1964) recommends the following procedure to form composite films of carbon and plastic:

Glass slides or freshly cleaved mica sheets are covered with a layer of evaporated carbon (see also page 38). A thickness of ~10 nm is recommended for extensive films. The carbon-covered slide or mica sheet is immersed in a solution of polyvinylformal (0.1%) or nitrocellulose (0.2%). For draining, the slides should be kept in a vertical position while drying. It is essential that the initial evaporation of solvent not take place in humid air, in order to avoid the formation of destabilizing holes or pseudoholes. Composite films with the carbon layer facing the glass or mica are easily stripped and floated onto a water surface, whereas plastic films on which carbon is evaporated usually firmly adhere to the support. Stripping can, if necessary, be facilitated by placing the slide coated with the double layer in a humidity box for several hours.

Films floating on the water surface are mounted on slot-type grids (Fig. 1.19)

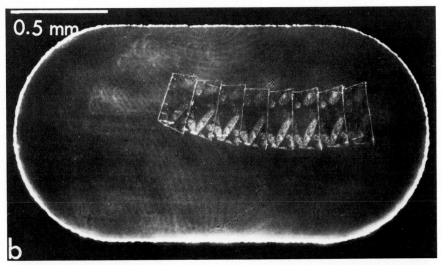

Fig. 1.19 A ribbon of serial sections on a thick (~100 nm) plastic coating a slot-type grid. *(From Stockem, 1970.)*

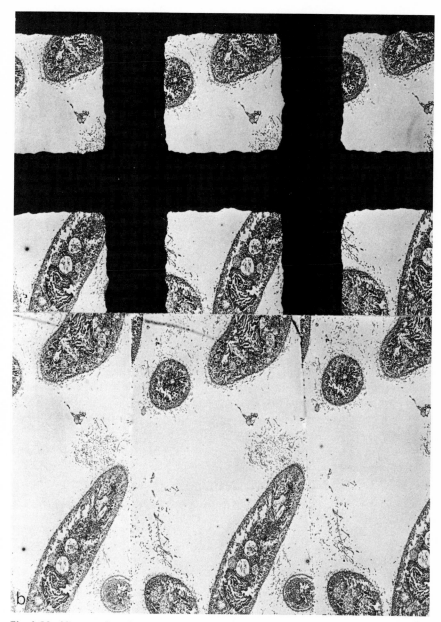

Fig. 1.20 Micrographs of three serial sections (a) mounted on a conventional specimen grid (mesh width ~90 µm) and (b) mounted on a slot-type grid. Supporting film: polyvinylformal. The advantage of unobstructed mounting is immediately apparent. (*From Stockem, 1970.*)

Fig. 1.21 Tripalmitin crystals transformed from Langmuir-Blodgett Layers deposited on a thin (20 nm) polystyrene film, spanning a 1-mm aperture. Note the high contrast of the thin (30 nm) unstained lipid crystals.

or on punched discs with hole diameters up to 1 mm. Mounting on grids is performed in the conventional way (see page 37), with either the carbon or the plastic film facing upwards. The latter type of mounting has the advantage that the plastic film adheres better to the grid. With specimen apertures or punched copper discs, a simple apparatus as described on page 30 (see also Fig. 1.15) is used. The covered discs are left on the carrier plate for several hours until they are completely dry. Excess water may be carefully drawn off with a filter paper. No attempt should be made to remove water from areas directly beneath the grids. Rapid drying will produce broken or wrinkled films (Dowell, 1959). The composite films are sufficiently stable to allow post-staining of sections if care is taken when removing excess stain (Reale and Luciano, 1965).

Stockem (1970) has more recently shown also that simple plastic films spanned over large-diameter holes may withstand the electron bombardment without significant shrinkage or movement. He used polyvinylformal (Pioloform F) films with an estimated thickness of 100 nm (Fig. 1.20). With more radiation-resistant polymers like polystyrene, considerably thinner films are expected to be sufficiently stable to be used for the unobstructed mounting of large-area specimens (Fig. 1.21) (Baumeister *et al.*, 1977).

Glass Specimen Supports

Thin glass layers as specimen supports for electron microscopy have only very rarely been employed, mainly for two reasons: (1) it is difficult to prepare sufficiently thin glass layers, and (2) most of the advantages of glass supports (i.e., temperature stability and resistance to organic solvents as well as to acid media) can nowadays also be met satisfactorily by a variety of other supports.

Möllenstedt (1947) and Ackermann (1947) described a technique to obtain sufficiently thin glass layers: a closed glass tube (e.g., Jenaer Geräte Glass 20) is heated in a gas burner and slowly blown up by means of compressed air from a cylinder equipped with a needle valve. The expanding glass tube is kept in a furnace, and the pressure in the glass sphere is gradually elevated until the interference colors disappear and black spots become visible. Such glass films have a thickness well below 100 nm. Electron diffraction proves (Ackermann, 1947) that glass supports are essentially amorphous.

Van Itterbeek et al. (1952) recommended blowing out the glass tube rapidly. Thus, very small glass flakes are dispersed in air. As the thinnest ones fall down most slowly, they can selectively be caught on the surface of water in a vessel. Glass layers must either be sandwiched between folding grids or fixed to the specimen grids by means of an adhesive (van Itterbeek et al., 1952), or be fused while being in the molten state (Ackermann, 1947).

Carbon Films

It has been found possible to evaporate carbon to form thin amorphous films (Bradley, 1954), which are most frequently used for high resolution work. An advantage of pure carbon films (especially as compared to most crystalline supports) is that they routinely can be made to cover relatively large areas in a uniform manner. This advantage has made them the most successful supports in spite of their pronounced granularity in phase contrast (Thon, 1965, 1966). Kakinoki et al. (1960) concluded from electron diffraction studies that the structure of an evaporated carbon film is a randon (glass-like) three- dimensional network formed by the recombination of atoms, molecules C_2, C_3,. . .), and ions. Graphite-like and diamond-like configurations, each occupying domains as small as several Ångstrom units, coexist without any systematic mutual orientation. The conditions of evaporation (vacuum, evaporation speed, direct-indirect evaporation, and target temperature), heat treatment, electron irradiation, and aging effects may influence the actual structure.

The density of carbon films, which is difficult to determine for such thin specimens with reasonable accuracy, is assumed to be in the range of 1.9 gcm^{-3} to 2.4 gcm^{-3} (Leder and Suddeth, 1960; Kakinoki et al., 1960). Stochastic spatial distribution of thickness causes the well-known phase contrast granularity of carbon films (Hahn, 1965). The presence of small (nanometersize) crystallites (especially in directly evaporated films) may additionally give rise to Bragg reflections, whose high intensity maxima may occasionally be seen in dark-field images, thus interfering with the specimen structure (Brakenhoff, 1974). Unlike the so-called 'amorphous carbons' being aggregates of very small graphite particles (Franklin, 1950), evaporated carbon films consist according to Kakinoki's model of a strongly interconnected three-dimensional network. This characteristic may account for their remarkable mechanical stability, even when as thin as 1-2 nm.

Another advantage of carbon films is that their good electric conductivity prevents accumulation of charges.

Carbon Evaporation. To evaporate carbon in a vacuum chamber, a current ($\sim$50 A, 20 V) is passed through two pointed carbon arc rods mounted in insulated holders. One rod is usually fixed and the other movable so that the points are pressed lightly together during the process. The pressure can be maintained either by means of springs or, if the assembly is mounted vertically, by gravity. Intense local heating occurs in the region of contact upon passing of the current and leads to carbon sublimation. The carbon rods should be of spectrographic grade and 4–7 mm in diameter. The pointed ends and their counterparts may be designed in different ways, with either one carbon rod sharpened to a point and the other flat or with both rods ground (Fig. 1.22). The latter design (Fig. 1.22c) has the advantage of yielding a more homogeneous cone of evaporation. The carbon evaporator unit may additionally be equipped with a simple device controlling the minimum distance between the carbon rod holders (governing the contact between the carbon rods), which allows one to evaporate a predetermined total amount of carbon (for details see Fig. 1.24 and the corresponding legend).

Bradley (1954, 1965) devised a simple means of monitoring the thickness of the evaporated carbon layer. A piece of white, glazed porcelain with a drop of vacuum oil on it is placed beside the target. The thickness is judged by the coloration of the porcelain indicator, which is only visible on the area not covered with oil (carbon does not form an absorbing layer on the oil surface). When a light brown color is barely detectable compared to the white of the oil-coverned area, the thickness of the film on the target is $\sim$5 nm, a satisfactory thickness for most purposes. For more accurate thickness estimation methods, see page 84. According to Reimer (1967) the brownish coloration is typical for a relatively poor vacuum. Grey carbon films, which are claimed to be more stable, are obtained in a good vacuum ($\sim$10^{-5} mbar) and, even more important, after carefully degassing the porous carbon rods prior to evaporation by preheating with a current of $\sim$20–25 Amp.

Whether or not a rapid evaporation has advantages over a slow evaporation is not yet quite clear. It should be considered, however, that the thermal energy transferred to a substrate during evaporation is composed of light quanta from the source absorbed in the target and the heat of condensation released at its surface. While the thermal emission power of the source is proportional to T^4 (T being temperature of the source), the evaporation rate is proportional to

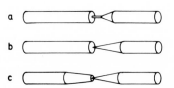

Fig. 1.22 Alternative methods of grinding carbon rods.

$\exp T$. Therefore, the evaporated mass per unit of emitted light energy increases with temperature, which would favor evaporation in short pulses at first glance. On the other hand, optical absorption in a target is generally higher for short wave quanta, whose abundance increases with temperature. In addition, the thermal power generated by condensation is proportional to the rate of deposition, whereas the resulting temperature rise in the target is dependent on its thermal conductivity.

In general, evaporation in short pulses but with large distances between source and target to keep the deposition rate low should be advantageous. For thermally sensitive targets the different propagation velocities of quanta and gas atoms could be utilized to separate them by means of a stroboscopic cylinder (Horn, 1962). It should also be mentioned that in the short pulse evaporation the development of a plasma bowl around the contacting tips of carbon rods enhances reproducibility because the influence of their microgeometry on the directional characteristic of the deposition rate is smoothed out. In the authors' laboratory a small commercial thyratron switching unit for welding purposes is used to energize the transformer of the evaporation unit for a predeterminable number of sine half-waves from the 50-Hz mains.

There is no doubt that indirect evaporation of carbon produces smoother and less grainy films than direct evaporation, because larger carbon clusters are prevented from reaching the target (Whiting and Ottensmeyer, 1972; Johansen, 1974; Kölbel, 1974). An indirect evaporation unit essentially consists of a stop positioned between the evaporation source and the target and a reflector. The design of such a ricochet cylinder should meet three conditions: (a) no direct sight from source to target; (b) no double reflection; (c) a large solid angle throughout. From this follows the geometry of the device (Fig. 1.23).

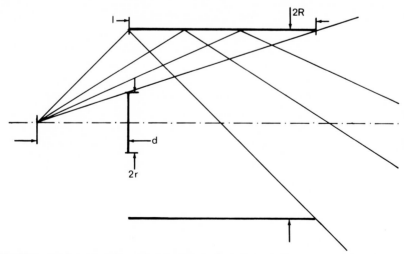

Fig. 1.23 Axial cross-section of "ricochet cylinder" for indirect evaporation of carbon showing optimum geometry. (For details see text.)

(a) $\dfrac{r}{d} = \dfrac{R}{l+d}$ (b) $2\,d = l$; hence $R = 3r$.

By choosing the stop radius r slightly smaller than $R/3$, a reinforced rim for easier handling is formed from the additional directly deposited carbon (see also Johansen's device described below). For a geometry according to (a) and (b) the maximum yield of the cylinder is obtained for $R/l \approx 0.9$ assuming an axially symmetric Lambert-law evaporation source. The inner surface of the cylinder should be very smooth i.e., well-polished metal or glass to prevent premature deposition.

A quite sophisticated evaporation unit (Fig. 1.24) for indirect deposition of carbon has been designed by Johansen (1974). The carbon rod holder consists of a horizontal rod, A, which allows the whole assembly to be moved vertically and laterally. The right-hand part of the holder is connected to the left by a bar fixed at B and insulated by means of mica at C. While the left carbon rod is fixed, the carbon rod on the right can be aligned by means of the adjustment screw D. The rods are pressed together simply by the leverage of the right-hand part of the holder operating through the hinge at B. The amount of carbon evaporated is regulated by two insulated (ceramic) distance screws (E, E). A 2,000-ml glass vessel (internal diameter 125 mm) cut to an open cylinder is used as a reflector (G) for indirect deposition of carbon on the circular target, $F(\sim 50$ mm in diameter).

In order to prevent direct deposition on the inner part of the target while at the same time allowing it to occur at the periphery, a circular stop, H, mounted on four thin metal supports is used (20 mm in diameter). Thus, the ultrathin carbon region deposited by indirect evaporation onto the central part of the target is automatically framed at the periphery. This framing facilitates observation of the film on the water surface and provides additional mechanical strength for handling. Johansen's evaporation unit is additionally equipped with an alignment jig (I), which is placed on top of the reflector to ease coaxial alignment of the carbon tips with the whole assembly before use. Williams and Glaeser (1972) obtained such "tab" regions, which can be used as guides for handling ultrathin carbon films by means of a rotary shutter (see also Fig. 1.25).

Targets for Carbon Evaporation. Many solid or liquid targets have been used for preparing carbon films: glass, mica, plastic films, organic glass, and glycerol. In the first two cases the evaporated carbon film is floated off onto a liquid surface; in the other cases the target is dissolved away after film formation. The advantages and disadvantages of the different targets are summarized in Table 1.6.

The easiest method to prepare carbon films is to deposit the layer on a thoroughly cleaned glass slide and to float the film off on a clean water surface by lowering the slide into the bath at a shallow angle (Bradley, 1965). While

Fig. 1.24 Carbon evaporator unit for indirect deposition (for detailed description see text): *A*, horizontal rod for adjusting the evaporation unit; *B*, hinge; *C*, mica insulators; *D*, adjustment screw; *E*, insulated screws for distance control; *F*, target; *G*, reflecting glass vessel; *H*, central stop; *I*, alignment jig. (*From Johansen, 1974.*)

glass is indeed a satisfactory target for the deposition of carbon films, at times difficulties arise when the carbon is to be released from it. Oil or detergent interlayers which have been recommended to aid floating off should be avoided, since they heavily contaminate the carbon layer. It should also be noted in this context that clean microscope slides as received from the manufacturers are already coated with a detergent layer (Fabergé, 1974), which is the reason why carbon films usually float off more easily from glass slides used without further cleaning.

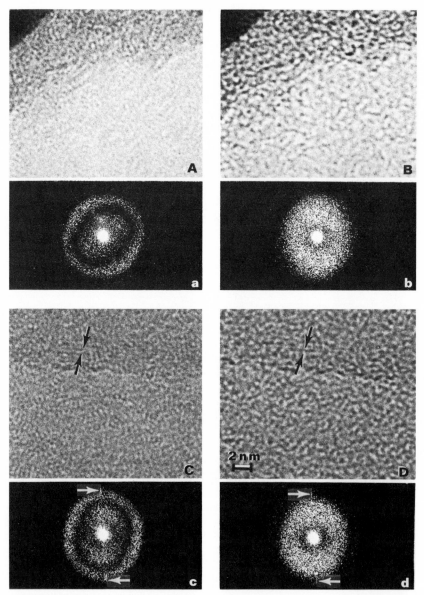

Fig. 1.25 Comparison between "ultrathin" carbon films and those of conventional thickness. *A* and *B* are micrographs of an "ultrathin" ($\sim$1.3 nm) carbon film at focus settings $d_z(A) \approx 0$ and $d_z(B) \approx \sqrt{C_s\lambda}$. *C* and *D* are micrographs of a carbon film about 17 nm thick at the above two foci, respectively. The defocus $d_z = 0$ leads to an image with minimum phase contrast granularity and the defocus $d_z = \sqrt{1.4\,C_s\lambda}$ yields a wide positive-phase-contrast main transfer interval, i.e., maximum overall phase granularity. The "ultrathin" film appears less granular than the thicker one at both foci. Note the "tab" regions in the upper halves of the pictures. (*From Johansen, 1974.*)

Table 1.6   Relative Merits of Different Targets for Carbon Film Preparation

Target	Advantages	Disadvantages	Aids to release the carbon film
Glass	Easily available.	Thorough cleaning necessary (detergent contamination!); floating off may be difficult; weak adhesion of carbon films to metal grids.	Detergent interlayer (–) (Bradley, 1965); oil interlayer (–) (Siddall, 1961); HF (1%) + acetone (10%) (+) (Münch, 1964); HF + detergent (?) (Bergeron and Pontefract, 1970).
Mica	Extremely smooth surface, free from detectable contaminations.	Weak adhesion of carbon films to metal grids.	Storage of freshly cleaved mica in humid atmosphere (+) (Johansen, 1974).
Plastic Films	Gentle deposition onto grids, high adhesion to metal grids (due to plastic remnants on grid bars).	Plastic remnants act as electric insulators; complete removal of plastic material is difficult to achieve; relatively coarse surface.	Target is dissolved by appropriate organic solvents (Bradley, 1954, 1965; Schober and Prokes, 1955; Reimer, 1967; Stolinski and Gross, 1969).
Glycerol	Extremely smooth surface, gentle deposition onto grids.	To prevent bubbling of the hygroscopic glycerol "inverse mounting" or cooling are necessary (see page 49).	Target is dissolved by water (Humphreys, 1963).
Organic glass (dibenzylamine tartrate)	Smooth surface, gentle deposition onto grids.	Preparation of the target material is tedious.	Target is dissolved by methanol (Fabergé, 1974).

(+) Recommendable.
(–) Not recommendable (contamination source).

The best procedure to aid release of the carbon films is that developed by Münch (1964): Subsequent to the evaporation, the carbon film is scratched into small squares adapted to the size of the grids. A mixture of 0.5-1% HF and ~10% acetone in water is prepared in a flat plastic trough. After removal of the carbon layer on the glass edges by scraping, the glass slide with the carbon film on top is immersed at a shallow angle to the liquid level. Acetone reduces the surface tension, and the HF solution creeps between glass surface and carbon layer and dissolves the top layer of the glass so that the carbon layer is lifted off. The floating carbon layer is picked up on grids from below.

Bergeron and Pontefract (1970) recommend a modified procedure: They use a double compartment Perspex trough; one compartment is filled with 1% HF and 0.025% of the nonionic surfactant Tween 80, the other compartment with 0.025% Tween 80 alone. Each compartment is carefully filled so that there is a slight positive meniscus around the edges. A Teflon bar is placed across the borders to prevent mixing. The carbon film is floated off on the HF-containing compartment in the conventional way, and then the film is carefully shifted to the second compartment. This step can be accomplished by placing another Teflon bar on the other side of the floating film and simultaneously sliding both bars, with the film between them. In this way the HF can be washed from the carbon film. Moreover, the presence of the surfactant reduces the surface tension, thus aiding the partition. It must be mentioned, however, that, irrespective of its relatively low concentration, the surfactant itself contaminates the carbon film, although the authors claim that the films are exceptionally clean. However, proof under high resolution conditions has not yet been presented.

Pure carbon films show a rather weak adhesion to metal grids, occasionally causing the film to come off the grid during further manipulations. There is usually no problem if the carbon film is deposited on grids precoated with perforated supports (which is necessary with ultrathin carbon films). Alternatively, the adhesion may be improved by treating the grids with an adhesive solution (e.g., polyisobutylene in toluene).

Although glass targets are quite convenient for preparing carbon films for general purposes (thickness 5-10 nm), ultrathin (<3 nm) carbon films required for high resolution work should be evaporated onto exceptionally smooth surfaces such as mica (Hall, 1956; Spencer, 1959; Whiting and Ottensmeyer, 1972; Williams and Glaeser, 1972; Johansen, 1974). Floating off usually involves no difficulties if the freshly cleaved mica sheets are kept for several hours under dust-protective conditions in a humidity box prior to evaporation. The thin water layer formed on the hydrophilic crystal surface facilitates the ingress of water between the substrate and carbon film (Johansen, 1974). A simple, but useful, tool for picking up floating carbon films has been devised by Johansen (1974) (Fig. 1.26).

Carbon deposition onto a plastic film, which subsequently is dissolved, requires less manipulation, and relatively fragile carbon films are gently deposited on the

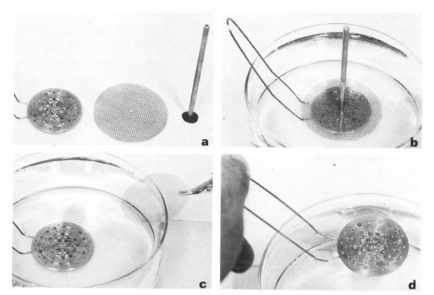

Fig. 1.26 Metal scoop for depositing carbon films on specimen grids. (a) Circular metal plate with several holes for accommodating the grids. Circular grooves facilitate handling with fine forceps. The plate is coated with silicon monoxide to make it hydrophilic. A stainless steel mesh secured by a rod screw keeps the grids in place during immersion of the whole assembly in water (b). After the screw is released, the mesh is removed with a pair of long pointed tweezers (c). The mica substrate, held by a clamp attached to a rack and a pinion is lowered towards the water until it just touches the surface at an angle of not more than 20°. It is kept in this position for several minutes to allow surface tension to aid the separation. By slowly turning the rack and the pinion, the mica is lowered into the water and the carbon film floated off. The mounting plate is carefully brought up underneath the ultrathin carbon film and removed from the water with the carbon film deposited on the grids (d). (*From Johansen, 1974.*)

grid, to which they firmly adhere owing to plastic remnants on the grid bars. It should be noted, however, that the carbon layer replicates the imperfect plastic surface, which causes variations of background scatter. Moreover, it is virtually impossible to remove completely the plastic target. In this respect, nitrocellulose is a better choice as a target material, because its very high mass loss upon electron irradiation automatically "cleans" the carbon of residual plastic material.

The procedure can be summarized as follows:

1. A number of grids are coated with plastic films by conventional methods. In addition to the usual plastic materials (nitrocellulose, polyvinylformal), methacrylates (Bradley, 1954; Schober and Prokes, 1955) and cellulose (Stolinski and Gross, 1969) have been tested.

2. The grids are placed in an evaporation unit, either mounted on specially designed holders (sandwiching brass plates with suitable perforations fitting the grid size and grooves to ease handling the grids with forceps) or simply on glass slides or on a gauze. The bell jar and hence the distance between the evaporating source and the target should be sufficiently great (~12 cm) to avoid thermic damage of the plastic target; otherwise, perforated or wrinkled carbon films may be obtained (Mahl and Möldner, 1972). As already stressed above in connection with thermosensitive plastic targets, a short pulse evaporation is advantageous.
3. To remove the plastic target, the grids (with carbon film upwards) are placed on a steel gauze, which is immersed in a suitable solvent in a Petri dish. The duration needed to remove the plastic depends upon the solubility of the plastic layer and its thickness. The grids are allowed to stay in the solvent overnight before they are dried on a stack of filter paper.

Theoretically, glycerol as used by Humphreys (1963) for preparing carbon films, and which is frequently used for preparing metal films (see page 47), should be an excellent target. It has an extremely smooth surface in common with mica, and the deposition on the evaporated layer on the grids is at least as easy and gentle as it is with plastic targets. It is difficult, however, to prevent the hygroscopic glycerol from "bubbling" upon evacuation; some possible solutions of this problem are discussed on page 49.

Fabergé (1970, 1974) proposed the use of a supercooled organic (dibenzylamine tartrate) glass as a target for carbon evaporation, which avoids some of the difficulties encountered in the glycerol method. When molten, grids are floated on it so that the target is automatically flush with the upper grid surface. After the carbon deposition, the target is dissolved in methanol. The drawback is that the dibenzylamine tartrate is not commercially available. Details of the preparation procedure are given by Fabergé (1974).

Silicon and Silicon Oxide Films

Silicon and silicon oxide (SiO, SiO_2) films show many of the desirable characteristic features of carbon films (chemical and thermal stability); nevertheless, they have only rarely been used as specimen supports, mainly because of difficulties in handling these rather brittle films. To our knowledge, the structure of these films, which might be distinctly different for various degrees of oxidation, has not yet been investigated systematically under high resolution conditions. There are, however, indications that silicon oxide films exhibit a pronounced inherent structure, especially upon electron irradiation (Reimer, 1967).

Silicon Films. Silicon films evaporated under "normal" (~$5 \cdot 10^{-6}$ mbar) vacuum conditions show a significant degree of oxidation (Pulker and Ritter, 1964). The oxygen content originates from oxidation during formation of the layer and from a superficial oxide layer formed when the film is exposed to air. For thin layers (<10 nm) it is assumed that complete oxidation takes place.

Silicon Monoxide Films. Hass and Kehler (1949) evaporated a stochiometric mixture of silicon and quartz at 1,200°C from a tantalum foil boat. At this temperature, which is lower than the volatilization temperature of the individual components, SiO escapes. Meanwhile, SiO formed by high-vacuum distillation is commercially available. Small pieces of SiO can easily be evaporated from tungsten wire baskets. Similar targets can be used for SiO-evaporation as described for carbon. Usually a slide is coated with a plastic film, upon which SiO is deposited. The film is cut into small squares and floated on a suitable solvent, which frees the SiO film from the plastic backing. The brittle films, which are difficult to see when floating on the solvent surface, are transferred on specimen grids or apertures in the usual way.

Silicon Dioxide. Silicon dioxide films can be obtained by vacuum evaporation of quartz at 2,230–2,590°C (Polivoda and Vinetskii, 1959). These authors directly evaporated the SiO_2 film on specimen grids precoated with nitrocellulose. The grids were transferred to a meshwork stage placed in a dish to which the solvent was added up to the level of the stage surface. Thus, the grids were in contact with the solvent only on the side carrying the plastic film. The dissolving process may last several hours (up to 24 hr). The advantage of this procedure is that the SiO_2-films are directly affixed to the grids, thus avoiding some delicate manipulations.

Polivoda and Vinetskii (1961) found that for the study of isolated membranes quartz films have some advantages over plastic supports. Erythrocyte ghosts deposited on quartz films appear to be completely spread out, while on nitrocellulose films the membrane shows multiple foldings. These authors also suggest that a more reliable representation of lipid depleted erythrocyte membranes can be obtained provided the extraction is carried out directly on the quartz film, thus avoiding artifacts which may arise during the desiccation of the preparation. Recently, Prestridge and Yates (1971) used aerosil (Cab-O-Sil), which essentially consists of SiO_2, as specimen supports for imaging clusters of rhodium atoms.

Metal and Metal Oxide Films

To overcome the practical thickness limits for plastic films set by their mechanical stability and radiation sensitivity, techniques to produce metal or metal oxide films of low mass thickness were developed during the forties. Beryllium ($Z = 4$, density 1.85), boron ($Z = 5$, density 2.54), and aluminum ($Z = 13$, density 2.7) have a reasonably low density and atomic number, yielding a scattering power which is approximately the same as for organic materials. Since very thin films of reasonable mechanical strength can be produced from these materials, they are obviously attractive as specimen supports.

Anodically Formed Aluminum Oxide Films. In 1941 Hass and Kehler as well as v. Ardenne and Friedrich-Freksa proposed the use of aluminum oxide films produced by anodizing aluminum foils. Walkenhorst (1947) developed an improved method to prepare aluminum oxide films of a defined thickness (below 10 nm). The starting material used was a 10-μm high purity aluminum foil, which was carefully degreased and then cleaned in dilute nitric acid. The electrolyte solution used was ammonium citrate (3%). With a carbon electrode as cathode the aluminum foil was anodized until the current was interrupted by the oxide layer. The final thickness of the Al_2O_3-layer proved to be proportional to the applied voltage (for small thicknesses). Walkenhorst obtained a thickness constant of 1.37 nm per volt, a value that was later substantiated by weighing, infrared reflection, and transmission measurements (Harris, 1955), and allowed the production of layers of predetermined thickness. To remove the oxide layer from the metallic aluminum, small pieces of the foil are floated on a 0.25% mercuric chloride (sublimate) solution. Thus, the aluminum is etched away, leaving the inert oxide layer on the surface of the solution. After deposition on a specimen grid, the aluminum oxide layer must be washed repeatedly with distilled water.

Varon *et al.* (1967) devised a modified procedure in which only selected areas of the metallic aluminum are removed by means of photo-lithography, leaving the remainder as a support grid. The solution used to etch away the aluminum, leaving the freely-suspended oxide over the holes, was two parts HCl to one part glycerol. The intimate contact between the aluminum oxide film and the grid structure, whose pattern can deliberately be chosen, increases the mechanical stability of the specimen support. It has been shown by electron diffraction that anodically formed aluminum oxide films are completely amorphous. They obviously exhibit, even at higher magnifications, only very little structure of their own.

Evaporated Aluminum Oxide and Beryllium Oxide Films. Hast (1947) proposed the evaporation of beryllium or aluminum onto the surface of plastic films and the subsequent dissolution of the latter with organic solvents. Cosslett (1948) described a technique for forming beryllium films directly on the specimen, i.e., to embed small specimens in the film. Hast (1948) devised quite an elegant alternative method for making thin metal films, which avoids the solution process (a complete removal of the plastic backing is extremely difficult to achieve, and the remainder creates a granular appearance similar to that of amorphous carbon). The metal film is prepared by evaporation onto a liquid surface. This method has an additional advantage in that liquids as targets provide extremely smooth surfaces.

Among the liquids tested (glycerol, vacuum oil, mercury), glycerol yielded the best results. Glycerol has a vapor pressure of less than 10^{-5} mbar at room temperature (10^{-9} mbar at $-20°C$; Gross, 1972) and is soluble in water. The glycerol, which is hygroscopic ought to be anhydrous, since even traces of water cause the formation of bubbles upon evacuation. This can be avoided by molecular distillation of the glycerol prior to use (Busch, 1954), by cooling the glycerol (Müller and Koller, 1972), or by using a specially designed holder which allows the grids to be suspended from glycerol films and coated from below so that bubbles can escape without disturbing the evaporated layer (Humphreys, 1963).

Müller et al. (1970, 1972) devised the following procedure for the preparation of very thin aluminum oxide films (Fig. 1.27): Grids covered with a perforated film and rendered hydrophilic by means of a glow discharge are placed onto a layer of glycerol spread over a sheet of filter paper and mounted on the cooling stage of a conventional freeze-etching apparatus. After cooling to $-50°$ to $-100°C$ to prevent bubbling of the glycerol during evacuation, the bell jar is usually evacuated until a vacuum of $1 \cdot 10^{-7}$ mbar is obtained. Pure oxygen is then injected through a needle valve and the pressure adjusted to $7 \cdot 10^{-4}$ mbar. Subsequently pure aluminum is evaporated from an electron gun at an angle of 90°

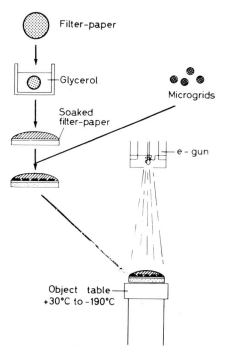

Fig. 1.27 Procedure for preparing aluminum oxide films. Aluminum is evaporated by electron bombardment under oxidizing conditions onto the surface of cold glycerol. (*From Müller and Koller, 1972.*)

onto the glycerol surface, where it condenses. It has been shown that only under such strongly oxidizing conditions (oxygen partial pressure $5 \cdot 10^{-4}$ -10^{-3} mbar) is it possible to obtain thin homogenous films (Gross, 1972). Films suitable for high resolution work should not exceed a thickness of 2.5–3.5 nm ($\approx 7 \cdot 10^{-7}$ gcm^{-2}), as estimated by means of a quartz coating monitor.

The cooling of the glycerol has no electron optically detectable effects on the structure of the evaporated aluminum oxide layers (Gross, 1972). It should, however, be mentioned that such cooling frequently leads to contamination of the glycerol surface, imparting to the otherwise smooth aluminum oxide films a carbon-like appearance (Vollenweider et al., 1973). This contamination can be reduced but not completely avoided by protecting liquid-nitrogen-cooled traps in the vicinity of the glycerol surface.

On completion of the evaporation the bell jar is flushed with dry nitrogen, while the glycerol and the cold trap are still cool. The aluminum oxide-coated preparation is placed on a pile of filter papers, which slowly absorbs the glycerol, thus gently depositing the film on the perforated carbon support. Finally, the grids are washed with distilled water and ethanol (Koller et al., 1973). Clean aluminum oxide films, which are essentially amorphous in contrast to polycrystalline films of pure aluminum, show in fact very little inherent structure (Fig. 1.28), making them suitable for high resolution work. It is difficult, however, to avoid contamination, which will strongly affect their smooth appearance. The main contamination sources are: (1) hydrocarbons condensing on the cooled glycerol surface during preparation; (2) residues from specimen mounting procedure; (3) carbonaceous deposits cracked from hydrocarbon sublimates by action of the electron beam.

The carbon deposits can be etched away from the inert aluminum oxide film at an increased oxygen partial pressure in the vicinity of the specimen during irradiation. The benefits of etching, however, are questionable, since a rapid destruction of the specimen is involved. Thus, rather than remove it by etching, it seems essential to avoid contamination, i.e., by keeping the glycerol at higher temperatures (see also Vollenweider et al., 1973), by keeping hydrocarbon partial pressures in the vicinity of the specimen as low as possible, and by improving the specimen preparation and mounting procedures.

Beryllium oxide films obtained under analogous conditions show crystal grains of ~5–10 nm in diameter (Vollenweider et al., 1973; see also Mihama et al., 1974). Although optical "noise" is extremely low within the crystalline domains, the coarse overall appearance makes BeO films unsuitable for most biological applications.

Aluminum-Beryllium Alloy Films. It has already been shown by Kaye (1949) that aluminum-beryllium alloy films evaporated onto smooth solid targets (glass or mica) are superior to films made from the respective pure metals, since the former are essentially amorphous. When Vollenweider et al. (1973) devised

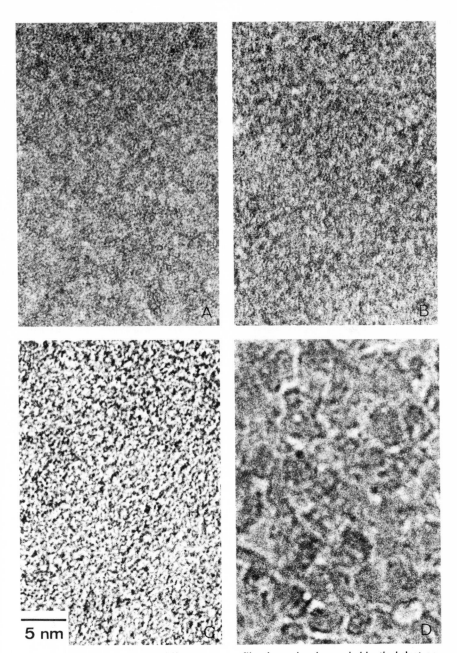

Fig. 1.28 Comparison of four different support films imaged under nearly identical electron-optical conditions: (a) aluminum-beryllium alloy film; (b) aluminum oxide film; (c) carbon film (condensed onto collodion); (d) beryllium film. (*From Vollenweider* et al., *1973.*)

a method for producing aluminum-beryllium alloy films by evaporation on a glycerol target (Fig. 1.29), their aims were (1) to avoid frequent contamination due to hydrocarbon condensation on the cooled glycerol surface and (2) to obtain films with a structure similar to that of aluminum oxide films but showing an improved wettability for aqueous solutions. Aluminum oxide films are, in fact, hydrophilic only immediately after preparation, and then gradually become hydrophobic, while the alloy films remain hydrophilic (Sogo et al., 1975).

An alloy containing 50% aluminum and 50% beryllium by weight is used for evaporation. Small pieces of wire ($\sim$15 mg of each metal) are squeezed together into a tungsten spiral mounted on an electron gun. For handling the poisonous beryllium, security precautions have to be followed. The metals are then melted under vacuum to give a uniform alloy. The evaporator (a freeze-etching apparatus can be used) is equipped with a shutter for controlling the amount of material deposited on the target, which if cooled may also serve as a cold trap in the immediate vicinity of the specimen. The hydrophilized perforated carbon supports are placed on a stainless steel mesh (mesh size $\sim$ 1 mm) wetted with redistilled glycerol, such that the perforated film is immersed in a glycerol layer. The mesh with the specimen grids mounted on top is placed on a holder with four small-size Teflon stoppers.

The whole assembly is then placed inside the evaporator which is pumped down to $\sim$5 $\cdot$ 10^{-6} mbar. The pressure is raised to 10^{-3} mbar by admitting oxygen through an adjustable needle valve. After the desired evaporation rate of 5 Hz/sec is reached (as monitored by a crystal oscillator), the shutter is opened and an aluminum-beryllium alloy layer corresponding to a frequency difference of $\sim$50 Hz on the quartz crystal coating monitor is condensed on the glycerol surface. To remove the glycerol and deposit the alloy film on the perforated support, the mesh with the grids is placed on a stack of filter paper and kept in an oven at 60°C overnight. The glycerol is less viscous at this temperature and thus easily absorbed by the filter paper. Finally, the grids are thoroughly washed with distilled water and baked in an oven for 30 to 45 min at 180°C.

As is shown in Fig. 1.28, the appearance of aluminum-beryllium alloy films is quite similar to that of aluminum oxide films. Both are suitable for high resolution work, although the improved hydrophilic character of the alloy film is probably an advantage. Ruben and Siegel (1978) used carbon-aluminum films, retaining good wettability over a long time.

Boron Film. Boron has only very rarely been used as a support film material. Cook and Kerecman (1962) evaporated boron, which has a melting point of $\sim$2,000°C, by means of an electron beam melting device. The films obtained show an inherent structure quite similar to that of carbon films; comparison of micrographs of both films indicate that with boron lower spatial frequencies are somewhat enhanced, i.e., the size of boron clusters appears to be somewhat larger.

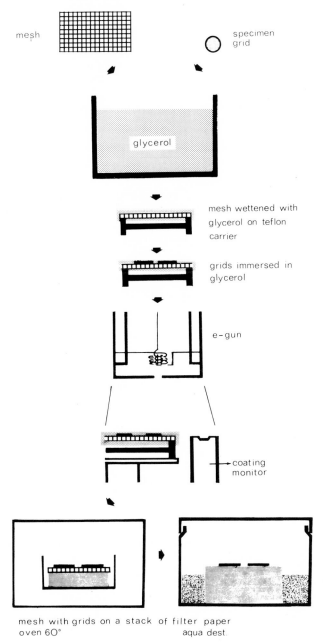

mesh

specimen grid

glycerol

mesh wettened with glycerol on teflon carrier

grids immersed in glycerol

e - gun

coating monitor

mesh with grids on a stack of filter paper
oven 60°   aqua dest.

Fig. 1.29 Procedure for making aluminum-beryllium alloy films according to Vollenweider *et al.* (1973). (For detailed description see text.)

Graphite Specimen Supports

Hexagonal graphite with its parallel and equidistant layered planes is a semiconductor showing satisfactory electrical and thermal conductivity along the a-axis, but not in the c-axis (Heidenreich et al., 1968). Graphite films suitable as specimen supports for electron microscopy are obtained by four basic techniques: (1) mechanical exfoliation, (2) oxidative exfoliation, (3) pyrolysis, and (4) thermal conversion of evaporated carbon films.

Mechanical Exfoliation. Mechanical stripping of natural or synthetic graphite, as originally suggested by Fernández-Morán (1960), is the easiest way to obtain graphite films. This technique involves stripping off surface layers by pseudoreplication or by means of Scotch tape, followed by dissolving of the plastic backbone and spreading on a liquid surface to select layers of adequate thickness by checking for the characteristic interference colors. Fernández-Morán (1960) reported that coherent single crystal films ~5–50 nm thick with several square millimeters of useful area can be obtained in this way.

Some modifications of this procedure for preparing graphite flakes were suggested by Beer and Highton (1962), who found it advantageous to use frozen ethanol as an adhesive for successive splittings, as it has been proved virtually impossible to completely remove the plastic adhesives. Hashimoto et al. (1973) prepared graphite films estimated to be only a few nanometers in thickness by cleaving natural graphite in an agate mortar filled with pure ethanol. Johansen (1975) prepared thin graphite flakes by ultrasonic exfoliation employing ethanol as an ultrasonic transfer medium. Relative dark field intensity measurements indicated that the exfoliated layers had a thickness between 2 and 5 nm; useful areas were up to 20 μm^2.

Heinemann (1974, personal communication) reported success using synthetic graphite exfoliated with the "gelatin method," similar to the approach of Palatnik (1970). A drop of 7–8% gelatin–water solution is applied to a graphite flake. During drying under an infrared lamp, a tangential stress develops along the cleavage plane of graphite, causing thin layers of graphite to peel off. After the gelatin is dissoved in water, the remaining graphite is picked up by a specimen grid.

Hines (1975) described a procedure to prepare large-area graphite films free of adhesive contamination. A crystal is glued to an adhesive-coated grid, and then the top graphite layers are cleaved away with the adhesive tape. The final graphite film areas in the centers of the grid openings do not come into contact with the adhesive during the preparation process. Graphite films obtained in this way are estimated to be a few tens of nanometers in thickness, and cover large areas in uniform thickness (Fig. 1.30). Recently, Iijima (1977) suggested combining mechanical exfoliation with adhesive tapes with oxidative etching in an electron beam. Small but extremely thin and noiseless areas are obtained (Fig. 1.31). For routine use, however, this method seems to be unsuitable.

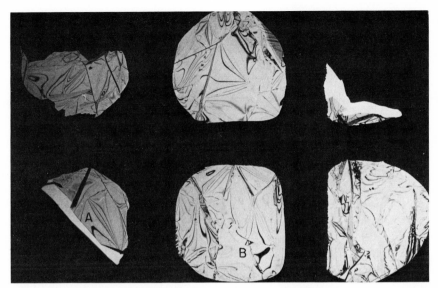

Fig. 1.30 Low magnification micrograph of a graphite crystal film, obtained by mechanical cleavage and mounted on a 200 mesh/in. specimen grid. (*From Hines, 1975.*)

Oxidative Exfoliation.   Dobelle and Beer (1968) have shown that graphite crystals can be cleaved into very thin layers by drastic oxidation. One gram of powdered graphite flakes (325 mesh) is thoroughly mixed with 0.5 gm sodium nitrate and 23 ml concentrated sulfuric acid. This dispersion is allowed to cool in an ice bath. Then 3 gm of potassium permanganate is added slowly with stirring, and the temperature is maintained below $20°C$ to avoid explosion hazards. After complete solubilization of the potassium permanganate, the temperature of the reaction mixture is increased to $35°C$ while the dispersion is stirred on a magnetic stirrer. An equal volume of distilled water is slowly added, and the temperature is maintained for 10–15 min at $98 ± 5°C$. After cooling, 10 ml of 3% hydrogen peroxide is added.

The dispersion is centrifuged at 500 rpm for 1 min, and the precipitate of larger, incompletely oxidized flakes is discarded. The supernatant is centrifuged again and the precipitate washed by several resuspension and sedimentation steps. Finally, the suspension is diluted in 10 volumes of ethanol and allowed to settle for 48 hr in a refrigerator. The yellow supernatant is suitable for spraying on perforated carbon films. The ethanol obviously wets the carbon films and aids stretching of the graphite oxide flakes across the holes.

Alternatively, the graphite oxide, which is composed of alternate layers of carbon and oxygen, can be reduced to pure graphite before it is used as a specimen support. For this purpose, 3 ml of the yellow-colored stock suspension is mixed in a test tube with an equal volume of 0.02 N sodium hydroxide. The suspension is allowed to stand for 10 min to permit the graphite oxide particles

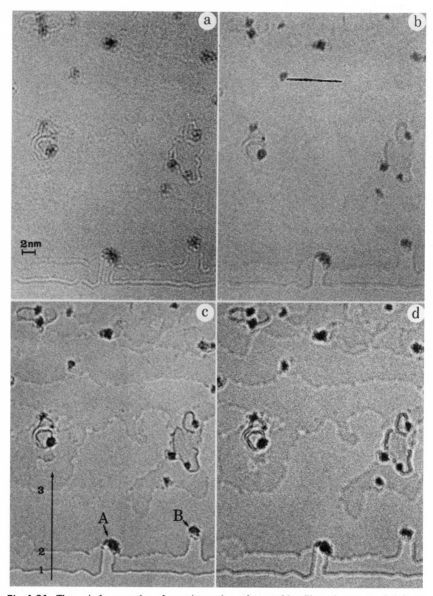

Fig. 1.31 Through focus series of an edge region of a graphite film. Amounts of defocus are (a) +42.5 nm, (b) −25 nm (c) −92.5 nm, and (d) −160 nm. Note the exceptionally smooth appearance of the graphite film through the whole focus series. The edge of the film is near the bottom of the images; contour lines running nearly horizontally are surface steps with a few atomic layers of carbon atoms. Dark particles are metallic tungsten, left on the film from etching in an electron beam in the presence of WO_3. (*Courtesy of Dr. Iijima, 1977.*)

to cleave into thin lamellae, a process that seems to be accelerated by ultrasonication. A few drops of hydrazine hydrate are added to the test tube and the contents heated to boiling for 2 min. During the reduction of the graphite oxide, the suspension becomes black. After cooling, the graphite lamellae are mounted on grids in the conventional way.

Pyrolysis. Karu and Beer (1966) developed a method for producing thin graphite films by pyrolysis of methane on hot single crystals of nickel. Strips of 0.0025-inch-thick highly polished reagent grade nickel foils were used as film substrates. These strips were successively cleaned with a detergent (acetone), 0.01 N hydrochloric acid, and quartz-distilled water. Strips of the foil were heated to $900°C$ for 8–10 hr in an oil-free ultrahigh-vacuum system (pressure below 10^{-7} mbar) by passing an appropriate current. The nickel substrate has a twofold function: it acts as a catalyst for the decomposition of methane, and the crystalline domains with a size range of a few microns formed during the preheating are the bases for epitaxial growth of the graphite.

Following the above pretreatment, the nickel strips are again heated to a temperature of $1,100°C$, and methane gas is admitted to the chamber through a bleeder valve to a pressure of $5 \cdot 10^{-3}$ mbar. The pyrolytic reaction is allowed to continue for 1 to 5 min, during which a dark layer grows on the nickel surface. Thickness and crystallinity of the graphite layer obviously depend on the pretreatment of the nickel strip, the partial pressure of methane (methane pressures below $1 \cdot 10^{-3}$ mbar and above $1 \cdot 10^{-2}$ mbar fail to yield graphitic films), substrate temperature, and the duration of the pyrolytic reaction. The graphite films formed by pyrolysis are removed from the nickel strips by dissolving the nickel in concentrated HCl or 5 N nitric acid. The graphite is transferred to distilled water and washed several times before being mounted on specimen grids.

Riddle and Siegel (1971) recommend using carbon monoxide instead of methane for the pyrolytic production of graphite films. The nickel serves as a catalyst for the decomposition of CO through the reaction

$$2\,CO \xrightarrow[\text{Ni-catalyst}]{} C + CO_2$$

Films obtained by the pyrolytic method, though relatively thick, show extremely smooth islands, i.e., fields of low stochastic intensity variations or structural noise in the submicron range embedded in areas with the characteristic granularity of amorphous carbon films.

It remains to be explored whether thinner graphite films with larger monocrystalline areas can be obtained by pyrolysis of other than gaseous materials. Edstrom and Lewis (1969) investigated the graphitization of several polynuclear aromatics after heating to $3,000°C$. Their results show that the nature of the final graphite is strongly dependent upon the chemical structure of the organic precursor. The majority of the aromatic hydrocarbons, especially planar condensed materials, yield well-ordered graphites. Certain advantages are also to be expected from the graphitization of aromatic polymer films, which, owing to preferential orientations of the chains, might yield graphites with well-ordered arrangements of the crystallites. An alternative approach, which remains to be explored, is the pyrolysis of glycerol, which is catalyzed in the interlayer position of natural monocrystalline vermiculite to yield extremely thin graphite layers (Walker, 1950).

Thermal Conversion of Thin Carbon Films. Jenkins *et al.* (1962) and Turnbull and Williamson (1963) have shown that carbon films prepared by evaporation in the normal way can be graphitized by high temperature heat treatment. White *et al.* (1971) adopted this method for the production of specimen supports and optimized it with respect to the crystallite size. To withstand the high temperatures necessary for the graphitization, specimen supports made of graphite must be used. These supports are produced by cleaving thin slices from graphitized carbon rods with the crystal *c*-axis running along the cylinder axis, and drilling holes into the graphite disk (Koller *et al.*, 1971). The supports are covered with a perforated carbon film, which again is covered with a standard carbon film produced according to the method of Bradley (1954).

The heat treatment is carried out under an argon atmosphere in a furnace permitting fast heat-up times at temperatures between 2,500° and 3,000°C. Temperatures of ~2,700°C yield crystallites of a few tenths of a micron in diameter, the size of the crystallites increasing with temperature and film thickness. White *et al.* (1971) recommend shielding the grids in the furnace with highly oriented stress-annealed graphite and placing graphite wool in their vicinity to minimize both oxidative etching and thickening through additional deposition. The effect of the ambient gas phase on the graphitization is not yet quite understood (Presland and White, 1969). While Noda and Inagaki (1964) claim that the presence of oxygen accelerates graphitization by preferential elimination of carbon atoms from criss-cross linkages, White *et al.* (1971) report that even small traces of oxygen lead to a loss of films. The relatively small size of monocrystalline domains and the abundant appearance of Moiré fringes in these films are still unsatisfactory.

In the present state of the art mechanical exfoliation seems to be the least tedious and most reliable method for producing graphite films of adequate thickness. Without any doubt the graphite films are less grainy than amorphous carbon films, owing to the narrowed spread of local thickness variations and a shift to less disturbing long-period variations (Fig. 1.32).

Molybdenite Specimen Supports

Molybdenite (MoS_2) has a layer structure in which the central molybdenum atoms are sandwiched between two sulfur layers. Six sulfur atoms are arranged around one molybdenum atom like a trigonal prism. The weak interactions

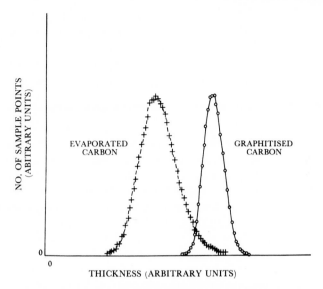

Fig. 1.32 Abundance distribution of mass-thickness values as measured by relative amount of elastically scattered electrons in a scanning beam. Although mean thickness of graphitized carbon (*right curve*) is higher, the spread (≈ standard deviation) is much smaller than for evaporated (amorphous) carbon. (*From Wiggins and Beer, 1972.*)

between the sulfur atoms ease the separation of layers by shear forces. Molybdenite has a density of 4.71; the electric conductivity is extremely low. Mechanically exfoliated molybdenite flakes have been used by Mihama *et al.* (1973) and Uyeda (1973, 1974) as specimen supports for extremely fine gold crystallites. The surface of molybdenite flakes was cleaned from contamination by heating in 10-Torr oxygen at 500°C for a few minutes, leading to a sublimation of molybdenum oxide (Mihama *et al.*, 1973). Noise appears to be quite low in cleaned areas; the remaining "islands," however, exhibit a noise pattern similar to that shown by amorphous carbon or metal films.

Layer Silicate Specimen Supports

The basic structure of layer silicates consists of an aluminum octahedral sheet sandwiched between two silicate tetrahedral sheets (Fig. 1.33). Some of the quadrivalent silicon ions are replaced by trivalent aluminum ions in the tetrahedral sheet, and ions of lower valency (magnesium, calcium) can replace aluminum in the octahedral sheet. The resulting negative charge of the silicate layers is neutralized by the presence of cations between the layers (interlayer position). In the case of micas, the interlayer ions are usually K^+, whereas in the vermiculites they are usually Mg^{2+} or Ca^{2+} and are associated with water molecules. Besides features in common with other crystalline supports (low stochastic variations of

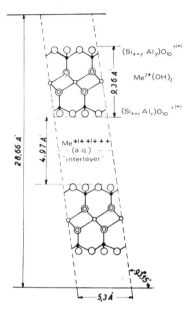

Fig. 1.33 Schematic representation of layer silicate structure.

density, atomically flat surfaces) layer silicate supports have the following advantages: (1) simple exfoliation procedures are available; (2) radiation sensitivity is low; (3) surface properties can be modified by ion exchange.

Mica. Fernández-Morán (1960) proposed the use of mechanically exfoliated mica, and Heinemann (1970) showed that useful mica films can be prepared either by direct mechanical stripping or by adhesive-tape cleavage. As a general drawback, all of these techniques are rather tedious, and the layers obtained are relatively thick. Layers thinned by means of ion beam etching (Poppa *et al.*, 1971) show rather rough surfaces, which makes it necessary to apply the ion bombardment from one side only.

Vermiculite. For vermiculite, which has essentially the same layer structure as mica, a number of relatively easy and reliable exfoliation techniques exist. These can roughly be divided into (1) chemical exfoliation and (2) thermic exfoliation (Table 1.7). Manley *et al.* (1971) used a chemical exfoliation technique as originally developed by Weiss *et al.* (1956) and Walker (1960); i.e., they introduced organic ions like butylammonium into the interlayer position. Single crystals of vermiculite containing such organic interlayers swell macroscopically when placed in water. Swelling is anisotropic and only occurs in a direction perpendicular to the plane of the silicate layers. When shearing forces are applied to swollen crystals (e.g., ultrasonication), the weak interlayer forces give way and the

Table 1.7   Methods of Vermiculite Exfoliation

	"Chemical" exfoliation	Thermal exfoliation	H_2O_2 exfoliation
Basic principle	Introduction of long chain organic ions into the interlayer position thus weakening the interlayer forces. Dissociation of layers by means of shearing forces	"Explosive" release of water from the interlayer position upon heating to $800°C$	Evolution of O_2 in the interlayer due to decomposition of H_2O_2
Time required for complete preparation procedure	~4 weeks	~2 hr	~2 days
Crystal structure preserved (+) or not (–)	+	+	+
Ion exchange capacity preserved (+) or not (–)	+	±	+
Drawbacks	Residual organic ions contaminate the crystal layers	Partial loss of ion-exchange capacity	?

From Baumeister and Hahn (1976b).

individual layers disperse. This technique is, however, rather time consuming, since the interstratification with organic ions may take several weeks, depending on the crystal size. Moreover, it is virtually impossible to remove the organic completely after exfoliation, giving the vermiculite a carbon-like appearance upon irradiation.

Baumeister and Hahn (1974a) exfoliated vermiculite by a simple heat treatment. The name "vermiculite" is derived from the Latin *vermiculus* ('little worm') and is due to the well-known phenomenon that elongate, curved, and twisted columns are formed when the crystals are suddenly heated to a high temperature (Walker, 1961). Owing to the explosive release of water upon sudden heating, the vermiculite crystals expand normally to the basal cleavage. Vermiculites with high water content (>12%) are suitable for thermic exfoliation. Thermo-gravimetric analysis and differential thermal analysis indicate that dehydration of vermiculite is almost complete at $900°C$.

The exfoliation procedure is as follow: A small quantity (~10 gm) of the plate-shaped crystals (average diameter 1 mm, thickness 0.05–0.2 mm) is heated in a muffle furnace to $800°C$ for 30 sec. The expanded crystals are suspended in distilled water. It takes ~12 hr for vermiculite columns to become disintegrated by means of a magnetic stirrer. The suspension is allowed to settle for 15 min, and afterwards the sediment of incompletely exfoliated vermiculite is discarded. The supernatant is treated with an ultrasonic disintegrator for several minutes. After centrifugation at ~10^4 g for 20-30 min, the sediment is discarded again and the supernatant diluted to a final concentration of 40-60 μg vermiculite/ml. A drop of this suspension is placed on a hydrophilized perforated gold film (hole diameter~0.5 μm). A satisfactory deposition and a reasonable density are obtained when the drop is allowed to remain on the perforated support for 10 min, and the excess suspension is removed. Suitably thin (2-4 nm) single crystal flakes may be of the order of tens of microns in lateral extent.

Baumeister and Hahn (1976a, 1976b) have recently shown that an even better yield of extremely thin crystals is obtained by exfoliation with hydrogen peroxide. The decomposition of H_2O_2 between the silicate layers, which may be catalyzed by the inorganic interlayer cations, and the evolution of oxygen yield quite satisfactory separations. The following procedure is recommended: A quantity of vermiculite (~10 gm) is placed in excess hydrogen peroxide (~30%). The dispersion is kept at 4°C for 24-48 hr. The vermiculite flakes are collected on a filter paper and placed in an oven heated to 100°C, which causes the crystals to begin to expand immediately. The bulk volume may increase by a factor of 30. The expanded crystals are dispersed in distilled water and treated with an ultrasonic disintegrator for several minutes. The further procedure is as described above for thermic exfoliation.

It is important that both methods, thermic and H_2O_2 exfoliation, avoid any organic contamination of the crystal layers. The yield of extremely thin layers is considerably better with the latter method. Moreover, in contrast to thermic exfoliation, the full ion exchange capacity is retained, which is of considerable importance, since it facilitates the modification of vermiculite supports so that they exhibit selective adsorption properties for various biomolecules (Baumeister and Hahn, 1974b) (see also page 69).

Vermiculite single crystal specimen supports show an extraordinarily smooth background structure. Especially after suppression of the lattice periods (Fig. 1.34), specimen details, which because of their faint contrasts are obscured by conventional amorphous support films, become visible. This holds true, at least to some extent, even after strong radiation damage has randomized the crystal. The appearance of strongly irradiated vermiculite resembles more or less that of aluminum films, showing lower optical noise as compared to carbon films. It should be pointed out, however, that vermiculite tolerates rather high electron doses without noticeable loss of crystallinity (see also page 80). Because of their perfectly smooth surface, vermiculite supports are well suited for negative staining. While with amorphous supports surface imperfections are exaggerated by negative stains and thus obscure the finest structural details, at least a very precise contour definition (Fig. 1.35) can be achieved with crystalline supports (Hahn and Baumeister, 1974; Johansen, 1975).

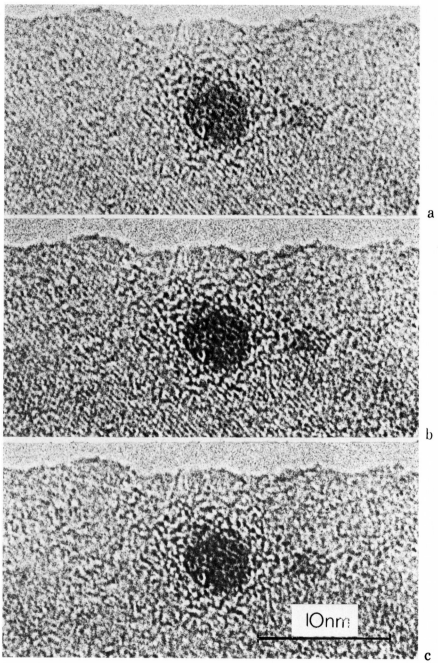

Fig. 1.34 Unstained ferritin molecule on a vermiculite single crystal support. (a) Print from original micrograph; (b) print from optical reconstruction of (a), without spatial frequency filter; (c) like (b) but vermiculite lattice pattern suppressed by spatial filter.

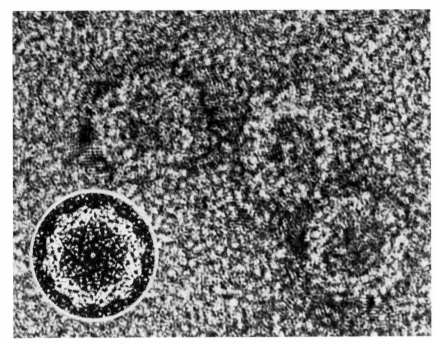

Fig. 1.35 Ferritin molecules, negatively stained with uranyl formate on a vermiculite support. The molecular contour is defined quite precisely owing to the smooth support surface. Inset: Ferritin after rotational averaging ($r = 6$).

Self-Supporting Specimens

It seems, at first glance, an attractive suggestion to circumvent the specimen support problem, especially in high resolution work, by freely suspending the specimen in the objective plane. This can, in principle, be accomplished by mounting the specimen over a hole of a perforated film or by adsorbing it onto a fibrous material, which itself spans over a hole. Some success has been achieved with chrysotile asbestos, which can be made to form a mesh across holes by ultrasonically dispersing some fibers in distilled water and allowing a drop of the suspension to dry on a grid or diaphragm (Fryer *et al.*, 1970). Specimens are deposited on the net by spraying techniques.

Fernández-Morán (1960) adsorbed protein molecules on asbestos fibers to be imaged without the interfering background structure of conventional supporting foils. That no significant improvement of image quality was achieved might have been due to charging hazards, which are a problem for isolated specimens. DNA as well as model membranes can be mounted on perforated films as self-supporting specimens. On irradiation, however, they tend to shrink or be rearranged so heavily that it is usually necessary to apply protective layers to

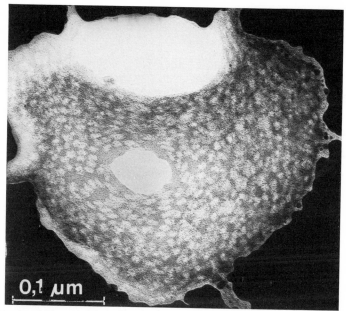

Fig. 1.36 Catalase molecules embedded in a thin self-supporting film or negative stain (uranyl formate).   Film and embedded particles are commonly deformed after rupture upon irradiation.

impede at least gross distortions (Baumeister and Hahn, 1973). The situation is essentially the same with thin, self-supporting layers of negative stains in which proteins or viruses are embedded.   Stain shrinkage upon irradiation (Unwin, 1974) usually entails a serious deformation of the particles (Fig. 1.36).   Moreover, with self-supporting specimens one must consider the effect of thermally excited vibrations, which might blur high resolution images.

Consider, for example, a DNA strand unsupported for the length of B base-pairs, spaced at $d = 3.4$ Å, and connected by four π-bonds over an effective area totaling $a = 20$ Å^2.   These bonds are considered to have the same shearing modulus as graphite ($G = 2.3 \cdot 10^9$ Nm^{-2}), which has similar bonds and distances between its layers. If we allow a lateral "blurring" rms-amplitude = 0.3 Å, only

$$B = \frac{6aGu^2}{kTd}$$

(Schnabl, 1971)

base-pairs spanning a hole can be observed without noticeable blur (k = Boltzmann's constant).   For the temperature $T = 300$ K we have $B = 1.8$, which shows this preparation to be completely impractical.   If, however, there were any hope of maintaining such a preparation at 5 K, during electron microscopic observation 106 base-pairs could be imaged without blurring due to thermal vibrations.

GENERAL SPECIMEN SUPPORT PROBLEMS

Surface Conditioning

"Surface conditioning" is a collective term for a multitude of surface modifications covering all aspects of wetting and adsorption. It is certainly beyond the scope of this chapter to give a complete survey of them or to provide a rationale for the concerted interplay of specimens and adsorbent surfaces. Surfaces nevertheless are of utmost importance because they govern final appearance, especially of isolated macromolecular specimens. This has been shown extensively for nucleic acids, whose configuration may drastically change upon even minor differences in surface properties of their underlying supports (Highton and Whitfield, 1974; Sogo *et al.*, 1975).

Wetting. Most specimen supports are usually regarded as hydrophobic, i.e., aqueous solutions fail to spread on their surface. Wetting of solid surfaces is a very complex phenomenon and difficult to assess quantitatively (contact angle measurements). Even traces of contamination and aging effects may drastically alter the hydrophobicity of a given support. Freshly evaporated carbon films are, for instance, hydrophobic, but upon aging, an increased oxygen content, which varies with conditions of atmosphere and temperature, leads to increased hydrophilicity. The increased polarity seems to be associated with the increased number of oxygen functional groups (Snoeyink and Weber, 1971). Aluminum oxide films seem to behave adversely: they become gradually hydrophobic upon aging (Sogo *et al.*, 1975).

Mahl and Möldner (1972) have shown that the degree of hydrophobicity of carbon films depends to a large extent upon the vacuum conditions during evaporation: with unbaffled oil-diffusion pumps extremely hydrophobic carbon films are obtained. The contaminating surface layer can be removed by means of organic solvents or a glow discharge in air at a low pressure. Carbon layers cleaned in this way show improved hydrophilicity.

Where a suspension containing small particulate specimens fails to spread spontaneously on a given support, a number of wetting aids can be employed. Basically, it is possible to modify the specimen support surface; for aqueous suspensions, this generally means increasing the number of surface polar groups. Alternatively, the surface tension of the liquid (specimen containing) phase can be lowered by means of suitable additives. The first approach includes methods as different as irradiation (ion-bombardment, UV) (Fig. 1.38) and surface coating with amphipatic molecules. Principally, preference should be given to the "irradiation methods" which create surface radicals and polar groups. A simple apparatus suitable for routine glow-discharge is shown in Fig. 1.37 (for details see legend). Similar designs are commercially available as part of a normal coating unit. Treating the specimen support with a detergent solution

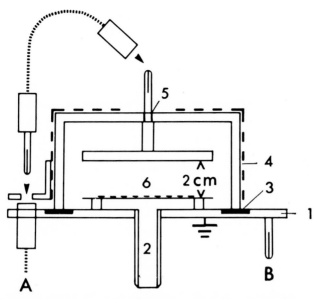

Fig. 1.37 Routine apparatus for glow-discharge. The bell is a 9-cm-wide and 5-cm-high crystallizing dish. A hole is pierced for the electric passage for one electrode and sealed by an epoxy resin. *1*, base plate; *2*, connection to pump, air and vapor reservoir; *3*, rubber gasket; *4*, protection against explosion; *5*, hole in the crystallizing dish for electric passage; *6*, specimen grids. (*From Dubochet* et al., *1971*.)

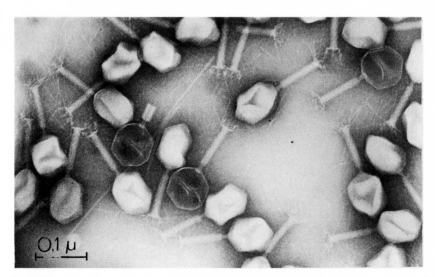

Fig. 1.38 Negatively stained T_2-bacteriophages on a polystyrene film hydrophilized by UV irradiation.

prior to use should be restricted to low resolution work, since it compromises most of the advantages of "low noise" supports.

The same restriction on detergents applies to the second approach to the spreading problem, i.e., adding a surface-active compound to the specimen-containing suspension. Generally, a satisfactory wetting agent, besides its ability to lower surface tension, should meet the following criteria:

1. It should be small enough not to be visualized, i.e., it should be of low molecular weight.
2. It should not aggregate (micelles, crystals) upon drying.
3. It should neither interact with the specimen to form heteromorphous complexes nor disintegrate it.
4. It should be compatible with negative stains if they are employed.

Various additives have so far been used: serum albumin (Valentine, 1961), sucrose (Misra and Ganguly, 1968), glycerol (Horne, 1965), and octadecanol (Gordon, 1972). Gregory and Pirie (1972, 1973), who thoroughly tested a number of conventional and potential compounds, found the dodecapeptide bacitracin to yield the most satisfactory results. Bacitracin, which is commercially available, has a relatively low molecular weight (1411), is effective in relatively low concentrations ($\sim$50 μg/ml), does not aggregate upon drying, and is biologically acceptable even with lipid-containing specimens. Further search for improved wetting agents should also consider compounds which are volatile under either normal or reduced pressure and thus do not interfere to any extent with the specimen structure.

Selective Adsorption. The somewhat ambiguous term selective adsorption here refers only to electrostatic interactions between specimens and charged specimen support surfaces. Sogo *et al.* (1975) have recently shown for DNA that the adsorption properties of different support films are well correlated with their surface charge and that the configuration of flexible macromolecules (such as DNA) depends to a great extent upon the kind of support film used. To compare the results of DNA adsorption with the sign of the surface charge of the support films, Sogo *et al.* (1975) measured the surface potential of various support films by means of a vibrating capacitor electrometer method and calculated the corresponding surface charge densities. The results are shown in Table 1.8.

Native DNA which behaves as an anion in low-ionic-strength buffer binds readily to positively charged films, while DNA intercalated with ethidium bromide, which is cationic under these conditions, perferentially binds to negatively charged films. As is expected, the charge of the support films should be opposite to the one on the macromolecules. This is also corroborated by the experiments of Gordon and Kleinschmidt (1970), who found that DNA is adsorbed from neutral aqueous solutions onto sheets of mica if the potassium ions on the cleavage surface (interlayer-cation) are replaced by aluminum ions

Table 1.8   Surface Potential and Charge Density of Different Supports

Material	Copper (reference)	Carbon	Aluminum	Aluminum-beryllium
Surface potential (V)	0	-0.1 (± 0.05)	$+0.5$ (± 0.2)	$+1.2$ (± 0.2)
Surface charge density (Cb/cm^2)	0	-3.10^{-8}	$+7.10^{-8}$	$+2.10^{-7}$

From Sogo et al., 1975.

through ion exchange. This means that mica, which in its natural potassium form is weakly negatively charged, is nonadsorptive for DNA, but aquires adsorptive properties when the potassium ions are replaced by the polyvalent aluminum ion. Al^{+++} can neutralize a negatively charged phosphate group of DNA while it still bears a positive charge to be retained on the mica. It is noteworthy in this context that the intercalated (cationic) DNA is readily adsorbed on native mica (Koller et al., 1974).

It is attractive to have specimen supports whose surface charge can be modified to accommodate particular adsorptive requirements. A simple but rather non-specific modification of surface charge patterns may be achieved by irradiating the support film; the charges generated depend on the material irradiated and on the environmental conditions. Dubochet et al. (1971) have, for example, shown that basic radicals are deposited on carbon films upon exposure to a glow discharge in 0.5 mbar amylamine vapor.

Quite an attractive way of manipulating surface charges and thus adsorption properties is to deposit a protein monolayer on a thin carbon film as described by Abermann and Salpeter (1974). Particles can be adsorbed from aqueous solutions on these rather smooth and homogeneous surfaces. Zingsheim (1977) has shown for ferritin that the particle density greatly depends upon the nature of the protein layer covering the carbon layer and on the pH of the aqueous solution (Fig. 1.39). This finding can be explained in terms of charge sign and density on the protein surface layer and on the proteins in solution; both depend on the respective isoelectric points of the proteins and on the pH of the electrolyte solution. It is interesting to note that the particle density may be manipulated by choosing the proper experimental conditions and that there may be a more than twentyfold difference between different surface layers under otherwise identical experimental conditions.

In a quite elegant way the surface charge can be modified with layer silicates. Here the negative charge of the silicate layers is compensated by exchangeable interlayer cations. An equimolecular but nonequivalent replacement of the resident interlayer cations modifies the surface charge. Exposure to solutions containing mono- or polyvalent cations yields crystals, which in the case of vermiculite with its very high exchange capacity (Weiss, 1958) may be homoionic in the interlayer position. The localized charges seem to be stochastically distributed on the silicate layer surface (Schön and Weiss, 1974) with a density of 0.55–0.66 e_0

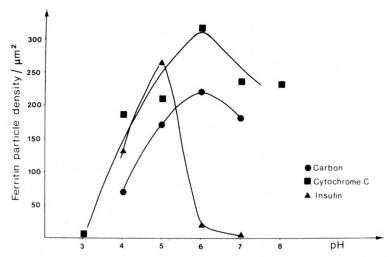

Fig. 1.39   Particle density of ferritin adsorbed to various surfaces as a function of pH. Ferritin concentration in 0.01 M Na acetate solutions: 10 mg/ml. Adsorption time: 30'. (*From Zingsheim, 1976*.)

per $(Si_{4-y} Al_y) O_{10}^{y(-)}$ unit.   In analogy to the micas, natural heteroionic vermiculite poorly adsorbs DNA, while positively charged Al-vermiculite shows excellent DNA adsorption properties.   It has also been shown that the charge pattern of vermiculite has important implications for the deposition and stability of model membranes (Baumeister and Hahn, 1974b, 1976a).

Since further progress in "molecular microscopy" depends on how far structural redundancies can be utilized, it is obviously attractive to bring biomolecules into two-dimensional periodic arrays or at least to achieve some kind of positional order.   Speciman supports with spatially defined charge patterns could serve as templates for a proper adsorption.   Protein layers provide a good starting point in this respect.   Zingsheim (1976) has recently shown that their stochastic charge pattern can be modified by high resolution electron beam writing.   Using the 0.5-nm electron spot of the STEM, parts of the protein layers are modified owing to the combined effect of radiation damage and the deposition of the contamination layers.   If such an experiment is carried out with an insulin-covered carbon support, adsorption takes place predominantly on the carbonized lines, i.e., the absorption of ferritin follows the two-dimensional charge pattern on the surface (Fig. 1.40).   Though this is only a first step towards the realization to two-dimensional periodic arrays, it is certainly a promising attempt.   Alternatively, one could design spatially defined charge patterns by monolayer techniques.   Especially charged homopolypeptides which self-assemble at liquid–air interfaces in well-defined configurations and with a considerable degree of long range order could provide suitable templates (Baumeister *et al.*, 1976).

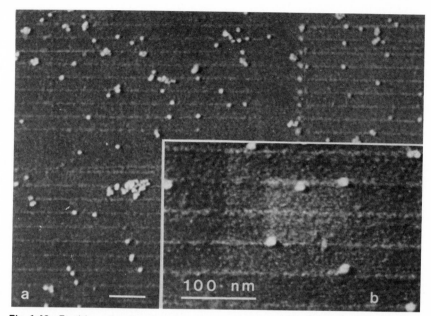

Fig. 1.40 Ferritin molecules adsorbed to insulin films, onto which 3-4-nm-wide lines were written by electron irradiation. Adsorption conditions similar to Fig. 1.39, at pH 6. Shadowed with tantalum-tungsten. Micrograph taken in the STEM (elastic) dark field mode, Shadows appear dark. Background granularity is caused mainly by the insulin layer, which is not as smooth as a pure carbon film. Note the preferential adsorption of ferritin in the vicinity of the irradiated lines. (*From Zingsheim, 1976.*)

Sag and Radial Slope

Under their own and the preparations' weight, thin foils sag and thus show a radially varying slope. Actually the sag itself is of minor importance because its amount is comparable to usual axial deviations of electron microscope specimen stages, which after larger horizontal displacements demand new focus setting anyway.

The slope, however, will lead to a defocus-gradient on a single image. If we assume the final image to be composed of $\sim 6,000^2$ picture elements of 10 μm diameter each, the maximum allowable slope $z'_{s\ max}$ can be estimated by $6,000\ \delta_0\ z'_{s\ max} \approx \delta_f$. Relating the depth of focus δ_f to resolution δ_0 and aperture α of the objective by $\delta_f = 2\delta_0/\alpha$, we obtain

$$z'_{s\ max} = \frac{1}{3,000\ \alpha}$$

and for a realistic value of $\alpha = 10^{-2}$ finally

$$z'_{s\ max}\ (10^{-2}) = \frac{1}{30}$$

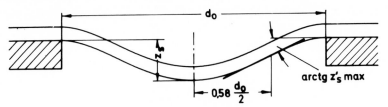

Fig. 1.41 Axial section trough support film, sagging under its own weight. The maximum radial slope angle arctg z'_s max is reached at 58% of its radius $r_0 = d_0/2$.

For a circular membrane of density ρ and Poisson number ν fixed over an aperture of diameter $d_0 = 2\,r_0$, the minimum thickness is given by

$$t_{\min} (z'_{s\,\max}) = \sqrt{\frac{\rho g r_0^3\,(1 - \nu^2)}{2\sqrt{3}\,Ez'_{s\,\max}}}$$

where $g = 9.81$ msec^{-2} (Fig. 1.41).

The maximum slope z'_s max is always reached at a distance of 0.577 r_0 from the center, while at the very center and at the periphery, we have $z'_s = 0$. Figure 1.42 shows $t_{\min}\left(\frac{1}{30}\right)$ for polystyrene and gold coated (perforated) plastic films; the weight of possible specimens on the supports is neglected. From the numer-

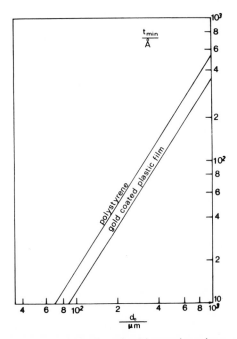

Fig. 1.42 Minimal thickness t_0 of plastic and gold coated specimen support films to keep radial slope angle below arctg (1/30) plotted over diameter $d_0 = 2r_0$.

ical situation it is clear that the required thickness of a support becomes limiting for optical performance only with bore diameters in the millimeter range as used for the unobstructed mounting of large specimen areas (see page 33). Just for completeness we finally state the sag $z_{s\ max}$ at the center:

$$z_{s\ max} = \frac{3g\rho r_0^4\ (1-\nu^2)}{16Et^2}$$

This sag stays within one micron even for supports spanning over large open areas ($\sim$1 mm) and hence does not cause a significant increase in magnification relative to the outer regions of the specimen.

Axial Oscillations

Although normally the amplitudes of Brownian motion in solids are far below the resolving power of electron microscopes, there are certain cases where thermally excited oscillations can really blur high power micrographs. The very thin specimen support films especially are capable of axial (out of plane) oscillations, whose amplitudes have to be kept within a fraction of the depth of focus to avoid noticeable loss of contrast in the finest image details. Here these limits are discussed for ultrathin supports suitable for high resolution work, mounted on perforated films, as well as for self-supporting liquid-condensed specimens (model and native membranes).

Because the smallest values of the spherical aberration constant C_s that can be realized in magnetic objective lenses (Riecke, 1972) vary inversely with the electron wavelength λ by $C_s\lambda \approx 1.9 \cdot 10^5$ Å^2, the depth of focus $\delta_f = \sqrt{C_s\lambda}/2$ is at least 300 Å for all electron microscopes operating at maximum resolving power ($\hat{=}$ optimum aperture). Allowing rms axial amplitudes $z_0 = 0.12\,\delta_f = 36$ Å, just causing a 10% contrast loss of very small ("point") objects, and using expressions given by Schnabl (1972), we obtain a relation for the minimum thickness t_0 of a specimen support with elastic modulus E for a given diameter d_h of the hole covered:

$$t_0 = \sqrt[3]{\frac{d_h^2 kT}{68\,E z_0^2}}$$

Figure 1.43 shows values of t_0 for $T = 300$ K and 5 K, (polystyrene). The considerable increase of E for low temperatures and high radiation doses was not taken into account because it cannot be generalized quantitatively, but it may widen the margin for t_0 in some cases. To make allowance for the axial oscillations of the supporting holey film with rms amplitudes z_s, we determine its minimum thickness t_s by demanding $z_t = \sqrt{z_0^2 + z_s^2} \leqslant 1.2\,z_0$, i.e., $z_s \leqslant 0.663\,z_0$. Figure 1.44 shows t_s numerically for plastic and gold holey supports spanning the 70-μm-diameter bore of a usual specimen aperture over temperature T. (To account for weakening by the holes, we attenuated the E values by 20%).

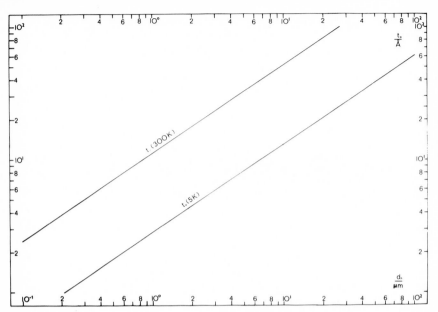

Fig. 1.43 Minimum thickness t_0 of polystyrene (or similar plastic) support films to keep thermally excited axial oscillation below 3.6 nm, plotted against film diameter d_0 and for film temperatures of 300 K and 5 K.

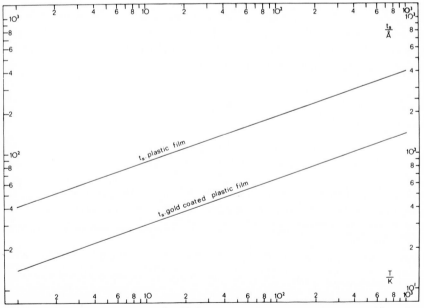

Fig. 1.44 Minimum thickness t_s of perforated plastic films with and without gold coating, to keep combined rms of thermally excited axial amplitudes of perforated and overlaying ultrathin supports below 4.3 nm, plotted against film temperature T.

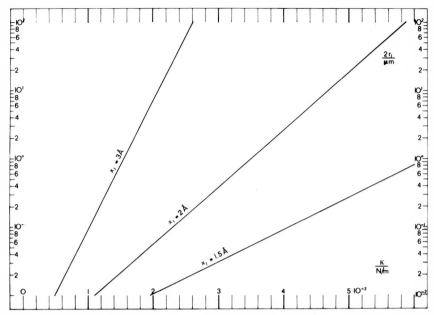

Fig. 1.45 Maximum diameter $2r_1$, of a self-supporting liquid-condensed membrane speci-men to keep rms of thermally excited lateral amplitudes below 0.3 nm, 0.2 nm, and 0.15 nm, plotted against surface tension κ for temperature T = 300 K.

Self-supported liquid-condensed membranes afford a somewhat different approach because they are kept stretched by surface tension instead of elastic forces.   A liquid with surface tension κ, spanning a circular hole of diameter $2\,r_1$, will perform axial oscillation with rms amplitudes

$$z_1 = \sqrt{\frac{kT}{20\kappa} \ln \frac{r_1}{d_a}}$$

if d_a is the mean distance of its independently moving elements. Even for the rather unfavorable values d_a = 6 Å and κ = 10^{-3} N/m, z_1 never exceeds 36 Å, at T = 300 K.   If we assume lateral movement of the molecules and set this equal to z_1, then Fig. 1.45 shows r_1 over κ for $x_1 = d_a/2; d_a/3; d_a/4; T$ = 300 K. It should be noted, however, that at the assumed resolution levels the applied electron doses will convert the fluid membrane into solid heteropolymer by crosslinking in a fraction of the exposure time needed even for the lowest usable signal/noise ratios.

Radiation Damage

Specimens as well as specimen supports in the electron microscope are inevitably exposed to electron irradiation. The fundamental processes involved in radiation damage and the effects of radiation on the physical and chemical properties of several types of matter have been reviewed extensively. The reader is referred to the work of Bahr *et al.* (1965), Reimer (1965c), Stenn and Bahr (1970), Grubb, (1974), Glaeser (1975), and Isaacson (1977). In general, some fraction of the mass is lost, the elementary composition changes, crystallinity degrades, and the mechanical as well as the electrical properties are altered.

Plastic Specimen Supports. It is a common phenomenon in electron microscopy that plastic films tend to move during observation and after heavy irradiation may finally break. Shrinkage of plastic films is a consequence of the considerable mass loss (Fig. 1.46), which is accompanied by changes in the elementary composition. Owing to the preferential loss of hydrogen, nitrogen, and oxygen-containing side groups the remaining mass is predominant, but not pure carbon, as indicated by its infrared spectra (Bahr *et al.*, 1965). Thus, the frequently used term carbonization inadequately describes the situation. Nevertheless, an obvious correlation exists between the original chemical composition of polymers, i.e., their carbon content and the remaining mass (Reimer, 1965c) (see also Table 1.3). The decrease in density upon irradiation (for nitrocellulose from 1.5 for nonirradiated to 1.03 for material irradiated with $2 \cdot 10^{-3}$ Cb/cm; Brockes, 1957) indicates that the polymer films assume a porous structure.

Generally, polymers can be divided into two main groups, those which crosslink and those which primarily degrade on irradiation. If bonds constituting the backbone of the molecules break, main chain scission results in degradation, i.e., the average molecular weight decreases. If chains of two different molecules join by free valence recombination crosslinking occurs. The molecular weight is observed to increase and the polymer is rendered insoluble, indicative of the formation of a three-dimensional network structure.

Polymers, such as polystyrene, which predominantly crosslink show improved mechanical properties at medium electron doses (10^{-2}–10^{-1} Cb/cm^2): the elastic modulus as well as the heat distortion temperature may increase by a factor of 3 to 5 (Henglein *et al.*, 1969). Degrading of polymers to be used as support film materials necessitates stabilization by surface coating with evaporated

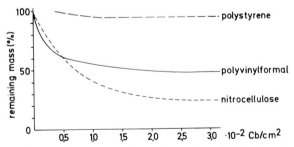

Fig. 1.46 Mass loss in different plastic specimen supports. (*Polystyrene and nitrocellulose from Brockes, 1957; polyvinylformal from Bahr* et al., *1965.*)

carbon. It is well established that aromatic compounds are much less sensitive to radiation than aliphatics. Phenyl groups can protect more sensitive groups over a distance of ~12 carbon atoms (Grubb, 1974). The relative insensitivity of aromatic compounds can be explained in terms of resonance delocalization (Spehr and Schnabl, 1973). Polystyrene, with its inherent aromatic stabilization, is therefore one of the most radiation-resistant polymers (Parkinson and Keyser, 1973).

Electrical conductivity is usually low in polymers, but may increase by several orders of magnitude upon irradiation. For polystyrene an increase in conductivity from 10^{-15} $(\Omega \, cm)^{-1}$ to 10^{-7} $(\Omega \, cm)^{-1}$ has been measured (Brockes, 1957). Polyvinylformal films exhibit, after an initial increase in conductivity, a decrease to very low values followed by an increase up to 10^{-6} $(\Omega \, cm)^{-1}$ at doses $>10^{-1}$ Cb/cm^2. Reimer (1965c) explains the decrease in a dose range around 10^{-2} Cb/cm^2, as due to the formation of electron traps in the conduction band, while the increase at higher doses is ascribed to the progressing carbonization. The observation of fluctuating charges in insulating plastic films which persist even after application of high electron doses (Mahl and Weitsch, 1960) indicates, however, that the electrical conductivity always remains lower than in evaporated carbon films.

Radiation-induced changes of specimen supports may also significantly influence the appearance of specimens. Kölbel (1972) has shown for a series of homologous sections that the image size and contrast of the specimen shows some correlation with the relative radiation resistance of their underlying supports (Fig. 1.47). As compared to pure carbon films, sections embedded in methacrylate may shrink in area on nitrocellulose supports by 20–25%. Upon stabilization with evaporated carbon layers, shrinkage is reduced to figures below 5%. As experiments with different embedding media indicate, specimen shrinkage is however not simply a function of the relative mass loss of the support; cooperative interactions between embedding medium and supporting film are obviously involved. Crosslinking between the embedding medium and the supporting polymer and the relative contraction of both may account for the effect. It might, at first glance, seem curious that the opposite effect has also been observed: polyethylene single crystals, which show quite a stable appearance on carbon films, expand by over 50% in area when supported on nitrocellulose films (Grubb, 1974). The stability of carbon supports is ascribed to their rigidity, which constrains the crystal to its original shape, while the soft nitrocellulose films allow the expansion of the crystal.

The possibility cannot be excluded that owing to intermolecular energy transfer processes specimen supports may increase or decrease the actual sensitivity of specimens (Hahn et al., 1976). Moreover, one has to be aware of the fact that volatile fragments released from degrading supports might participate in the degradation of the specimen (Dubochet, 1975a). Further development of improved specimen supports will have to consider these aspects carefully. Despite

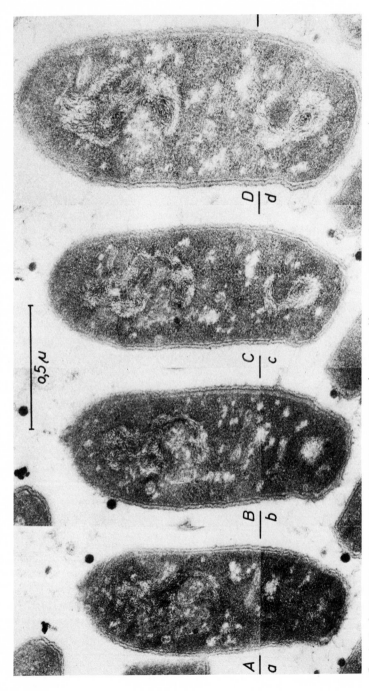

Fig. 1.47 Comparison of image size of homologous sections (specimen: *Mycobact. vaccae* infected with phage Bo 5) on different supports demonstrates the concomitant shrinkage of radiation-sensitive plastic films and thin sections mounted on them. *A, a* = nitrocellulose; *B, b* = nitrocellulose + silicon oil; *C, c* = polyvinylformal; *D, d* = carbon. (*From Kölbel, 1972.*)

its importance, the complex interplay between the specimen and its supporting film during irradiation has yet scarcely been investigated.

Carbon and Metal Films. Carbon, metal, and metal oxide films are relatively radiation-resistant. They do not noticeably change their overall appearance even at high electron doses. This does not however mean that atomic transpositions, surface migration, and etching, as well as deposition of new material (contamination), which may balance each other do not occur. The conditions under which neither contamination (due to carbonization of condensed hydro-

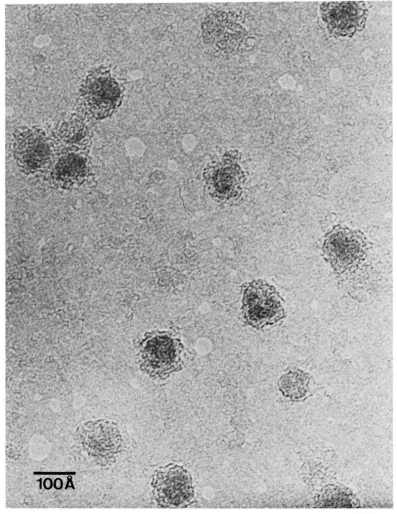

100 Å

Fig. 1.48 Ferritin molecules encapsulated into an aluminum oxide support. (*From Müller* et al., *1975*.)

carbons) nor removal of carbon (mainly due to radiation-induced changes in the proportion of water and subsequent CO formation) takes place are very difficult to realize experimentally (Heide, 1962, 1963; Fourie, 1975). Only with carefully designed cold chambers operating at temperatures below $-130°C$ these undesirable radiation-induced effects can be minimized (Heide, 1963). The relative radiation resistance of metal and metal oxide films make it attractive to encapsulate sensitive molecules into these supports (Cosslett, 1948; Müller et al., 1975) (Fig. 1.48). Such conducting coatings reduce heating and charging effects and favor recombination, and electrons tunneling into the organic specimen may act as scavengers, neutralizing the ionization produced by the electron beam (Salih and Cosslett, 1974).

Layer Silicate Supports. One might suspect that during electron irradiation atom positions in crystalline supports are randomized, thus creating new random pictorial noise, which owing to its "white" frequency spectrum can no longer be discriminated from specimen structures. Thus the most important advantage of crystalline over amorphous specimen supports would in practice not be maintained. As is to be expected, a degradation of crystallinity is in fact observed. Relating beam damage in vermiculite supports to the decay of distinct diffraction spots yields the dose-response function shown in Fig. 1.49.

Comparing the characteristic damaging dose for vermiculite

$$D_{1/e} \approx 10^6 \ e^-/Å^2 = 1.6 \cdot 10^3 \ Cb/cm^2$$

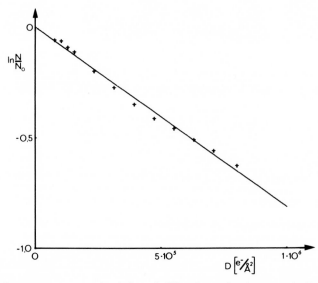

Fig. 1.49 Dose-response curve for fading of diffraction orders from vermiculite single crystal specimen supports.

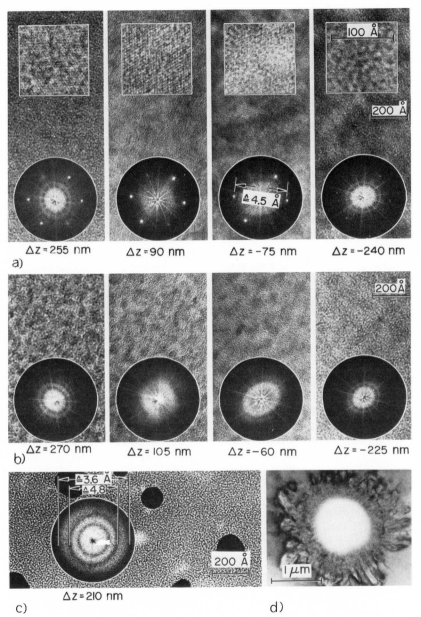

Fig. 1.50 Radiation damage in a mica support film. (a) Defocus series of a thin mica layer, prepared by mechanical exfoliation. (b) Same area after irradiation by 100-keV electrons with about 500 $Cb \cdot cm^{-2}$; disappearance of diffraction spots reflects loss of crystallinity. (c) Carbon film (with evaporated gold crystals) taken and photographically processed under the same conditions. (d) Mica irradiated by an electron beam of 1 μm diameter with about 1,000 $Cb \cdot cm^{-2}$. Note the relatively high amount of phase contrast granularity as compared to the irradiated mica. (*From Heinemann, 1970.*)

with the dose tolerated by biomolecular specimens, it becomes immediately apparent that the radiation sensitivity of vermiculite will certainly not be a serious problem. No significant changes of crystalline order are to be expected for doses which already exceed the steady state doses of almost all organic specimens by several orders of magnitude. Even the doses that must be applied for the detection of single heavy atoms within an organic matrix (Baumeister and Hahn, 1976a) would not significantly degrade the vermiculite lattice. Heinemann (1970) monitored radiation damage in mica supports by means of high resolution imaging and subsequent light optical diffraction (Fig. 1.50): the doses destroying crystallinity in mica are obviously of the same order of magnitude ($\sim 3 \cdot 10^5 \ e^-/\text{Å}^2$) as those measured for vermiculite. It is remarkable in this context that even after complete destruction of the crystalline structure at extremely high electron doses ($D > 10^6 \ e^-/\text{Å}^2$) the amorphous layer silicate supports show a relatively low amount of background phase contrast, especially as compared to carbon (Fig. 1.50c). Amorphous vermiculite resembles more or less the structure of low noise aluminum oxide or aluminum-beryllium alloy films.

Temperature and Charging Effects

Because only a very small fraction of the kinetic energy of the transmitted electrons is lost in the specimen, temperature effects are negligibly small in most cases. Temperature rise estimated on the basis of inelastic scattering cross sections by Isaacson (1977) come out a factor of two lower than the data obtained with coefficients derived from the Bethe stopping power theory by Kanaya (1955) put into the expressions of Leisegang (1956). Although measurements of Reimer and Christenhuss (1965) showed energy exchange coefficients only slightly smaller than predicted by Kanaya, only Isaacson's inelastic cross sections were directly measured on very thin specimens, excluding any risk of multiple scattering and exaggerated linear energy transfer due to slowed and secondary electrons. In general, temperatures are hardly elevated, so energy dissipation is due only to thermal conduction in the support, while radiative dissipation can be neglected.

For the thin specimen supports considered here, the linear energy transfer is independent of thickness, and the dependences of molecular number densities and inelastic scattering cross sections of atomic number nearly cancel out. Therefore, only the number N_b of electrons hitting the specimen per second and its thermal conductivity k (in $\text{Jcm}^{-1} \ \text{sec}^{-1} \ \text{K}^{-1}$) govern the temperature rise ΔT for a given primary electron velocity $v = \beta c$ (c = velocity of light) and the fraction of conduction radius r_c over the radius r_b of a homogenously irradiated circular field (Isaacson, 1977):

$$\Delta T \approx 5 \cdot 10^{-14} \ \text{Jcm}^{-1} \ \frac{N_b}{k\beta^2} \left(\ln \frac{r_c}{r_b} + \frac{1}{2} \right)$$

In some cases, the current density distribution has a gaussian shape, which leads to more complicated expressions for the center temperature increase (Gale and Hale, 1961; Reimer and Christenhuss, 1962).

About 1 electron/μm^2 is required in the image plane for a proper electron microscope image, and a circular patch of 20 cm radius has to be illuminated to achieve sufficiently even current density distribution over a recording area of ~10 cm diagonal width. Further assuming that only 50% of the electrons reach the image plane and a 2-sec exposure time, we get $N_b = 3.14 \cdot 10^{10}$ e^-/sec to hit the specimen independently of magnification. Figure 1.51 shows ΔT over r_c/r_b for support material of low thermal conductivity ($k_{plastic} = 2 \cdot 10^{-3}$ Jcm^{-1} sec^{-1} K^{-1}; $k_{Al_2O_3} = 13 \cdot 10^{-3}$ Jcm^{-1} sec^{-1} K^{-1} and 100 keV primary energy. With gold coated perforated supports specimen temperature decreases further owing to much higher conductivity ($k_{Au} = 3.1$ Jcm^{-1} K^{-1}) and because $r_c/r_b \approx 1$ is easily achievable.

Because practically no beam electrons really get stuck in a specimen, the build-up of charge patterns on a preparation is governed by local secondary emission coefficients ϵ. These, in turn, depend much more on the material and distribution of the specimen than on the support itself. Moreover, surface reactions with the environmental residual gases as well as chemically manifest radiation damage tend to alter these values in a way that can hardly be generalized or predicted.

If we nevertheless suppose the secondary emission coefficient to be proportional to the thickness t of (thin) support films and define $\epsilon/t = \epsilon'$, the irradiated

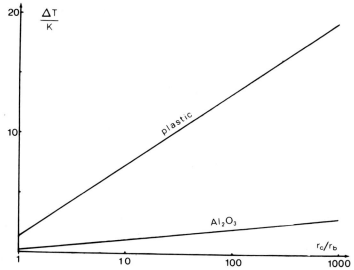

Fig. 1.51 Temperature rise ΔT in the center of a homogeneously irradiated circular area of radius r_b in plastic and Al_2O_3 support films plotted over the fraction radius r_c of heat sink/ r_b.

volume represents a line source yielding $eN_b\epsilon'$ A/m. Analogous to the considerations of thermal conduction, we find the stationary potential U_s of the irradiated patch (radius r_b) against the grounded aperture (radius r_c) to be

$$U_s = \frac{eN_b\epsilon'\rho_R \ln r_c/r_b}{2\pi}$$

with ρ_R = specific resistance of the support material and e = electron charge.

For $\epsilon' = 10^{-3}$ nm^{-1}, $eN_b = 5$ nA, and ρ_R (carbon) $= 7 \cdot 10^{-5}$ Ωm, we find $U_s = 55$ mV $\cdot \ln r_c/r_b$.

For insulating support materials, e.g., ρ_R (bulk polystyrene) $\approx 10^{16}$ Ωm, $e \cdot U$ soon reaches the mean energy of secondary electrons, thus influencing ϵ and, by establishing a feedback, leading to an equilibrium of a few volts (Reimer, 1965b). Although considerable field strengths ($\approx 10^8$ Vm^{-1}), which could cause a mechanical disruption of the support (Isaacson, 1977), can build up, image distortions were not found in high resolution micrographs (Baumeister and Hahn, 1974a) of vermiculite ($\rho_R \approx 10^{15}$ Ωm). Only in projection images of very low magnification was a fast fluctuating "granularity" observed by Mahl and Weitsch (1960).

Methods for Thickness Measurement

A variety of methods to measure or estimate the geometrical thickness t or the mass thickness $\rho \cdot t = m_t$ (ρ = density) of a preparation have been developed for or applied in electron microscopy. The following survey will show which method can be used to monitor (m) film deposition during evaporation and which are only applicable to calibrate (c) samples or test preparations (Table 1.9).

It is evident that in electron microscopy the best-suited specimen support has the lowest possible mass thickness. This is true because the amount of mass per unit area is responsible for the degree of contamination of the imaging beam with inelastically scattered electrons; and, owing to the chromatic aberration of the imaging lenses and the delocalized nature of the inelastic scattering process (Isaacson et $al.$, 1974a; Rose, 1976), such electrons attenuate the overall contrast and blur contours on electron microscope images. Nevertheless, the geometrical thickness is of importance because it limits the sag and radial slope of the support film (see page 77).

1. Weighing is often used to calibrate other methods of thickness determination because microbalances usually cannot be operated in a vacuum bell jar to monitor the evaporation process. To obtain a feeling of sensitivity and accuracy of a microbalance needed we refer to De (1962), who used glass cover strips of 3.24 cm^2 assumed to be 0.17 mm thick and recorded mass thicknesses down to $7 \cdot 10^{-7}$ gm/cm^2 to calibrate the optical absorption of carbon layers versus mass thickness. His balance (Sartorius MPR -5) therefore had to show mass differences $\geqslant 2$ μg in a total of 0.15 gm with still reasonable accuracy. Johansen (1974)

Table 1.9   Thickness Measurement

No.	Method	Measured quantity	Application
1	Weighing	m_t	c
2	Vibrating quartz	m_t	m
3	Contrast in e.m. (light or dark field)	m_t	c
4	Optical absorption	m_t	m
5	Ellipsometry	t	m
6	Optical interference (n = refractive index)	$n \cdot t$	c
7	Transmission e.m. (sections, shadowing, bent edges)	t	c
8	Electrical resistance	m_t	m
9	Radioactive tracer	m_t	(m?)

c, calibration
m, monitoring
m_t, mass thickness
t, geometrical thickness
$n \cdot t$, optical thickness

recorded mass differences $\geqslant 8$ μg $\pm 15\%$ with a Mettler M 5 SA microbalance. Williams and Glaeser (1972) weighed aluminum foils with a total carbon deposit of 155 μg to determine the density of carbon layers.

2. A second method yielding mass thicknesses and, in contrast to weighing, capable of "on line" monitoring of vacuum evaporation processes is the vibrating quartz coating monitor. In most commercial units the difference between the frequencies of natural oscillations of a measuring quartz crystal and a (covered) reference crystal, a difference developing from the frequency decrease due to the mass of the deposit on the measuring crystal, is used as signal.

Because the measuring crystal obviously cannot be at the same position relative to the evaporation source as the substrate for the actual film to be generated, and because the extension of homogeneously covered fields strongly depends on the geometry of the source as well as surrounding surfaces (Moore et al., 1971), quartz monitors have to be calibrated in the actual environment used for evaporation. In addition, the sticking coefficients for condensation on the actual target and the quartz crystal are not necessarily equal and will also depend on the kind and amount of deposits already covering the surfaces of the substrate and the quartz at the beginning of measurement (Schwarz, 1966; Dubochet, 1975b).

3. Apparently the most realistic method of mass thickness determination should be the measurement of electron optical contrast of a given support in the microscope or on micrographs. Indeed, relatively accurate and reliable data for electron optical contrast of light element specimens have been worked out

for mass determination of interesting items in biological preparations (Zeitler and Bahr, 1962; Burge and Silvester, 1960; Lippert, 1970).

If I_0 and I are the intensities in transmission images of open areas and of areas with mass thickness m_t, imaged with aperture ϑ (radian) and wavelength λ, dependent on acceleration voltage U by

$$\lambda(U) = \frac{2.425 \cdot 10^{-12}\ \text{m}}{\sqrt{\dfrac{U}{511.3\ \text{kV}}}\left(2 + \dfrac{U}{511.3\ \text{kV}}\right)}$$

we obtain

$$m_t(I_0/I) = \frac{84\,(1 + \varphi^2)\ln(I_0/I)}{h(U)\,2.5 + (1 + \varphi^2)\ln\left(\dfrac{1 + \varphi^2}{\varphi^2}\right)}\mu\text{g/cm}^2$$

with the scaling functions

$$\varphi(\vartheta, U) = \frac{1.52 \cdot 10^{-10}\,\text{m}}{\lambda(U)}\,\vartheta$$

and

$$h(U) = \frac{\lambda(U)\left(1 + \dfrac{U}{511.3\ \text{kV}}\right)}{5.436 \cdot 10^{-12}\ \text{m}}$$

Numerical factors in these formulas are fitted to measurements on carbon at 60 keV. It is obvious that small U and ϑ values enhance the contrast and increase the sensitivity of this method. Especially for $U = 60$ kV and $\vartheta = 3.6 \cdot 10^{-3}$ (i.e., $h(U) = 1$; $\varphi(U) = 0.1125$) contrast is practically independent of the atomic number Z (Reimer, 1961), and we obtain for small mass thicknesses:

$$m_t(I/I_0) = 12.26\,(1 - I/I_0)\ \mu\text{g/cm}^2$$

In this context it should be mentioned that a proper consideration of the influence of the neighboring atom screening potential allows the description of scattering contrast experiments with fair accuracy in terms of atomic properties only (Reimer, 1969; Schwertfeger, 1974).

The imaging aperture ϑ in most cases cannot simply be determined by dividing the radius of the objective aperture by focal length. Because the focus of electron lenses of high excitation lies well within the field, the ray paths are still curved after intersecting the axis, and the effective aperture can be 20-50% larger than its nominal value. Results apparently different from those mentioned previously (Sarkar, 1966) might be due to the neglect of this consideration. If electron image contrast is measured by the built-in insulated screen, its voltage-dependent efficiency (Grubb, 1971) has to be known. Measurements of image contrast on photographic emulsions can benefit from the nearly linear dose–optical density relationship (Frieser and Klein, 1958). But effects of delayed processing have to be considered for actual measurement as well as intermittency effects for calibration work on emulsions as detectors (Lippert, 1969).

If we assume a bright field contrast of $1 - I/I_0 = 3\%$ as just sufficient for a thickness measurement of reasonable accuracy, the limit of this method is ~.36 $\mu g/cm^2$, corresponding to 1.8 nm carbon foil of density 2 gcm^{-3} for the above-described imaging conditions, yielding Z-independent contrast. Williams and Glaeser (1972) as well as Johansen (1974) used the intensity of dark field images for mass thickness determination of very thin evaporated carbon films (Fig. 1.52). In both cases the calibration was based on weighing, but to convert measured mass thicknesses into geometrical thicknesses, Williams and Glaeser (1972) determined the density of a 20-nm film to be 3.1 gcm^{-3} and, applying this value to ultrathin films, found an 0.8-nm film ($\hat{=}$ .25 $\mu g/cm^2$) to yield still 5% of the bright field intensity of a hole. Johansen assumed a density of 1.9 gcm^{-3}, which was found by Leder and Suddeth (1960), calibrating energy loss measurements with thicknesses determined by optical interference (Tolansky method).

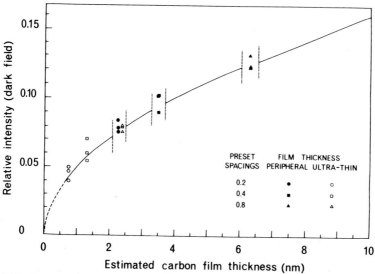

Fig. 1.52 Relative measurements of dark field intensities for carbon films of different thickness. The solid points refer to the thickness of the peripheral zones ("tab-area") and the open ones to the ultrathin part. The vertical broken lines indicate the standard deviation of the thickness measurements, and the points refer to the pre-set spacings of the carbon rods.(*From Johansen, 1974.*)

The higher sensitivity of dark field thickness measurements is possibly due to the utilization of inelastically scattered electrons, which for light elements are at least two times more abundant than elastically scattered ones at the usual voltages and apertures. The measured relative dark field intensity I_d/I_0 depends on the geometrical thickness t of the carbon film by

$$\frac{I_d}{I_0}(t) = \left(\frac{t}{t_0}\right)^{3/4}$$

with $t_0 = 1.06 \cdot 10^3$ Å in the work of Williams and Glaeser (1972) ($U = 80$ kV, $\rho = 3.1$ gcm^{-3}) and $t_0 = 0.632 \cdot 10^3$ Å for Johansen's plot ($U = 100$ kV, $\rho = 1.9$ gcm^{-3}) (Fig. 1.52). The different t_0 values might be largely due to different scaling values $\varphi(\vartheta, U)$ and $h(U)$.

Williams and Glaeser (1972) ascribed the nonlinearity of $I_d(t)$ to the transition from the lower scattering amplitudes of bound atoms in bulk material to larger scattering amplitudes assumed valid for very thin layers. This, however, is hardly consistent with the assumption of thickness-independent density. Additionally, the slope of $I_d(t)$ should not rise more than $\approx 50\%$, because the forward scattering amplitudes $f(0)$ of free atoms are approx. $\frac{3}{2}$ of the amplitudes of bound atoms (Schwertfeger, 1974). (For the physical meaning of scattering amplitudes $f(\theta/\lambda)$, θ = scattering angle, $\theta/\lambda = q$ = spatial frequency, see Zeitler, 1965).

4. Optical absorption as a thickness-measuring method has the advantage of simplicity as well as monitoring capability. In its simplest form it is used to estimate visually the thickness of an evaporated carbon layer. For this purpose a piece of white plastic is partially covered with vacuum-oil, on which no carbon layer forms. By observing the contrast developing between the clean and the covered area during evaporation, the thickness already attained can be estimated (Bradley, 1954). For higher accuracy, the simultaneous display of a calibrated sample is convenient.

Quantitative measurements of optical absorption for thickness determination of thin carbon layers were first published by Agar (1957). Early controversies concerning the lack of numerical conformity of optical density versus thickness plots published by different authors were largely muted after Graff (1961) found carbon evaporation to yield reproducible values only for vacua $\leqslant 10^{-4}$ mbar and a source distance $\geqslant 10$ cm. He also measured how for layers of different thicknesses the optical properties (absorption constant, refractive index) changed with the measuring wavelength, λ_m, for 0.4 μm $\leqslant \lambda_m \leqslant 1.4$ μm.

Deviations from Beer's law (i.e., the linear relationship of optical density over thickness) reported by some authors for very thin layers were discovered by comparison with theory (Mayer, 1950) to be due to interference effects and were noticeable up to thicknesses of $\lambda_m/8$ by Graff. Recent measurements of De Boer and Brakenhoff (1974) are within 10% of Graff's values.

Defining optical density S conventionally as

$$S = \log \frac{L_0}{L}$$

with L/L_0 the relative light intensity transmission, we obtain from De Boer and Brakenhoff (1974) the numerical relation for the geometrical thickness t as a function of S for $\lambda_m = 550$ nm:

$$t = 86.45S \ [\text{nm}]$$

While De Boer and Brakenhoff (1974) calibrated their measurements by the Tolansky (1948) light-interference method, Holl et al. (1974) used an oscillating quartz crystal. From their plot of $\ln(L_0/L)$ over the mass thickness m_t we derived:

$$m_t = 9.785S - 1 \ [\mu\text{g/cm}^2]$$

for $m_t > 1$ μg/cm^2 and $\lambda_m = 213.9$ nm.

Using a scaling function obtained from a plot of De Boer and Brakenhoff (1974), for the transmission of an 18-nm-thick carbon film:

$$\frac{L}{L_0}(\lambda_m) = 0.248 + 0.5045 \frac{\lambda_m}{\mu\text{m}}$$

we compared both measurements, and found them numerically equal for a density $\rho = 1.88$ gcm^{-3}. This practically coincides with the findings of Leder and Suddeth (1960), derived from energy losses. These facts justify some confidence in the monochromatic optical absorption method, as well as in the calibration methods mentioned, although it seems restricted to strongly absorbing carbonaceous material on a transparent backing.

5. The multiple-beam interference method of Tolansky (1948) is the most sensitive of a variety of methods using phase shifts of light beams for length measurements. Thickness differences down to ~5 nm can be determined, if several single measurements are averaged. The accuracy depends on the flatness of the surfaces and is in the range of less than 1 nm, i.e., $\approx \lambda_m/500$. It is, of course, necessary to furnish the surface to be measured with a reflective coating of ~5% residual transparency. Obviously, this evaporated silver layer has to be of very uniform thickness to secure resemblance to the original surface. A small slope with respect to an opposite mirror also of some percent transparency gives rise to interference fringes (distance d_f) in case of sufficient spatial and temporal coherence of illumination.

The thickness-step t denoting the thickness of a support film is proportional to the fringe shift s_f occurring at the step by (Fig. 1.53)

$$t = \frac{s_f \lambda_m}{2d_f}$$

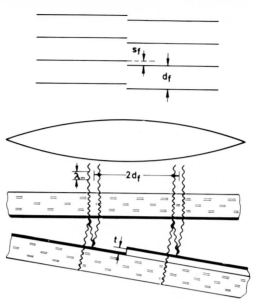

Fig. 1.53 Tolansky method to measure geometrical thickness t by the shift s_f of the fringe patterns (fringes distance d_f) along a thickness-step. In reality, the pitch of the lower plate, bearing the step to be measured, has to be rotated by 90° around the optical axis to show the fringe pattern drawn above the lens symbol. Wavelength λ_m, thickness t, and tilt angle of the lower plate are drawn much larger than to scale.

This method was also used to calibrate early thickness measurements based on electron optical transmission by Weber and v. Fragstein (1954), and for the same purpose in the recent investigation of optical absorption of carbon films by De Boer and Brakenhoff (1974). King *et al.* (1972) compared the Tolansky method (see also Bennett and Bennett, 1967) with shearing interferometry (Dyson, 1970) and an electromechanical scanning stylus method, and found all three methods to yield an approximately equal ultimate accuracy of less than 1 nm in geometrical thickness.

6. For thickness measurements in surface-film science the ellipsometer (Rothen, 1968) is a well known tool, capable of detecting differences of fractions of nanometer units in optical thickness $n \cdot t$ (n = refractive index) of sufficiently transparent coatings on plane mirrors. Therefore, small changes of magnitude and orientation of the electrical vector of polarized light are to be detected by measurement of analyzer rotation necessary to restore minimum intensity after exchanging mirrors of known and unknown coating thickness.

The method seems well suited for transparent plastic support films where optical absorption does not yield a sufficient signal, although it lacks the monitoring capability of the latter method. Moreover, the considerable cost and complexity of modern ellipsometers may prevent its widespread use only for thickness measurements in electron microscopy. Nevertheless, it was used by Kleinschmidt and Vasquez (1969) for support film thickness determination.

7. Three different procedures are available for obtaining geometrical thickness by transmission electron microscopy:

a. A support film is mounted on a plane cut slab of polymerized embedding material. The mount is topped with fresh monomer and, after polymerization, sectioned perpendicularly to the film plane. Examination of these sections for electron micrographs of at least $\gtrsim 5.10^4$ magnification and searching for parts of the film really appearing in a projection parallel to their plane yields thickness values down to 2 nm with accuracies not far from the resolution of the instrument (~ 0.5 nm) (Moretz et al., 1968).

b. While the above method depends upon a difference in density between the film and the embedding polymer, which differences become quite small in case of plastic films, it is also possible to gain thickness values from the projected intensity profiles of the bent edge of a protruding flap, occasionally observable with a deliberately ruptured support film. Figure 1.54 shows how accurately the geometrical thickness can be deduced from a density scan over the image of an edge, which was assumed to belong to a 2-nm carbon film bent with a 20-nm radius and observed at 100 kV. The primary intensity profile $I(x)/I_0$ was assumed to be imaged at zero defocus (gaussian focus) with a spherical aberration constant $C_s = 1.3$ mm, yielding the image intensity profile $I'(x)/I_0$.

c. Finally, a common method of determining the surface profile of a preparation is also applicable to thickness measurement: Heavy metal shadowing allows determination of the thickness t of a film from the width w_s of the shadow if the elevation angle α_s of the (small) source is known. It is:

$$t = w_s tg\alpha_s$$

if the direction of evaporation is perpendicularly pointing from the film to its edge.

To achieve a noticeable contrast with a small additional thickness, heavy metal as evaporant is necessary. An angle $\alpha_s \approx 40°$ is a good compromise between low sensitivity and exaggeration of striated contrasts duel to surface roughness and build-up of heavy metal micro-crystals. A Pt layer of 1–2 μgcm^{-2} under usual conditions yields sufficient contrast for measurable shadows. For all three methods mentioned it is advisable to select micrographs taken at or near zero defocus for the proper evaluation of distances and contrasts due to differences in mass-thickness. Otherwise, Fresnel fringes and phase contrasts may render the location of the real edge position a tedious job.

8. To monitor evaporation of conducting material (e.g., carbon) Wiesenberger (1963) used a glass rod containing two platinum wires. One end of this rod was polished flat and put near the target, facing the source. When the resistance between the two leads, decreasing from very high values upon deposition, reached a predetermined value, the power supply of the evaporator was cut off

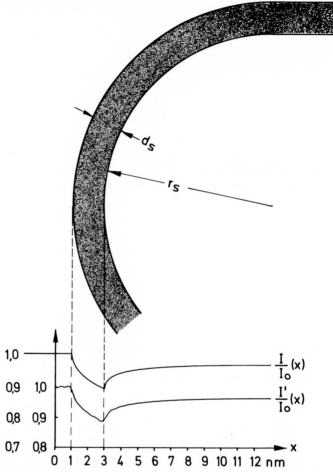

Fig. 1.54   Relative electron intensity $I(x)/I_0$ over image coordinate x in projection of bent edge of 20-Å carbon film for 100 keV electrons. Upper curve: Intensity distribution for aperture $\vartheta = 7.3 \cdot 10^{-3}$ according to the relations given in this chapter. Lower curve: Intensity profile in image, at gaussian focus ($\triangleq$ zero defocus) with spherical aberration constant $C_s = 1.3$ mm and objective aperture $\vartheta = 7.3 \cdot 10^{-3} = \sqrt[4]{\lambda/C_s}$.

via an electronic sensing circuit. Assuming a distance a from the rod axis for both wires, each having a radius r_w and covered with a material of specific resistance ρ_R over an axial thickness t (the wires should protrude from the glass over the distance t at least), we measure

$$R = \frac{\rho_R \ln(a/r_w)}{\pi t} \quad \text{if} \quad r_w \ll a$$

The values are:

$$\rho_R = 70 \cdot 10^{-6} \ \Omega\text{m (bulk carbon)}$$
$$a = 5 \text{ mm and } r_w = a/20 = 0.25 \text{ mm}$$
$$\text{yield } R = 33 \text{ k}\Omega \text{ for } t = 20 \text{ Å}$$

9. Radioactive tracer methods are principally suited for measuring mass-thicknesses of thin films, as was shown by Beischer (1950), and for mass-losses due to radiation damage, shown by Thach and Thach (1971). The possibility of contamination hazards in handling radioactive isotopes and the connected security and legal problems, and, moreover, the feasibility of the many alternative methods mentioned above, may have prevented the wider use of tracer methods for support film thickness measurement.

Electron Optical Noise

Because specimens have to be of fairly low mass-thickness, if the capability of molecular resolution of the electron microscope is really utilized, the influence of the molecular structure of the support films on the image becomes more pronounced with higher resolutions and magnifications.

It is well known that evaporated thin carbon films without specimens are successfully used as test objects to gain information about the electron optical performance of electron microscopes. This is possible because they impose a nearly stochastic spatial distribution of phase delays on the passing electron wave, thus representing nearly the same spatial spectrum as a hypothetical phase-point, but with much higher intensity. High resolution phase-contrast images of specimens on such films therefore have a 'white' noise-like appearance ('granularity') and are further deteriorated by the shot noise of the electrons and the recording process.

To assess quality parameters for specimen support materials, the disturbing "structural noise" must be quantified independently of mass thickness as well as of imaging and recording parameters. Measurements of image intensity variations by microdensitometry (Hashimoto et al., 1974), although not corrected for equal mass thickness, revealed a decrease of "structural noise" intensity for the sequence amorphous carbon $> Al_2 O_3 >$ graphite single crystal. If one makes the very probable assumption that the natural graphite flakes used by Hashimoto et al. (1974) had a higher mass-thickness than the amorphous, evaporated films, the advantage of the crystalline support may be still more pronounced than can be read from these "noise" figures.

Kübler and Downing (1977) used the peak values of spatial correlation functions, which are proportional to the mean square image intensity fluctuations, to assess the amount of noise generated by different support films. To ascertain the same defocus for different pictures, they tilted the specimen stage and thus

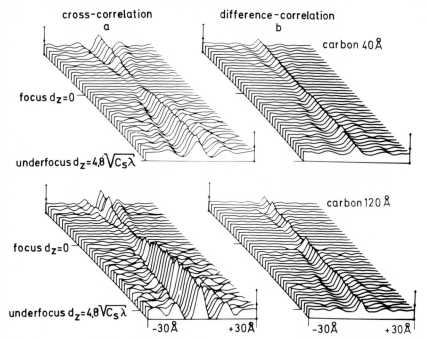

Fig. 1.55 Cross- (a) and difference- (b) correlation parallel to the tilt axis of micrographs of evaporated carbon films. Upper row: thickness $t \approx 40$ Å, lower row: $t \approx 120$ Å. Original micrographs taken at $2 \cdot 10^5$ magnification, $C_s = 1.3$ mm, $\lambda = 0.037$ Å ($\hat{=}$ 100 kV), tilt angle 60°. Heights of functions are proportional to rms noise over correlation distances marked on horizontal scale. (*Modified after Kübler and Downing, 1978.*)

established a linear relationship of defocus over the image coordinate perpendicular to the tilt axis. Auto-, cross-, and difference-correlations were performed on digitized densities from scans parallel to the tilt axis, each ~ 0.21 $\sqrt{C_s\lambda}$ apart. The dependence of cross-correlations of defocus (Fig. 1.55) shows that enhanced granularities in certain spatial frequency domains are mainly due to the phase-contrast transfer properties of the objective-lens, while the spatial spectrum of the carbon film is practically "white," within the resolution range of the electron microscope (≈ 0.3 nm). The same conclusions can be read from the defocus dependence of the signal/noise ratio, measured by Frank and Al-Ali (1975).

Assuming stochastic spatial distribution of scattering atoms, and a geometrical thickness t well within the depth of focus, δ_f, the relative spectral density (i.e., normalized power spectrum) of scattered intensity was given by Rose (1976) to be

$$P^2(q) = [f(q)/f(0)]^2 \left[1 - \frac{2J_1(2\pi qb)}{2\pi qb} \right]$$

(J_1 = Bessel function of 1st order)

where $f(q)$ is the scattering amplitude of the atoms constituting the film, q is the spatial frequency (see also p. 98), and b is the mean atomic distance.

The absolute magnitude of the rms noise contrast $\sqrt{g_s^2}$ due to the support film is obtained from the mean square noise contrast:

$$\overline{g_s^2} = \frac{m_t \lambda^2 f^2(0)}{A \cdot m_H} \, 2\pi \int_0^{q \, max} P^2(q) u^2(q) q \, dq$$

with $m_t/A m_H$ = mass thickness/(atomic weight · hydrogen atom mass) representing the number of atoms per unit area. The term $u(q)$ is the phase-contrast transfer function of the electron microscope and q_{max} is the highest q where $u(q)$ has a sensible magnitude. For phase contrast in a coherently illuminated bright field conventional electron microscopy (CEM), $u(q)$ was given by Hanssen and Morgenstern (1965) as:

$$u_{CEM}(q) = -2 \sin \left[2\pi \left(C_s \lambda^3 \frac{q^4}{4} - d_z \lambda \frac{q^2}{2} \right) \right]$$

Phase-contrast transfer functions for the scanning transmission electron microscope (STEM) were given by Rose (1975), and for proper defocus look roughly similar to

$$u_{STEM}(q) = \frac{1}{\pi} \{ \text{arc} \cos Q - Q \sqrt{1 - Q^2} \} \quad \text{with } Q = \frac{\sqrt[4]{2 C_s \lambda^3}}{4} q$$

known as Duffieux-integral. To plot $P^2(q)$ we used the method of Schwertfeger (1974) to calculate the scattering amplitudes $f(q)$ of carbon for different densities (Figs. 1.56 and 1.57). The consideration of nonstochastic spatial distribution of atoms, e.g., microcrystalline clusters, will lead to less peaked spectra for reasonably wide size distributions of clusters.

Experimental evidence for the dependence of the noise contrast on mass thickness can be deduced from the increasing maxima of the cross-correlations for the 4-nm and 12-nm carbon films measured by Kübler and Downing (1978) (Fig. 1.55a) as well as from measurements in dark field of Isaacson et al.(1974b) It cannot yet be clearly assigned to the theoretical prediction (Hahn, 1965; Rose, 1975) that rms noise contrast is proportional to the mean standard deviation of mass thickness for a given spatial frequency. This might be due to a dependence of cluster size and abundance on mass thickness; the different correlations (Fig. 1.55b) only show a very weak dependence of mass thickness, suggesting that the radiation damage between the two exposures acts merely in the vicinity of the surfaces. A similar behavior of crystalline vermiculite supports was reported by Baumeister and Hahn (1976b) and was ascribed to a caging effect. Only two ways can be named to ease the noise problem for supports of a given mass thickness: For amorphous materials the greatest density will lead to the lowest mean atomic distances, i.e., the peak of spectral density $P(q)$ is shifted to higher spatial frequencies. For crystalline supports a low vacancy and

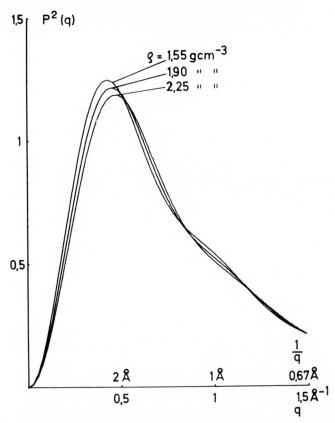

Fig. 1.56 Normalized power spectrum of carbon support film over spatial frequency q, assuming stochastic spatially distribution of atoms and geometrical thickness $t \ll$ depth of focus δ_f. Influence of density ρ on mean atomic distance b as well as on forward scattering amplitude $f(0)$ was considered.

dislocation density must be obtained for sufficiently large areas with little or no increase upon irradiation.

For obvious reasons most interesting biological structures are not built up of identical units, and in some cases not even the previously mentioned structural features can be specified in real space. If in these cases the power spectrum of the object, $P_o^2(q)$, and that of the support, $P_s^2(q)$, are known, a spatial filter function

$$F_{os}(q) = [1 + P_s^2/P_o^2]^{-1}$$

acting similarly to the well-known noise-suppressing Wiener-filter, will allow enhanced readability (Welton, 1971). While P_s^2 can be measured by electron diffraction, the knowledge of P_o^2 will in most practical cases be too crude to

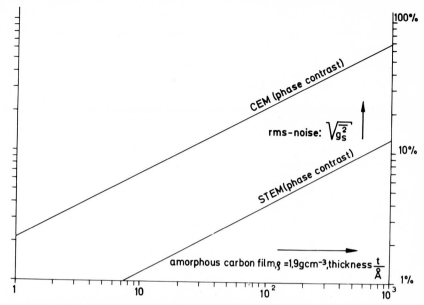

Fig. 1.57 Plot of rms phase contrast image intensity noise $\sqrt{\overline{g_s^2}}$ due to an amorphous carbon film of density $\rho = 1.9$ gcm^{-3} over thickness t. Imaging was assumed with a conventional transmission e.m. (CEM), $C_s = 0.51$ mm, at a defocus of $d_z = \sqrt{(1.46 + 2N)C_s\lambda}$, $N = \sigma$ and a maximum spatial frequency

$$q_{max} = \{(C_s\lambda^3)^{-1/2} [1.46 + d_z/\sqrt{C_s\lambda}]\}^{1/2} \quad (*)$$

as well as with a scanning transmission electron microscope (STEM) operated with same C_s and λ and a phase contrast transfer function $u_{STEM}(q)$ as given in the text with

$$q_{max}(STEM) = 4(2C_s\lambda^3)^{-1/4}$$

In general, the overall noise of the recorded image is $g_r = \sqrt{\left(g_s^2 + \dfrac{1}{n_i}\right)(DQE)^{-1}}$ with g_s = noise due to substrate, n_i = number of electrons contributing to a picture element, and DQE = detective quantum efficiency of the recording process. (*)These values yield maximum positive phase contrast of small objects for even , negative contrast for odd N.

justify spatial filtering alone. But filter implementation for other reasons, e.g., correction of contrast transfer function and suppression of shot noise (Hahn and Baumeister, 1974), allows inclusion of terms like $F_{os}(q)$ in the shaping of the filters used.

Support Structure Suppression

Obviously the support image can be separated from the specimen image only to the extent that they show known differences between whatever spatial features are involved.

Usually a priori knowledge of the specimen will be expressible in terms of real space: e.g., relative location, shape, and density. Because a suitable support film will not show the same features as the object in comparable abundance, different pattern-recognition techniques could be applied: rotational (Markham et al., 1963) or oriented translational (Valentine, 1964) repetition and photographic superposition, arraying after visual orientation and spatial band-pass filtering of repeated structures (Ottensmeyer, 1976), digital searching routines (Kübler and Koller, 1976), and cross-correlation with known or supposed structures by optical (Vander Lugt, 1964) or digital (Langer et al., 1970) means. Great care must, however, be applied to secure objective results with techniques utilizing subjective visual detection and orientation.

Depositing specimens with broad spatial spectra on crystalline supports allows us to get rid of their periodic pattern, which is superimposed on the actual specimen structure in the image (Baumeister and Hahn, 1974a). Electron optical, light optical, or digital frequency filtering can be used to avoid or punch out the narrow peaks of the support spectrum without noticeable loss of object information.

If all relevant object structures are already exhibited in a repeating array as in crystalline specimens on amorphous supports, spatial band-pass filtering will yield a fairly good object/support separation (Berger et al., 1972). Also photographic (or digital) superposition of repeating unit cells will step up the signal/noise ratio according to the number of cells used.

From light-optical diffractograms of electron micrographs (Thon, 1965) as well as from the theory of electron microscope image formation by phase-contrast (Hanssen and Morgenstern, 1965) it is known that transfer-gaps (i.e., zero-crossings of the contrast-transfer functions) cut out certain spatial frequencies $\vec{q}$ (a vector, oriented perpendicularly to the lines in the specimen plane, where magnitude denotes the number of line pairs per unit length) from being imaged. Measuring all spatial frequency vectors in normalized units defined by $\vec{R} = \vec{q}\sqrt[4]{C_s\lambda^3}$ and defining a normalized defocus by $\Delta = d_z/\sqrt{C_s\lambda}$, respectively (Hanssen, 1966), we find the defocus Δ_s necessary to suppress a certain frequency R_s by putting a transfer gap over it:

$$\Delta_s = \frac{n}{R_s^2} + \frac{R_s^2}{2}; \quad n = 0, 1, 2 \ldots.$$

Although this method offers the advantage of working without any further processing of the micrographs, it has some drawbacks: The transfer-function belonging to Δ_s and the position of further zero-crossings are not always favorable for proper imaging of the wanted information. Besides that, suppression acts for all azimuthal directions, although only a few spatial frequency vectors really are disturbing.

If special modifications are performed in the condenser or objective aperture plane as for dark field imaging or correction purposes, tilting of the crystalline

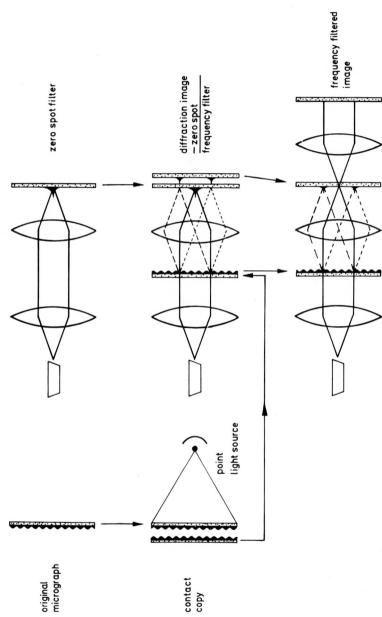

Fig. 1.58 Suppression of periodic patterns in images by optical spatial frequency filtering. A negative contact copy of the original micrograph is put into an optical diffractograph. A transparent diffraction picture is exposed, in which the zero-spot is blocked off by a prefabricated mask. The processed diffractograph-transparency acts as a matched spatial frequency filter for the filtered reconstruction of the contact copy.

original
micrograph

contact
copy

point
light source

zero spot filter

diffraction image
−zero spot
frequency filter

frequency filtered
image

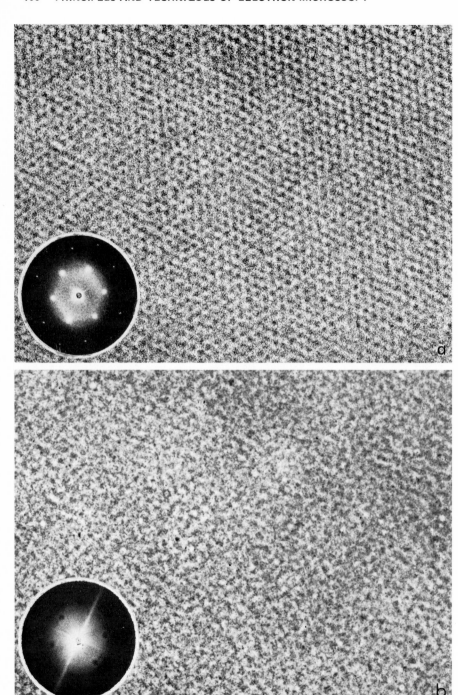

support can locate its diffraction spots over blocked regions of the aperture plane. Hashimoto *et al.* (1971) used this method for the suppression of lattice patterns of a graphite single crystal support. Schnabl (1972) investigated the choice of materials for single crystal supports for optimum conditions with tilted illumination instead of a tilted crystal.

A posteriori light-optical frequency filtering (Maréchal and Croce, 1956; Klug and de Rosier, 1966) can, among many other features, be used to dispose of unwanted periodic patterns at any defocus. For this purpose the appropriate diffraction spots are blocked off by matched filters in the Fourier-($\triangleq$diffraction-) plane of an optical processor (Hahn, 1972). The appropriate masks can either be exposed by the light of the actual diffraction image, while the zero spot is blocked off by a prefabricated mask (Fig.1.58), or be made from a drawing, reduced by a proper factor. The first method is advisable if the diffraction spots to be blocked outweigh other features of the diffraction pattern far enough to yield a negative with only these spots following the use of a steep gradient (high γ) emulsion and processing (Fig. 1.58).

After removal of the periodic pattern only stochastically distributed contrast remains, which might be due to a partial randomization of the original atom positions after irradiation and contamination (Fig. 1.59). The diffractogram of a filtered image shows dark spots where the filter (in this case made from a drawing to create spots of visible size) acted on the Fourier-transform of the original image.

Obviously digital filtering (Burge and Scott, 1975; Burge, 1976) is also suited for the same purpose, although digitizing, computational processing, and pictorial display of the several million picture elements for even one image is only justified in case of excellent original micrographs (Kübler *et al.*, 1978).

It is a pleasure to express our thanks to all colleagues who generously provided us with published or unpublished material for this chapter, and especially to Dr. F. P. Ottensmeyer for critical reading of the manuscript. We also wish to thank Dipl.-Phys. Seredynski for numerous calculations and plots, Mr. Schneider for drawings, Mrs. Ludolph and Mrs. Werner for bibliographic work and typing the manuscript, and Misses Türken, Wenzel, and Konz for photographic work and preparing the figures.

References

Abermann, R., and Salpeter, M. M. (1974). Visualization of deoxyribonucleic acid molecules by protein film adsorption and tantalum-tungsten shadowing. *J. Histochem. Cytochem.* **22**, 845.

Ackermann, I. (1947). Silikatglas als Trägerfolie für elektronenoptische Anordnungen. *Optik* **2**, 280.

Fig. 1.59 Identical fields of a vermiculite single crystal before (a) and after (b) suppression of lattice periods. The remaining stochastic "granularity" stems from residual contamination and radiation-induced deterioration of the crystal. ×6,000,000. (*From Baumeister and Hahn, 1976.*)

Adam, N. K. (1941). *Physics and Chemistry of Surfaces*, p. 209. Oxford University Press, New York.

Agar, A. W. (1957). The measurement of the thickness of thin carbon films. *Brit. J. Appl. Phys.* 8, 35.

v. Ardenne, M., and Friedrich-Freksa, H. (1941). Die Auskeimung der Sporen von *Bacillus vulgatus* nach vorheriger Abbildung im 200-kV-Universal-Elektronenmikroskop. *Naturwissenschaften* 29, 523.

Bahr, G. F., Johnson, F. B., and Zeitler, E. (1965). The elementary composition of organic objects after electron irradiation. *Lab. Invest.* 14, 1115.

Baumeister, W.. and Hahn, M. (1973). Elektronenmikroskopische Untersuchungen bei atomarar Auflösung an Modellmembranen. *Cytobiologie* 7, 244.

Baumeister, W., and Hahn, M. H. (1974a). Suppression of lattice periods in vermiculite single crystal specimen supports for high resolution electron microscopy. *J. Microscopy* 101, 111.

Baumeister, W., and Hahn, M. (1974b). Surface charge modification by ion exchange in single crystal specimen supports. *Proc. 8th Int. Congr. Electron Micros.* (Canberra) 2, 176.

Baumeister, W., and Hahn, M. (1975). Radiation resistant plastic specimen supports. *Naturwissenschaften* 62, 527.

Baumeister, W., and Hahn, M. (1976a). Prospects for atomic resolution electron microscopy in membranology. In: *Progress in Surface and Membrane Science*, Vol. 11 (Danielli, J. F., and Cadenhead, D., eds.), p. 227. Academic Press, New York.

Baumeister, W., and Hahn, M. (1976b). An improved method for the preparation of single crystal specimen supports, *Micron* 7, 247.

Baumeister, W., and Seredynski, J. (1976). Preparation of perforated films with pre-determinable hole size distributions. *Micron* 7, 49.

Baumeister, W., Hahn, M., and Fringeli, U. P. (1976). Electron beam induced conformational changes in polypeptide layers: an infrared study. *Z. Naturforsch.* 31c, 746.

Baumeister, W., Hahn, M., and Seredynski, J. (1978). Polystyrene specimen supports. To be published.

Beer, M., and Highton, P. J. (1962). A simple preparation of graphite-coated grids for high resolution electron microscopy. *J. Cell Biol.* 14, 499.

Beischer, D. E. (1950). Electronic radiography by transmission using radioactive monolayers. *Science.* 112, 535.

Bennett, H. E., and Bennett, J. M. (1967). *Physics of Thin Films*, Vol. 4. Academic Press, New York.

Berger, J. E., Taylor, C. A., Shechtman, D., and Lipson, H. (1972). Miscellaneous applications. In: *Optical Transforms* (Lipson, H., ed.), p. 401. Academic Press, London and New York.

Bergeron, G., and Pontefract, R. D. (1970). Production of carbon films for electron microscopy: hydrofluoric detachment from glass. *Stain Technol.* 45, 53.

Boersch, H., Hamisch, H., and Löffler, K. H. (1959). Elektronenoptische Herstellung freitragender Mikrogitter. *Naturwissenschaften* 46, 596.

Bradley, D. E. (1953). A new method of making electron microscope specimen support films. *Nature* 171, 1076.

Bradley, D. E. (1954). Evaporated carbon films for use in electron microscopy. *Brit. J. Appl. Phys.* 5, 65.

Bradley, D. E. (1965). The preparation of specimen support films. In: *Techniques for Electron Microscopy*, p. 58, F. A. Davis Corp., Philadelphia.

Brakenhoff, G. J. (1974). On the sub-nanometer structure visible in high-resolution dark-field electron microscopy. *J. Microscopy* 100, 283.

Brockes, A. (1957). Über Veränderungen des Aufbaus organischer Folien durch Elektronenbestrahlung. *Z. Phys.* 149, 353.

Burge, R. E. (1976). Binary filters for high resolution electron microscopy. II. *Optik* **44**, 159.

Burge, R. E., and Scott, R. F. (1975). Binary filters for high resolution electron microscopy. I. *Optik* **43**, 53.

Burge, R. E., and Silvester, N. R. (1960). The measurement of mass thickness and density in the electron microscope. *J. Biophys. Biochem. Cytol.* **8**, 1.

Busch, H. (1954). Dünnste Aufdampfungschichten auf Flüssigkeitsoberflächen (Glycerin). *Z. wiss. Mikros.* **62**, 152.

Cook, C. F., and Kerecman, A. J. (1962). Preparation and investigation of boron as a substrate material. *Proc. 5th Congr. Electron Micros.* (Philadelphia), EE-8.

Cosslett, V. E. (1948). Beryllium films as object supports in the electron microscopy of biological specimens. *Biochim. Biophys. Acta* **2**, 239.

Crisp, D. J. (1949). Surface films of polymers. Pt. 1. Films of the fluid type; Pt. 2. Films of the coherent and semi-crystalline type. *J. Colloid Sci.* **1**, 49, 161.

De, M. L. (1962). Considerations of image contrast in electron microscopy of objects composed of low atomic number. *Z. Naturforsch.* **17b**, 728.

De, M. L., and Sarkar, N. H. (1962). Optical density of thin carbon films. *Naturwissenschaften* **49**, 55.

De Boer, J., and Brakenhoff, G. J. (1974). A simple method for carbon film thickness determination. *J. Ultrastruct. Res.* **49**, 224.

DeSorbo, W., and Cline, H. E. (1970). Metal membranes with uniform submicron-size pores. *J. Appl. Phys.* **41**, 2099.

Dobelle, W. H., and Beer, M. (1968). Chemically cleaved graphite films for electron microscopy. *J. Cell Biol.* **39**, 733.

Dowell, W. C. T. (1957). Carbon-stabilized collodion substrates for electron microscopy. *J. Appl. Phys.* **28**, 634.

Dowell, W. C. T. (1959). Unobstructed mounting of serial sections. *J. Ultrastruct. Res.* **2**, 388.

Dowell, W. C. T. (1964). Die Entwicklung geeigneter Folien für elektronenmikroskopische Präparatträger großen Durchlaßbereichs und ihre Verwendung zur Untersuchung von Kristallen. *Optik* **21**, 47.

Dowell, W. C. T. (1970). The rapid production of holey carbon-formvar supporting films. *Septième Congr. Int. Micros. Electron.* (Grenoble) **I**, 321.

Drahos, V., and Delong, A. (1960). A simple method for obtaining perforated supporting membranes for electron microscopy. *Nature* **186**, 104.

Dubochet, J. (1975a). Carbon loss during irradiation of T4 bacteriophages and *E. coli* bacteria in electron microscopes. *J. Ultrastruct. Res.* **52**, 276.

Dubochet, J. (1975b). Personal communication.

Dubochet, J., Ducommun, M., Zollinger, M., and Kellenberger, E. (1971). A new preparation method for dark-field electron microscopy of biomacromolecules. *J. Ultrastruct. Res.* **35**, 147.

Dyson, J. (1970). *Interferometry as a Measuring Tool.* The Machinery Publishing Co., London.

Edstrom, T., and Lewis, I. C. (1969). Chemical structure and graphitization: X-ray diffraction studies of graphite derived from polynuclear aromatics. *Carbon* **7**, 85.

Fabergé, A. C. (1970). Benzylamine tartrate as an organic glass substrate for carbon films by evaporation. *Proc. 28th Ann. EMSA Meeting*, p. 484. Claitor's Pub. Division, Baton Rouge, La.

Fabergé, A. C. (1974). Carbon support films for electron microscopy by deposition on an organic glass. *J. Phys. E: Sci. Instr.* **7**, 95.

Fernández-Morán, H. (1960). Single-crystals of graphite and of mica as specimen supports for electron microscopy. *J. Appl. Phys.* **31**, 1840.

Fourie, F. T. (1975). The controlling parameter in contamination of specimens in electron microscopes. *Optik* **44**, 111.

Frank, J., and Al-Ali, L. (1975). Signal-to-noise ratio of electron micrographs obtained by cross correlation. *Nature* **256**, 376.

Franklin, R. E. (1950). The interpretation of diffuse x-ray diagrams of carbon. *Acta Cryst.* **3**, 107.

Frieser, H., and Klein, E. (1958). Die Eigenschaften photographischer Schichten bei Elektronenbestrahlung. *Z. angew. Phys.* **10**, 337.

Fryer, J. R., Hutchson, J. L., and Paterson, R. (1970). An electron microscopic study of the hydrolysis products of zirconyl chloride. *J. Coll. Interface Sci.* **34**, 238.

Fukami, A., Adachi, K., and Katoh, M. (1972). Micro grid techniques and their contribution to specimen preparation techniques for high resolution work. *J. Electronmicroscopy* **21**, 99.

Gale, B., and Hale, K. F. (1961). Heating of metallic foils in an electron microscope. *Brit. J. Appl. Phys.* **12**, 115.

Glaeser, R. M. (1975). In: *Physical Aspects of Electron Microscopy and Microbeam Analysis* (Siegel, B., and Beaman, D. R., eds.). John Wiley and Sons, New York.

Gordon, C. N. (1972). The use of octadecanol monolayers as wetting agents in the negative staining techique. *J. Ultrastruct. Res.* **39**, 173.

Gordon, C. N., and Kleinschmidt, A. K. (1970). Electron microscopic observation of DNA adsorbed on aluminum-mica. *J. Coll. Interface Sci.* **34**, 131.

Graff, K. (1961). Optische Dichte und Dicke von aufgedampften Kohlenstoffschichten. *Optik* **18**, 120.

Gregory, D. W., and Pirie, B. J. S. (1970). Wetting agents for electron microscopy of biological specimens. *Proc. 5th Europ. Congr. Electron Micros.* (Manchester), p. 234.

Gregory, D. W., and Pirie, B. J. S. (1973). Wetting agents for biological electron microscopy. I. General considerations and negative staining. *J. Microscopy* **99**, 261.

Gross, H. (1972). *Untersuchungen über das Aufdampfen von Trägerschichten für die Elektronenmikroskopie im Ultrahochvakuum.* Diplomarbeit ETH, Zürich.

Grubb, D. T. (1971). The calibration of beam measurement devices in various electron microscopes, using an efficient Faraday cup. *J. Phys. E: Sci. Instr.* **4**, 222.

Grubb, D. T. (1974). Radiation damage and electron microscopy of organic polymers. *J. Mat. Sci.* **9**, 1715.

Hahn, M. H. (1965). Zur Deutung der "Granulation" elektronenmikroskopischer Bilder hoher Vergrößerung. *Z. Naturforsch.* **20a**, 487.

Hahn, M. H. (1972). Eine optische Ortsfrequenzfilter- und Korrelationsanlage für elektronenmikroskopische Aufnahmen. *Optik* **35**, 326.

Hahn, M., and Baumeister, W. (1974). High resolution negative staining of ferritin molecules on vermiculite single crystal supports. *Biochim. Biophys. Acta* **371**, 267.

Hahn, M., and Baumeister, W. (1976). New specimen supports for high and ultrahigh resolution. EMAG 75th Meeting (Bristol). *Developments in Electron Microscopy and Analysis.* Academic Press, London.

Hahn, M., Seredynski, J., and Baumeister, W. (1976). Inactivation of catalase monolayers by 100 keV electrons. *Proc. Nat. Acad. Sci.* **73**, 823.

Hall, C. E. (1951). Scattering phenomena in electron microscope image formation. *J. Appl. Phys.* **22**, 655.

Hall, C. E. (1953). *Introduction to Electron Microscopy.* McGraw-Hill, New York, London, and Toronto.

Hall, C. E. (1956). Visualization of indvidual macromolecules with the electron microscope. *Proc. Nat. Acad. Sci. USA* **42**, 801.

Hall, D. M. (1957). The deterioration of nitro-cellulose films used for electron microscopy. *Brit. J. Appl. Phys.* 8, 380.

Hanssen, K.-J. (1966). Generalisierte Angaben über die Phasenkontrast- und Amplitudenkontrast-Übertragungsfunktionen für elektronemikroskopische Objekte. *Z. angew. Phys.* 20, 427.

Hanssen, K.-J., and Morgenstern, B. (1965). Die Phasenkontrast- und Amplitudenkontrast-Übertragungsfunktionen für elektronenmikroskopische Objeke. *Z. angew. Phys.* 19, 215.

Harris, L. (1955). Preparation and infrared properties of aluminum oxide films. *J. Opt. Soc. Amer.* 45, 27.

Harris, W. J. (1962). Holey films for electron microscopy. *Nature* 196, 499.

Hashimoto, H., Kumao, A., Endoh, H., and Ono, A. (1974). Image contrast of atoms and substrate films in bright and dark field technique. *Proc. 8th Int. Congr. Electron Micros..* (Canberra). I, 244.

Hashimoto, H., Kumao, A., Hino, K., Endoh, H., Yotsumoto, H., and Ono, A. (1973). Visualization of single atoms in molecules and crystals by dark field electron microscopy. *J. Electron Microscopy* 22, 123.

Hashimoto, H., Kumao, A., Hino, K., Yotsumoto, H., and Ono, A. (1971). Image of thorium atoms in transmission electron microscopy. *Jap. J. Appl. Phys.* 10, 1115.

Hass, G., and Kehler, H. (1941). Über eine temperaturbeständige und haltbare Trägerschicht für Elektroneninterferenzaufnahmen und Übermikroskopische Untersuchungen. *Kolloid-Z.* 95, 26.

Hass, G., and Kehler, H. (1949). Über ein Verfahren zur Untersuchung unzusammenhängender dünner Schichten im Übermikroskop und zur Herstellung von Abdruckfilmen und Trägerfolien mittels Aufdampfen von Silizium-Monoxyd. *Optik* 5, 48.

Hast, N. (1947). Preparation of thin specimen films. *Nature* 159, 170.

Hast, N. (1948). Production of extremely thin metal films by evaporation onto liquid surfaces. *Nature* 162, 892.

Hayat, M. A. (1970). *Principles and Techniques of Electron Microscopy: Biological Applications,* Vol. 1. Van Nostrand Reinhold Company, New York and London.

Heide, H. G. (1962). The prevention of contamination without beam damge to the specimen. *5th Int. Congr. Electron Micros.* (Philadelphia) 1, A4 Academic Press., New York and London.

Heide, H. G. (1963). Die Objektverschmutzung im Elektronenmikroskop und das Problem der Strahlenschädigung durch Kohlenstoffabbau. *Z. angew. Phys.* 15, 116.

Heidenreich, R. D., Hess, W. M., and Ban, L. L. (1968). A test object and criteria for high resolution electron microscopy. *J. Appl. Cryst.* 1, 1.

Heinemann, K. (1970). A comment on mica as electron microscope specimen support film. *Proc. 28th Ann. EMSA Meeting* (Arcenaux, C. J., ed.) p. 526. Claitor's Pub. Division, Baton Rouge, La.

Heinemann, K. (1974). Personal communication.

Henglein, A., Schnabel, W., and Wendenburg, J. (1969). *Einführung in die Strahlenchemie.* Verlag Chemie GmbH, Weinheim/Bergstr.

Highton, P. J., and Whitfield, M. (1974). The control of the configuration of nucleic acid molecules deposited for electron microscopy, by ionic bombardment of carbon films. *J. Microscopy* 100, 299.

Hines, R. L. (1975). Graphite crystal film preparation by cleavage. *J. Microscopy* 104, 257.

Hoelke, G. W. (1975). Preparation and use of holey carbon microgrids in high resolution electron microscopy. *Micron* 5, 307.

Holl, M., Haas, D., Berger, R., Schürmann, D., and v. Buttlar, H. (1974). Preparation of self-supporting carbon foils of uniform thickness below 2 μg/cm^2 density. *J. Appl. Phys.* 45, 3069.

Horn, H. R. F. (1962). Thermische Bedampfung ohne Licht- und Wärmestrahlungsbelastung des Objektes. *5th Int. Congr. Electron Micros.* (Philadelphia) 1, A9. Academic Press, New York and London.

Horne, R. W. (1965). Negative staining methods. In: *Techniques for Electron Microscopy* (Kay, D. H., ed.), p. 328. Blackwell Scientific Publications, Oxford.

Humphreys, W. J. (1963). Preparation of carbon films by use of a glycerin substrate. *J. Cell Biol.* 19, 634.

Iijima, S. (1977). Observation of single and clusters of atoms in bright field electron microscopy. *Optik* 48, 193.

Isaacson, M. S. (1977). Specimen damage in the electron microscope. In: *Principles and Techniques of Electron Microscopy: Biological Applications, Vol. 7* (Hayat, M. A., ed.) Van Nostrand Reinhold Company, New York and London.

Isaacson, M., Langmore, J. P., and Rose, H. (1974a). Determination of the non-localization of the inelastic scattering of electrons by electron microscopy. *Optik* 41, 92.

Isaacson, M., Langmore, J., and Wall, J. (1974b). The preparation and observation of biological specimens for the scanning transmission electron microscope. In: *Scanning Electron Microscopy* (Johari, O., and Corvin, I., eds.) p. 19. IITRI, Chicago.

van Itterbeek, A., de Greve, L., van Veelen, G. F., and Tuynman, C. A. F. (1952). Glass layers as supports for electron microscopy. *Nature* 170, 795.

Jaffe, M. S. (1948). Auxiliary supporting nets for fragile electron microscope specimens. *J. Appl. Phys.* 19, 1191.

Jenkins, G. M., Turnbull, J. A., and Williamson, G. K. (1962). Graphitization deformation in carbon films. *Proc. 5th Int. Congr. Electron Micros.* (Philadelphia) 1, GG-11.

Johansen, B. V. (1974). Bright field electron microscopy of biological specimens. II. Preparation of ultra-thin carbon support films. *Micron* 5, 209.

Johansen, B. V. (1975). Bright field electron microscopy of biological specimens. IV. Ultrasonic exfoliated graphite as "low-noise" support films. *Micron* 6, 165.

Johansen, B. V. (1976). Bright field electron microscopy of biological specimens. VI. Signal-to-noise ratio in specimens prepared on amorphous carbon and graphite crystal supports. *Micron*, 7, 57.

Johnston, H. S., and Reid, O. (1971). An improved method for preparing perforated carbon films for electron microscopy using ultrasonic vibration. *J. Microscopy* 94, 283.

Kakinoki, J., Katada, K., Hanawa, T., and Ino, T. (1960). Electron diffraction study of evaporated carbon films. *Acta Cryst.* 13, 171.

Kanaya, K. (1955). The temperature distribution of specimens on thin substrates supported on a circular opening in the electron microscope. *J. Electronmicroscopy* 3,1.

Karu, A. E., and Beer, M. (1966). Pyrolytic formation of highly crystalline graphite films. *J. Appl. Phys.* 37, 2179.

Kausche, G. A., Pfankuch, E., and Ruska, H. (1939). Die Sichtbarmachung von pflanzlichem Virus im Übermikroskop. *Naturwissenschaften* 27, 292.

Kaye, W. (1949). An aluminum-beryllium alloy for substrate and replica preparation in electron microscopy. *J. Appl. Phys.* 20, 1209.

King, R. J., Downs, M. J., Clapham, P. B., Raine, K. W., and Talim, S. P. (1972). Comparison of methods for accurate film thickness measurement. *J. Phys. E: Sci. Instr.* 5, 445.

Kirchner, F. (1930). Interferenzapparat für Demonstration und Strukturuntersuchungen. *Physikal. Z.* 31, 772.

Kleinschmidt, A. K., and Vasquez, C. (1969). Electron microscopy of unstained biomacromolecules. *Proc. 27th Ann. EMSA Meeting* (Arcenaux, C. J., ed.), p. 264. Claitor's Pub. Division, Baton Rouge, La.

Klug A., and de Rosier, D. J. (1966). Optical filtering of electron micrographs: reconstruction of one-sided images. *Nature* 212, 29.

Kölbel, H. (1965/66). Die Stabilisierung von Trägerfilmen für die Elektronenmikroskopie mit Hilfe von Silikonen (Organopolysiloxanen). *Z. wiss. Mikros.* **67**, 171.

Kölbel, H. K. (1972). Der Einfluß verschiedenartiger Trägerfilme auf Bildgröße und -kontrast von Dünnschnitten biologischer Objekte im Elektronenmikroskop. *Mikroskopie* **28**, 202.

Kölbel, H. K. (1974). Indirect evaporation of carbon onto films and other substrates. *8th Int. Congr. Electron Micros.* (Canberra) **1**, 404.

Koller, T., Beer, M., Müller, M., and Mühlethaler, K. (1971). Electron microscopy of selectively stained molecules. *Cytobiol.* **4**, 369.

Koller, T., Beer, M., Müller, M., and Mühlethaler, K. (1973). Electron microscopy of selectively stained molecules. In: *Principles and Techniques of Electron Microscopy: Biological Applications*, Vol. 3 (Hayat, M. A., ed.), p. 53. Van Nostrand Reinhold Company, New York and London.

Koller, T., Sogo, J. M., and Bujard, H. (1974). An electron microscopic method for studying nucleic acid–protein complexes. Visualization of RNA polymerase bound to the DNA of bacteriophages T7 and T3. *Biopolymers* **13**, 995.

Kübler, O., Hahn, M. and Seredynski, J. (1978). Digital and optical spatial frequency filtering of electron micrographs. I. *Optik*, in press.

Kübler, O., and Downing, K. H. (1978). Quantitative Untersuchung von Trägerfilmen. In preparation.

Kübler, O., and Koller, Th. (1976). Automatic tracing of nucleic acid molecules. *6th Europ. Reg. Conf. Electron Micros.* (Jerusalem) **II**, 552.

Lange, R. H. (1972). Korrelative Licht- und Elektronenmikroskopie unter Berücksichtigung der Histochemie. *Mikroskopie* **28**, 193.

Langer, R., Frank, J., Feltynowski, A., and Hoppe, W. (1970). Anwendung des Bilddifferenzverfahrens auf die Untersuchung von Strukturveränderungen dünner Kohlefolien bei Elektronenbestrahlung. *Ber. Bunsenges. Phys. Chem.* **74**, 1120.

Leder, L. B., and Suddeth, J. A. (1960). Characteristic energy losses of electrons in carbon. *J. Appl. Phys.* **31**, 1422.

Leisegang, S. (1956). Elektronenmikroskope. In: *Handbuch der Physik* **33**, 396 (Flügge, S., ed.). Springer-Verlag, Berlin, Göttingen, Heidelberg.

Lickfeld, K. G., and Menge, B. (1968). Ein neues Verfahren zur Herstellung von Lochfolien für die Elektronenmikroskopie. *Z. wiss. Mikros.* **69**, 18.

Lippert, W. (1969). Erfahrungen mit der photographischen Methode bei der Massendickenbestimmung im Elektronenmikroskop. *Optik* **29**, 372.

Lippert, W. (1970). Quantitative Auswertung von Helligkeitsunterschieden im elektronenmikroskopischen Bild (elektronenmikroskopische Massendickenbestimmung). In: *Methodensammlung der Elektronenmikroskopie.* S. 1 (Schimmel, G., Vogell, W., Hrsg.). Wissenschaftl. Verlagsges. mbH, Stuttgart.

Loeffler, K. H. (1964). Erzeugung freitragender Mikroobjekte durch elektronenstrahlaktivierten Kohlefolienabbau. Dissertation D 83, Techn. Univ. Berlin.

Mahl, H. and Möldner, K. (1972). Beobachtungen an aufgedampften C-Schichten. *Mikroskopie* **28**, 139.

Mahl, H., and Weitsch, W. (1960). Nachweis von fluktuierenden Ladungen in isolierenden Filmenbú Elektronenbestrahlung. *Optik* **17**, 107.

Manley, J. H., Williams, D. L. and Okinaka, R. (1971). Vermiculite lamellae as substrate for transmission electron microscopy. *J. Microscopy* **94**, 73.

Maréchal, A., and Croce, P. (1956). Amélioration de la perception des détails des images par filtrage optique des fréquences spatiales. In: *Problems in Contemporary Optics*, p. 76. Istituto Nazional di Ottica, Firenze.

Markham, R., Frey, S., and Hill, G. J. (1963). Methods for the enhancement of image detail and accentuation of structure in electron microscopy. *Virology* **20**, 696.

Martin, J. P., and Speidel, R. (1972). Herstellung freitragender Mikrogitter im Elektronen-Rastermikroskop "Steroscan Mk II." *Optik* **36**, 13.

Marton, L. (1935). La microscopie électronique des objets biologiques (4. communication). *Bull. Acad. Roy. Belg. (Classe des Sciences)* **21**, 606.

Marton, L. (1936). La microscopie électronique des objets biologiques (4. communication). *Bull. Acad. Roy. Belg. (Classe des Sciences)* **22**, 1336.

Marton, L. (1937). La microscopie électronique des objets biologiques (5. communication). *Bull. Acad. Roy. Belg. (Classe des Sciences)* **23**, 672.

Mayer, H. (1950). *Physik dünner Schichten I.* Wissenschaftl. Verlagsges. mbH, Stuttgart.

Mihama, K., Horata, A., and Uyeda, R. (1973). Observation of extremely fine gold crystallites grown on molybdenite films by high resolution electron microscopy. *Jap. J. Appl. Phys.* **12**, 746.

Mihama, K., Shima, S., and Uyeda, R. (1974). BeO supporting films for high resolution electron microscopy. *Jap. J. Appl. Phys.* **13**, 377.

Misra, D. N., and Ganguly, P. (1968). Negative staining of human hemoglobin molecules with uranyl acetate. *Arch. Biochem. Biophys.* **124**, 349.

Moharir, A. V., and Prakash, N. (1975). Formvar holey films and nets for electron microscopy. *J. Phys. E: Sci. Instr.* **8**, 289.

Möldner, K. (1965). Ein einfaches Verfahren zur Herstellung von Lochfolien. *Naturwissenschaften* **52**, 449.

Möllenstedt, G. (1947). Silikatglas als haltbare, temperaturbeständige und säurefeste Trägerfolie für Elektroneninterferenzen und Elektronenmikroskopie. *Optik* **2**, 276.

Moore, N. T., Chatterji, S., and Jeffery, J. W. (1971). Dependence of film thickness on the heater geometry in vacuum coating technique. *J. Phys. E: Sci. Instr.* **4**, 1075.

Moretz, R. C., Johnson, H. M., and Parsons, D. F. (1968). Thickness estimation of carbon films by electron microscopy of transverse sections and optical density measurements. *J. Appl. Phys.* **39**, 5421.

Müller, F. H. (1954). Untersuchungen an Oberflächenfilmen hochpolymer Substanzen. *Kolloid-Z.* **136**, 127.

Müller, K.-H. (1970). Micro-recording by use of the Elmiskop 101. *Septième Congr. Int. Micros. Electron.* (Grenoble) **I**, 183.

Müller, M., Downing, K. H., Kübler, O., and Koller, T. (1975). Encapsulation of macromolecules into "low noise" supports. *J. Microscopie Biol. Cell.* **23**, 117.

Müller, M., and Koller, Th. (1972). Preparation of aluminum oxide films for high resolution electron microscopy. *Optik* **35**, 287.

Müller, M., Koller, Th., Moor, H. (1970). Preparation and use of aluminum films for high resolution electron microscopy of macromolecules. *Proc. 7th Int. Congr. Electron Micros.* (Grenoble) **1**, 633.

Münch, G. (1964). Simplified preparation method for carbon replicas and carbon films for specimen support in electron microscopy. *Rev. Sci. Instr.* **35**, 524.

Myers, G. E., and Montet, G. L. (1966). Optical density and thickness of graphite lamellae. *J. Appl. Phys.* **37**, 4196.

Noda, T., and Inagaki, M. (1964). Effect of gas phase on graphitization of carbon. *Carbon* **2**, 127.

Ottensmeyer, F. P. (1976). Personal communication.

Palatnik, L. S. (1970). *Sov. Phys.* **11**, 2086. (In Russian)

Parkinson, W. W., and Keyser, R. M. (1973). Polystyrene and related polymers. In: *The Radiation Chemistry of Macromolecules* 57 **II**, (Dole, M., ed.). Academic Press, New York and London.

Pease, D. C. (1975). Micronets for electron microscopy. *Micron* **6**, 85.

Polivoda, A. I., and Vinetskii, Y. P. (1959). A method of preparing quartz films for electron microscopy in studies of the fine structure of erythrocytes. *Biofizika* **4**, 599.

Polivoda, A. I., and Vinetskii, Y. P. (1961). Electron microscopic study of erythrocytes on quartz and collodion films. *Biofizika* 6, 128.

Poppa, H., Heinemann, K., and Elliot, A. G. (1971). Epitaxial orientation studies of gold on uhv-cleaved mica during early stages of nucleation and growth. *J. Vacuum Sci. Technol.* 8, 471.

Presland, A. E. B., and White, J. R. (1969). Graphitization of evaporated carbon films. *Carbon* 7, 77.

Presland, A. E. B., and White, J. R. (1970). An electron diffraction study of graphitization in evaporated carbon films. *Micron* 2, 73.

Prestridge, E. B., and Yates, D. J. C. (1971). Imaging the rhodium atom with a conventional high resolution electron microscope. *Nature* 234, 345.

Pulker, H., and Ritter, E. (1964). Optische Eigenschaften von Siliciumaufdampfschichten im Bereich 500-800 Millimikron. *Optik* 21, 21.

Reale, E., and Luciano, L. (1965). Die Anwendung der Dowell'schen Präparatträger in der Histologie. *J. Microscopie* 4, 405.

Reimer, L. (1961). Messung der Abhängigkeit des elektronenmikroskopischen Bildkontrastes von Ordnungszahl, Stromspannung und Aperturblende. *Z. angew. Phys.* 13, 432.

Reimer, L. (1965a). Contrast in amorphous and crystalline objects. *Lab. Invest.* 14, 939.

Reimer, L. (1965b). Aufladung kleiner Teilchen im Elektr. mikr. *Z. Naturforsch.* 20a, 151.

Reimer, L. (1965c). Irradiation changes in organic and inorganic objects. *Lab. Invest.* 14, 1082.

Reimer, L. (1967). *Elektronenmikroskopische Untersuchungs- und Präparationsmethoden.* 2. Aufl. Springer-Verlag, Berlin, Heidelberg, New York.

Reimer, L. (1969). Elektronenoptischer Phasenkontrast. II. Berechnung mit komplexen Atomstreuamplituden für Atome und Atomgruppen. *Z. Naturforsch.* 24a, 377.

Reimer, L., and Christenhuss, R. (1962). Experimenteller Beitrag zur Objekterwärmung im Elektronenmikroskop. *Z. angew. Phys.* 14, 601.

Reimer, L., and Christenhuss, R. (1965). Determination of specimen temperature. Proc. Symp. Quant. EM (Washington, 1964). *Lab. Invest.* 14, 1158.

Revell, R. S. M., Agar, A. W., and Lee, A. M. (1955). The preparation of uniform plastic films. *Brit. J. Appl. Phys.* 6, 23.

Riddle, G. H. N., and Siegel, B. M. (1971). Thin pyrolytic graphite films for electron microscope substrates. *Proc. 29th Ann. EMSA Meeting* (Arceneaux, C. J., ed). Claitor's Pub. Division, Baton Rouge, La.

Riecke, W. D. (1972). Objective lens design for transmission electron microscopes–a review of the present state of the art. *Proc. 5th Europ. Congr. Electron Micros.* (Manchester), p. 98.

Rose, H. (1975). Zur Theorie der Bildentstehung im Elektronenmikroskop I. *Optik* 42, 217.

Rose, H. (1976). Image formation by inelastically scattered electrons in electron microscopy. I. *Optik* 45, 139. II. *Optik* 45, 187.

Rothen, A. (1968). Surface film techniques. In: *Physical Techniques in Biological Research,* 2nd ed., Vol. II, p. 217, Pt. A., "Physical Chemical Techniques," Academic Press, New York and London.

Ruben, G., and Siegel, B. M. (1975). Thin carbon-aluminum films for dark field electron microscopy of DNA. 33rd Ann. proc. E.M.S.A., p. 658. Las Vegas, Nev., G. W. Bailey (ed).

Ruska, E. (1934). Über ein magnetisches Objektiv für das Elektronenmikroskop. *Z. Phys.* 89, 90.

Ruska, H. (1939). Übermikroskopische Untersuchungstechnik. *Naturwissenschaften* 27, 287.

Sakata, S. (1958). Preparation of micro-grids for the electron microscopic observation with high magnification. *J. Electronmicroscopy* 6, 75.

Salih, S. M., and Cosslett, V. E. (1974). Reduction in electron irradiation damage to organic compounds by conducting coatings. *Philos. Mag.* 30, 225.

Sarkar, N. H. (1966). Mass-scattering cross section of thin carbon films. *J. Appl. Phys.* 37, 4389

Savdir, S. (1963). Eine einfache Methode der Bildung von Formvarfilmen für elektronenmikropische Schnitte auf Gittern. *Sci. Tools* 10, 12.

Schaefer, V. J., and Harker, D. (1942). Surface replicas for use in the electron microscope. *J. Appl. Phys.* 13, 427.

Schnabl, H. (1971). Thermische Schwingungen elektronenmikroskopischer Objekte. *Z. angew. Phys.* 31, 214.

Schnabl, H. (1972). Verringerung des Störkontrasts im Elektronenmikroskop durch Neigen der kristallinen Objektträger-Folie. *Optik* 36, 37.

Schober, B., and Prokes, V. (1955). Supporting films of carbon for use in electron microscopy. *Folia Biol.,* Ceskoslovenska akademie ved. Praha, 1, 316.

Schön, G., and Weiss, A. (1974). Ein statistisches Modell der Ladungsverteilung in quellungsfähigen glimmerartigen Schichtsilicaten mit beidellitischer Ladungsverteilung. *Z. Naturforsch.* 29b, 44.

Schuster, M. C., and Fullam, E. F. (1946). Preparation of powdered materials for electron microscopy. *Ind. Eng. Chem.* 18, 653.

Schwarz, H. (1966). Evaporation and deposition rates, sticking probability, and background pressure during the production of thin films in vacuum. *J. Appl. Phys.* 37, 4341.

Schwertfeger, W. (1974). Zur Kleinwinkelstreuung von mittelschnellen Elektronen beim Durchgang durch amorphe Festkörperschichten. Dissertation, Tübingen.

Siddall, G. (1961). A method of stripping carbon films from glass. *Proc. Europ. Reg. Conf. Electron Micros.* (Delft) 1, 584.

Sjöstrand, F. S. (1956). An improved method to prepare Formvar nets for mounting thin sections for electron microscopy. In: *Stockholm Conf. Electron Microscopy.* Proc. 120.

Snoeyink, V. L., and Weber, W. J., Jr. (1971). Surface functional groups on carbon and silica. In: *Progress in Surface and Membrane Science* (Danielli, J. F., and Cadenhead, D., eds.) p. 327. Academic Press, New York.

Sogo, J. M., Portmann, R., Kaufmann, P., and Koller, T. (1975). Adsorption of DNA molecules to different support films. *J. Micorscopy* 104, 187.

Spehr, R., and Schnabl, H. (1973). Zur Deutung der unterschiedlichen Strahlen-Empfindlichkeit organischer Moleküle. *Z. Naturforsch.* 28a, 1729.

Spencer, M. (1959). The preparation of carbon films for electron microscopy. *J. Biophys. Biochem. Cytol.* 6, 125.

Stenn, K., and Bahr, G. F. (1970). Specimen damage caused by the beam of the transmission electron microscope, a correlative reconsideration. *J. Ultrastruct. Res.* 31, 526.

Stockem, W. (1970). Die Eignung von pioloform F für die Herstellung elektronenmikroskopischer Trägerfilme. *Mikroskopie* 26, 185.

Stolinski, C., and Gross, M. (1969). A method for making thin, large surface area carbon supporting films for use in electron microscopy. *Micron* 1, 340.

Tanaka, M., Higashi-Fujime, S., and Uyeda, R. (1974). Ultrafine grids for specimen supporting media in high resolution electron microscopy. *8th Int. Congr. Electron Micros.* (Canberra). II, 180.

Tesche, B. (1973). Herstellung dünner Lochfolien mit einheitlichen Lochdurchmessern. *Naturwissenschaften* 60, 549

Thach, R. E., and Thach, S. S. (1971). Damage to biological samples caused by the electron beam during electron microscopy. *Biophys. J.* 11, 204.

Thon, F. (1965). Elektronenmikroskopische Untersuchungen an dünnen Kohlefolien. *Z. Naturforsch.* 20a, 154.

Thon, F. (1966). Zur Defokussierungsabhängigkeit des Phasenkontrastes bei der elektronenmikroskopischen Abbildung. *Z. Naturforsch.* 21a, 476.

Thon, F., and Willasch, D. (1970). Hochauflösungs-Elektronenmikroskopie mit Spezialaperturblenden und Phasenplatten. *Septième Congr. Int. Micros. Electron.* (Grenoble) **I**, 3.

Tolansky, S. (1948). *Multiple Beam Interferometry of Surface and Films.* Oxford Univ. Press, London and New York.

Trenktrog, W. (1923). Messungen an sehr weichen Röntgenstrahlen. Dissertation, Kiel.

Turnbull, J. A., and Williamson, G. K. (1963). Graphitization as observed in thin carbon films, and its relation to the mechanical properties of polycrystalline graphite. *Trans. Brit. Ceram. Soc.* **62**, 807.

Unwin, P. N. T. (1974). Electron microscopy of the stacked disk aggregate of tobacco mosaic virus protein. II. The influence of electron irradiation on the stain distribution. *J. Mol. Biol.* **87**, 657.

Uyeda, R. (1973). An attempt to visualize single atoms by means of electron microscopy [in Japanese]. *Physics* **28**, 378.

Uyeda, R. (1974). Innovations in specimen supporting media for high resolution. *8th Int. Congr. Electron Micros.* (Canberra) **I**, 246.

Valentine, R. C. (1961). Contrast enhancement in the electron microscopy of viruses. *Adv. Virus Res.* **8**, 287.

Valentine, R. C. (1964). Fundamental difficulties in obtaining very high resolution of biological specimens. *Proc. 3rd Europ. Reg. Conf. Electron Micros.* (Prague) **B**, 23.

Vander, Lugt, B. A. (1964). Signal detection for complex spatial filtering. *IEEE Trans. Inform. Theory* IT-10, No. 2, 139.

Varon, J., Schiavo, J., and Janus, T. P. (1967). Ultrathin Al_2O_3 support films for electron microscopy. *Rev. Sci. Instr.* **38**, 690.

Vollenweider, H.-J., Koller, T., and Kübler, O. (1973). Aluminum-beryllium alloy films as specimen supports for high resolution electron microscopy. *J. Microscopie* **16**, 247.

Walkenhorst, W. (1947). Ein einfaches Verfahren zur Herstellung strukturloser Trägerschichten aus Aluminiumoxyd. *Naturwissenschaften* **12**, 373.

Walker, G. F. (1950). Vermiculite-organic complexes. *Nature* **166**, 695.

Walker, G. F. (1960). Crystallography. Macroscopic swelling of vermiculite crystals in water. *Nature* **187**, 312.

Walker, G. F. (1961). *Vermiculite Minerals in the X-Ray Identification and Crystal Structures of Clay Minerals,* (Brown, G., ed.), p. 297. Mineralogical Soc., London.

Watts, J. (1949). Perforated object supporting films for the electron microscope. *J. Sci. Instr. Phys. Industry.* **26**, 158.

Weber, K., and v. Fragstein, C. (1954). Über die Durchlässigkeit von Trägerfolien für den Elektronenstrahl im Übermikroskop. *Optik* **11**, 511.

Weichan, C. (1970). Hochauflösungs-Testpräparate. In: *Methodensammlung der Elektronenmikroskopie* (Schimmel, G., und Vogell, W., eds.), Nr. 3. 1. 1. 1.

Weiss, A. (1958). Der Kationenaustausch bei den Mineralen der Glimmer-, Vermikulit- und Montmorillonitgruppe. *Z. anorg. all. Chemie* **297**, 257.

Weiss, A., Mehler, A., and Hofmann, U. (1956). Zur Kenntnis von organophilem Vermiculit. *Z. Naturforsch.* **11b**, 431.

Welton, T. A. (1971). Electron optical factors limiting resolution in transmission electron microscopes. *Proc. Workshop Conf. Microscopy of Cluster Nuclei in Detected Crystals,* Chalk River Nuclear Labs. CRNL-622-1, p. 125. Canada.

White, J. R., Beer, M., and Wiggins, J. W. (1971). Preparation of smooth graphite support films for high resolution electron microscopy. *Micron* **2**, 412.

Whiting, R. F., and Ottensmeyer, F. P. (1972). Heavy atoms in model compounds and nucleic acids imaged by dark field transmission electron microscopy. *J. Mol. Biol.* **67**, 173.

Wiesenberger, E. (1963). Personal communication.

Wiggins, J. W., and Beer, M. (1972). Prospects for molecular microscopy. *Anal. Chem.* **44**, 77A.

Williams, R C., and Glaeser, R. M. (1972). Ultrathin carbon support films for electron microscopy. *Science* **175**, 1000.

Yang, G. C. H., and Shea, S. M. (1974). The precise measurement of the thickness of ultrathin sections by a "resectioned section" technique. *J. Microscopy* **103**, 385.

Zeitler, E. (1965). Theory of elastic scattering of electrons. *Lab. Invest.* **14**, 36; *Proc. Symp. Quant. Electron Micros.* (Washington, 1964), p. 774.

Zeitler, E., and Bahr, G. F. (1962). A photometric procedure for weight determination of submicroscopic particles. *J. Appl. Phys.* **33**, 847.

Zingsheim, H. P. (1976). The physical deposition of information on a molecular scale. *Ber. Bunsenges. Phys. Chem.*, **80**, 1185.

2. PREPARATION AND ANALYSIS OF SERIAL SECTIONS IN ELECTRON MICROSCOPY

Robert L. Knobler, Jerome G. Stempak, and
Mary Laurencin

Departments of Neurology, Anatomy and Cell Biology, State University of New York,
Downstate Medical Center, Brooklyn, New York

INTRODUCTION

Each thin section analyzed in electron microscopy yields a two-dimensional image. A third dimension may be added to individual sections by stereomicroscopy (Helmcke, 1965; Peachey, 1965; Gray and Willis, 1968); for a cell or tissue a third dimension may be inferred from stereological methods (Weibel, 1969; Elias et al., 1971; Weibel and Bolender, 1973) or be determined from micrographs of a series of thin sections by reconstruction. Serial section analysis and reconstruction, although the most accurate method for determining three-dimensional form and relationships, is not a commonly used technique, owing in part to the technical difficulty of collecting and analyzing consecutive series of sections, and the tedious, time-consuming nature of making reconstructions from micrographs.

Despite these obstacles, there has been a growth of interest in applying the technique in recent years, especially to studies of development. Recent studies have attempted to elucidate the structure and development of intracellular organelles and complex cellular arrangements, particularly in tissues of the nervous system. Many of the studies are included in Table 2.1.

Table 2.1   Some Serial Section Electron Microscopical Studies

Autoradiography Nadelschaft and Rose (1974)	*Neurocytology, myelination-CNS* Knobler and Stempak (1973)
Histochemistry Zotikov and Bernhard (1970)	Knobler *et al.* (1974, 1976) Stempak and Knobler (1972)
Microorganisms Donelli *et al.* (1970) Gillies (1972) Keddie and Barajas (1969, 1972) O'Leary *et al.* (1973) Sotelo *et al.* (1973) Suganuma (1968) Vivier and Petitprez (1972) Ward *et al.* (1975)	*Neurocytology, myelination-PNS* Martin and Webster (1973) Webster (1971) Webster *et al.* (1973) *Neurocytology, neurocellular* Conradi (1969a,b,c) Conradi and Skoglund (1969a,b) Elfvin (1963a, 1971a) Famiglietti and Peters (1972)
Mitochondria Berger (1973) Hoffman and Avers (1973) Stempak (1967) Stempak and Laurencin (1976)	Hinds and Hinds (1972, 1974) Hinds and Ruffet (1971, 1973) Karlsson (1966a,b) Kishi (1972) Pinching and Powell (1971a)
Red cell differentiation Campbell (1968)	Poritsky (1969) Price and Powell (1970a,d) Rakic (1972) Vaughn and Peters (1973)
Heart, cardiovascular De Kock and Dunn (1964) Thaemert (1969, 1970, 1973)	*Neurocytology, synaptic organization* Elfvin (1963b, 1971b,c) Hinds and Hinds (1976a,b) Jones and Brearley (1973)
Kidney Barajas (1970) Barajas and Muller (1973) Shimamura and Sorenson (1965)	Pinching and Powell (1971b,c) Price and Powell (1970b,c) Sjöstrand (1958, 1974) Špaček *et al.* (1973)
Neurocytology, fiber systems Andersson-Cedergren and Karlsson (1966) Hildebrand (1971a,b,c) LoPresti *et al.* (1973, 1974) Macagno *et al.* (1973) Reiter (1966)	Špaček and Lieberman (1974) White (1972, 1973, 1976) *Muscle* Andersson-Cedergren (1959) Karlsson *et al.* (1966) Pachter *et al.* (1973, 1974)
Neurocytology, glial cells Špaček (1971) Stensaas and Stensaas (1968)	Swash and Fox (1975) Thaemert (1966)

Detailed reconstructions also provide information of value in interpreting physiological and biochemical parameters of cell function. This information includes the distribution of organelles within a cell, and their relative numbers. Quantitative analyses of normal structures are of particular interest with respect to the interpretation of pathological changes; three-dimensional analyses can be used to check the reliability of quantitative information obtained by the analysis of single thin sections.

Serial section electron microscopical analysis provides the only means of investigating cellular arrangements below the resolution of the light microscope, yet too complex to be worked out from single section electron micrographs. Serial section analysis can define a cytoplasmic profile by demonstrating continuity with its cell of origin, as well as defining its course, shape, and extent. Thus, serial section electron microscopical analysis and three-dimensional reconstruction can extend the limits of single section microscopy.

Since the first demonstration of the practicality and usefulness of the reconstruction technique by Sjöstrand (1958), considerable technical knowledge applicable to serial section electron microscopical analysis has accumulated. The methods of collecting and photographing serial sections are a routine matter with sufficient practice. Newer automated methods of analysis and reconstruction already in use (Hillman *et al.*, 1977; Llinás and Hillman, 1975; Ware and LoPresti, 1975) are leading to more widespread interest in the technique. It is our aim to discuss the most practical methods of performing serial section electron microscopical analysis.

Serial section electron microscopical analysis makes use of all the routine aspects of electron microscopic technique, such as fixation, embedding, staining, sectioning, and support films, which have been covered in great detail in Volume 1 of this series (Hayat, 1970). Thus, a knowledge of basic electron microscopic techniques is presumed, and only those steps relevant to serial section electron microscopy will be emphasized. It should be pointed out that most laboratories presently involved in serial section electron microscopical analysis have devised their own methods and preferences of section collection and reconstruction. We will present the various steps as practiced in our laboratory, and will point out, where they are known and thought to be significant, variations from the procedures which we follow. The technique may be subdivided into the following components: tissue fixation, staining, embedding, grid preparation, sectioning, section collection, microscopy, analysis, and reconstruction. This will be the order of discussion.

TISSUE FIXATION

Tissues fixed by osmium tetroxide were recognized as well-fixed if they exhibited "a large area of intact tissue, in which clefts, shrinkage or retraction spaces, explosions, vacuoles, disrupted limiting membranes and dehiscences are all absent" (Palay *et al.*, 1962). Introduction of glutaraldehyde fixation (Sabatini *et al.*, 1963) made it clear that considerable extraction occurred in tissues fixed by osmium tetroxide alone, and any current definition of fixation must include a statement concerning satisfactory retention of all cell components.

Further, it has become clear that differences exist in the quality of fixation that may be obtained in different tissues with the same fixative, or in the same tissue at different stages in development. Tennyson (personal communication,

1973) stated: "I have been working on embryonic tissue for 15 years and I still have not found the 'perfect fixative.' . . . Each embryonic stage is a different problem and will require a different fixative . . . I usually fix any particular embryonic stage with a number of different fixatives and use the specimen which is best preserved . . . good results are often more by luck than design."

Serial section analysis of developing nervous tissue appearing adequately fixed by single section analysis reveals that components fixed well at one level may contain artifacts at subsequent levels. In addition, not all components at a given level are fixed to the same extent (Knobler and Stempak, 1976). This reemphasizes the point that no currently known fixation procedure is ideal for every component in labile tissues.

Pilot fixation experiments, as suggested by Tennyson (above), are advised for determining the most satisfactory method. A good starting point would be the methods described in the literature that provide adequate fixation for the particular tissue and stage of development to be studied. Starting with these methods, modifications can be made within a much narrower range than if one is starting fresh.

Although many recipes have been suggested as providing adequate fixation of labile tissues such as developing nervous tissue, we cannot confirm any of them, but we can offer some guides. In our opinion, double fixation is a requirement, and the first fixative should be an aldehyde solution, preferably a combination of at least two, one of which should be glutaraldehyde. The second fixative should be a heavy metal-containing fixative; we prefer osmium. The aldehyde fixative should be graduated (if possible) from hypo- or isotonic to hypertonic, the specific tonicity to be determined for each tissue (Sumi, 1969, Schultz and Willey, 1973).

Other factors that may influence the quality of fixation are: choice of buffer and its ionic composition, pH, temperature, method, and duration of fixation. Finally, specific experimental requirements such as enzyme histochemistry or autoradiography may require a given recipe to be modified to avoid adverse effects. These parameters are discussed more thoroughly in Hayat (1970).

After testing more than 30 recipes, we found that the most uniform fixation of developing nervous tissue in the rat was by perfusion with two concentrations of glutaraldehyde and formaldehyde in 0.1 M sodium phosphate buffer, administered after exsanguination of the animal with buffer. The procedure was accomplished at room temperature.

The buffer is made by mixing ~7 parts sodium diphosphate to 3 parts sodium monophosphate, each a 0.1 M solution. The fixatives are used in two dilutions, as described by Peters (1970). The *initial fixative* contains 1.25% glutaraldehyde and 1% formaldehyde. The formaldehyde is methanol-free and made fresh from paraformaldehyde powder by dissolving the desired weight per volume in 0.1 M sodium diphosphate buffer, which is heated on a hot plate–stirrer to 60°C. A 4% solution of formaldehyde is usually made this way, and a portion is diluted to the 1% concentration. The *final fixative* contains 5% glutaraldehyde and 4% formaldehyde. For both fixative combinations, the glutaraldehyde may be diluted directly in the formaldehyde solution. The final pH of these solutions is adjusted to between 7.0 and 7.2 with small quantities of the more acidic sodium monophosphate or more basic sodium diphosphate as needed.

The method of fixation must be tailored to the accessibility of the tissue to be studied. Most major organs such as kidney, liver, and brain can be fixed by vascular perfusion. This method maintains a more physiological environment until the fixative arrives, as well as providing a more rapid and uniform penetration of the tissue by fixative than by simple immersion. Details of an elaborate perfusion arrangement are provided in Palay *et al.* (1962). We have been using a simpler apparatus in our laboratory (Fig. 2.1A, 2.1B). The major component is a three-stopcock manifold from the Becton Dickinson Company, to which three 60-cc syringes containing buffer rinse solutions and fixatives can be attached; or, if larger

Fig. 2.1 (A) Perfusion apparatus. Note manifold (*M*) with three stopcocks, into which 60-cc syringes are inserted. In our protocol, the syringe at the far right contains only buffer (*B*), the syringe in the center contains the low concentration (*LC*) aldehydes, and the syringe on the left contains the high concentration (*HC*) aldehyde fixatives. Instead of syringes, plastic tubing can be used to connect reservoir bottles to the manifold. The manifold is supported by additional stopcocks, which are connected to a syringe supported by a ringstand clamp (*C*). The tubing at the left of the manifold (*T*) is part of a scalp vein needle (23-gauge for 5-day-old rats). There are four needles (*N*) inserted into the wax block (*W*), which are used to immobilize the animal after anesthesia for perfusion. The block and ringstand are situated in a tray to catch excess fluid drainage.

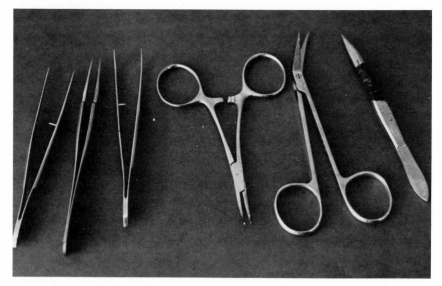

Fig. 2.1 (B) Dissection instruments. There are three pairs of fine forceps, one mosquito clamp, one angled scissor, and one scalpel blade mounted in a forceps for greater control. These instruments are used to perform the thoracotomy for intracardiac perfusion, on small immature animals. Larger instruments are necessary for operating on adult animals.

volumes are necessary, plastic tubing connected to reservoir bottles can be attached instead. The manifold has a single outlet to which a needle can be attached, to deliver solutions to the animal via an intravascular or intracardiac route. In some cases, such as biopsy, autopsy, or tissue culture material, or specimens with poorly accessible vascular routes, such as distal segments of peripheral nerves, perfusion is not possible. However, immersion with fixative *in situ* is preferable, where possible, to isolation and immersion of tissue fragments.

Perfusion Procedure:
1. Anesthetize the animal by exposure to ether vapors in a closed chamber or by intramuscular injection of Nembutal (0.9 mg/kg).
2. When it is anesthetized, but still breathing spontaneously, immobilize the animal.
3. Cannulate the trachea if the procedure is anticipated to be prolonged, as described in Palay *et al.* (1962).
4. To expose the heart for intracardiac perfusion, make a transabdominal incision just below the ribs. This will expose the diaphragm, which is then severed from the anterior abdominal wall. A flap is created of the anterior thoracic wall by incising the ribs bilaterally at the anterior axillary lines. The mosquito forceps is utilized to clamp the internal mammary arteries at the tip of the sternum (xyphoid).

5. The pericardium is incised, and the scalp vein needle is placed and held securely in the left ventricle.

6. The right atrium is incised to allow exsanguination.

7. Fixation is accomplished by administering buffer rinse and then fixatives.

The quantity of fixative to be used in perfusion is determined empirically by determining the amount of buffer required to wash out the animal's blood in a total body perfusion, ~10 cc in young rats, or to blanch the organ or region of interest in localized perfusion. For young rats, administration of the buffer rinse is usually completed within 20 sec, allowing rapid introduction of the fixative. Approximately three times as much fixative solution is needed as a minimum quantity to produce good results.

We excise the tissue and immerse it in the second aldehyde fixative overnight, although some workers have not found this necessary. Peters (1970) has left the perfused animal in a plastic bag overnight in a refrigerator without adverse effect, and Mugnaini (1965) recommends leaving the perfused animal intact in the second aldehyde to reduce the presence of artifacts such as dark cells.

Once the tissue is isolated and the desired components are dissected free, they are placed in individually labeled specimen bottles and rinsed briefly with buffer prior to postfixation in osmium tetroxide. The tissue blocks should have at least one dimension restricted to 1-1.5 mm thickness, which is important in permitting adequate penetration of the osmium, and subsequent solutions such as stains and dehydration and embedding agents. Care should be taken in handling the specimens, since they are fragile and easily distorted. Penetration of fluids into the specimens is enhanced by mounting the jars containing them on a variable speed rotator, which provides constant but gentle agitation.

STAINING

Serial sections may be stained by any of the methods described in the literature and discussed by Hayat (1975), and summarized in Table 2.2. Our initial efforts in serial section microscopy were on short ribbons (under 100 sections) which could be included on a few grids. We stained them in a 20% solution of methanolic uranyl acetate for 5 min (Stempak and Ward, 1964) and occasionally followed it with lead citrate (Reynolds, 1963) for very short intervals (from 3 sec to a few minutes). The restriction on time is essential, since the extent of lead stain in a tissue seems proportional to the amount of uranyl taken up by the tissue, and much more is taken up from the highly concentrated methanolic solution than from aqueous solutions.

For some years the research of interest in our laboratory required very long ribbons on many grids. The occasional loss of a group of sections due to film breakage on a grid during staining was too costly a loss in very long ribbons. We tested several en bloc stains (Locke et al., 1971), and formulated our own method.

The details of the en bloc staining technique have been described by Locke et al.

Table 2.2   Staining Methods

I. *En bloc*
    A. Prior to embedment
        1. Aqueous solutions
        2. Alcoholic solutions
    B. Following embedment
        1. Alcoholic solutions
II. Section staining
    A. Prior to final collection
        1. On loop or unfilmed 1 × 2 mm slot grid
    B. Following final collection
        1. On filmed loop
        2. On filmed 1 × 2 mm slot grid
        3. On mesh grid
    C. Multiple grid staining

(1971). Following fixation and postfixation excess buffer is washed out with several changes of distilled water. If staining is to be done with an aqueous solution of uranyl acetate, a 1–2% solution is prepared by dissolving 0.25-0.50 gm uranyl acetate in 25 cc distilled water with mechanical agitation. Staining is carried out in tightly stoppered vials, in an oven heated to 40-60°C, for 8-16 hr. Staining intensity is usually proportional to time and temperature as well as the concentration of the stain; ideal parameters must be determined for each tissue. When staining is complete, remove the excess stain in several changes of distilled water. Dehydration and embedding are then carried out as usual.

If alcoholic uranyl acetate is preferred, a 1–2% solution in 95% or absolute ethanol can be prepared. The tissues must be dehydrated through a graded series of alcohol through 95% ethanol. We use 10-min periods with gentle agitation in 50%, 70%, 80%, 90%, and 95% ethanol, after rinsing the tissue in distilled water following fixation, to remove excess buffer. As with the aqueous stain, staining should be carried out in tightly stoppered vials, at 40-60°C, for 8-16 hr. Alcoholic uranyl acetate is light-sensitive and is best freshly prepared with about 1 drop of glacial acetic acid in every 10 ml of stain (Locke et al., 1971). When staining is complete, remove the excess stain in several changes of 95% or absolute ethanol. Hot stain should not be allowed to dry out on the specimen. Infiltration and embedment may then be carried on as usual.

We have stained blocks of tissue with a 2% solution of methanolic uranyl acetate during the dehydration sequence. After dehydrating to 95%, we immerse the tissue in a 2% solution of uranyl acetate in methanol for 1-2 hr at room temperature. Dehydration is continued by two changes of 100% alcohol for 10 min each; the remainder of the sequence is normal. The intensity of the stain is not great, but the short staining time at room temperature minimizes possible extraction of cellular material.

Another method of *en bloc* staining was reported by Karnovsky (1971). In his technique potassium ferrocyanide is slowly added to, and dissolved in, 1% osmium tetroxide in water to achieve a final concentration of 1.5% potassium ferrocyanide. This solution is used for postfixation at room temperature or 4°C for 1-2 hr. The pH is ~10.5, but it can be buffered to a lower pH with little effect on the staining. Decreasing the concentration of ferrocyanide decreases the intensity of staining. Tissue so treated shows greatly increased contrast, without additional staining, as well as improved staining with uranyl acetate and increased intensity of certain cytochemical reactions involving peroxidase and oxidase methods.

There is a method available for *en bloc* staining of blocks following embedment (Locke and Krishnan, 1971), which can be useful if serial sections are desired from old blocks that were not stained prior to embedment. The authors caution that blocks be fully polymerized; otherwise they may be softened by the hot alcohol. The blocks should be trimmed and then placed in a tightly stoppered vial of 1-2% uranyl acetate in 95% ethanol at 60°C for 8-12 hr. Remember that alcoholic uranyl acetate is light-sensitive and should have one drop of glacial acetic acid in every 10 ml of stain. Following staining, wash the blocks in 95% ethanol to remove the excess stain. If the blocks soften, dry them and place them in the oven to harden again. Cut sections from the exposed face, discarding the first few sections. The stain should penetrate for many microns into the block, allowing many adequately stained thin sections to be cut.

Sections may conveniently be stained prior to their being mounted on a grid by being picked up from the trough with a plastic ring (Barbolini *et al.*, 1974), a wire loop (Anderson and Brenner, 1971), or an unfilmed 1 X 2 mm slot grid (Galey and Nilsson, 1966), and transferred to either a Petri dish or a wax plate containing a droplet of stain. Sections carefully handled in this way will successfully float, remain visible, and readily survive transfer from solution to solution through mounting on grids.

Sections that have been picked up on a filmed loop or filmed grid are handled in the same fashion as any of the single section staining techniques routinely performed. The risk of film breakage with section loss, as well as contamination and unevenness of staining, is great here, but the method continues to be practiced. Furthermore, it is very time-consuming and tedious to stain many grids in a serial section study one by one. A time-saving apparatus which was developed primarily for electron microscopic autoradiography allows staining of many grids stimulataneously. These multiple grid holders are available commercially.

DEHYDRATION AND EMBEDDING

Dehydration procedures are in part determined by the embedding medium used. Currently, the most popular embedding medium is Epon used according to the method of Luft (1961). Among the advantages of Epon-embedding are its adequate penetration of the tissue, controlled variability of hardness, suitable sectioning qualities, suitable staining properties for both light and electron microscopy, and stability in the electron beam. Epon embedments are made by combining the following components:

Mixture A (Epon 812, 62 ml; dodecenyl succinic anhydride—DDSA, 100 ml)
Mixture B (Epon 812, 100 ml; methyl nadic anhydride—MNA, 89 ml)
Accelerator (DMP-30 which is 2,4,6-tri dimethylaminomethyl phenol, 1.5-2% volume/volume)

Mixtures A and B should be made separately, without accelerator, and can be stored in a refrigerator as such for as long as six months. Before use, the mixtures should be allowed to come to room temperature, to avoid the collection of water vapor. Mixture A alone provides a soft embedment, while mixture B alone is very hard. We usually use a 50% A-50% B, or 40% A-60% B combination, which provides well-penetrated blocks of sufficient hardness for serial thin sectioning. Care should be taken to avoid direct handling of all of the component compounds, since they are potentially toxic to the skin. For mixing the complete embedding medium, it is most convenient to use a jar into which the desired proportions and needed quantities of mixtures A and B can be mixed, with 2% of the total volume provided by the accelerator—that is, 2 ml DMP-30 for 100 ml of 40% A-60% B combined, added while the mixture is being gently agitated by hand or a mechanical stirrer and taking care to avoid the introduction of air bubbles.

There are three mixtures of this embedding medium routinely used in our laboratory:

a. 1:1 propylene oxide: complete medium. This is needed for tissue infiltration and making a 1:3 propylene oxide: complete medium mixture. It is made by mixing equal parts of propylene oxide and complete embedding medium.

b. 1:3 propylene oxide: complete medium. This is used for tissue infiltration. It is made by mixing 1 part of the 1:1 mixture with 2 parts of the complete embedding medium.

c. Undiluted complete medium for final embedding.

The propylene oxide should be refrigerated to reduce its high vapor pressure. It should be brought to room temperature for use.

Embedding is accomplished by dehydrating the tissue following fixation, postfixation, and *en bloc* staining. This is done by 10-min rinses of the tissue in 50%, 70%, 80%, and 95% ethanol and two changes of 100% ethanol. This is followed by two changes of propylene oxide, 10 min each, 1 hr in 1:1 propylene oxide: complete medium, and 1-12 hr in 1:3 propylene oxide: complete medium. Gentle but constant agitation aids the infiltration of the tissue, but is generally not necessary if at least one dimension of the block is 1 mm or less. Final embedding is then carried out when most convenient.

Flat embedding molds have allowed more accurate orientation of blocks than previous techniques utilizing capsules, although other embedments can be cut

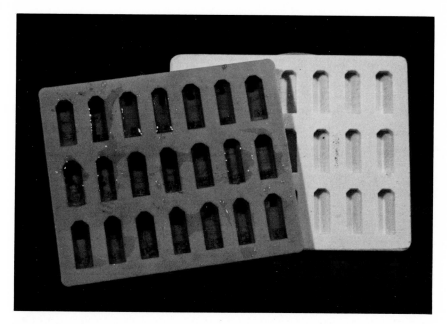

Fig. 2.2 Embedding molds. These rubber molds contain 21 compartments, each measuring 13 X 5 X 2 mm. The specimens are usually placed at the narrowed end, and the labels are placed at the opposite end.

into suitable shape, oriented as necessary, and glued to a support. It is best to fill the compartments of the embedding mold (Fig. 2.2) with complete embedding medium and then place labels, written in pencil, at one end. Care should be taken to evacuate air bubbles that form around the specimen as it is introduced and oriented, since they may be retained and weaken the embedment. A dissecting probe or toothpick is most useful for this purpose. A cool initial incubation period (35-37°C) is useful to evaporate remaining traces of propylene oxide without boiling and favors the establishment of linear polymerization of the Epon rather than crosslinkages. The crosslinked polymers are more difficult to cut (Luft, 1961). Gradual heating of the blocks for a day at 35°C, a day at 45°C, and a day at 60°C is suggested to improve the cutting quality, but curing for a day or two at 60°C has provided easily sectioned blocks as well.

GRID PREPARATION

Grids with single holes of various sizes and shapes can be obtained from any of the electron microscopy supply houses. The decision on which type of grid to use depends on two considerations, the viewing area sought and the preparation time available. Grids with large unobstructed viewing areas, such as the 1 X 2 mm single slot serial section grids, require preparation of a support film and carbon reinforcement, but are able to carry 1-40 adjacent sections, depending upon block face size and section placement. On the other hand, there are those who may choose to collect each section on an individual grid, for which a wire mesh grid without either support film or carbon reinforcement can be used. The latter method, although saving some time in grid preparation, and allowing slightly increased contrast, requires accurate placement of each section.

We have routinely used 1 X 2 mm slot serial section grids which have been ultrasonically cleaned and coated on their dull side with a Formvar film and then with several layers of carbon evaporated onto the Formvar surface. These grids allow the greatest number of serial sections to be mounted per grid, while providing a completely unobstructed viewing area. In addition, we can collect ribbon segments instead of having to separate ribbons into individual sections for collection.

Commercially available grids should be sonicated in at least 2-3 changes of absolute ethanol to be certain contaminants are removed. Following this cleaning, we prepare a 0.5% solution of Formvar in ethylene dichloride, from which films can be made and secured to the grid by one of two methods. The film is prepared by dipping a glass slide into the Formvar solution and allowing the slide to air-dry in a dust-free area. When dry, the film is removed from the slide by scraping the four edges of the film with a sharp tool and submerging the slide in a water bath, slowly and steadily, at an angle of 45°. This usually allows the film to float off onto the surface of the water bath, although it does not work every time, and several slides may have to be discarded. The floating film can be seen by its reflection of light.

Grids can be placed dull side down on the Formvar film (Fig. 2.3A). If placed closely, 20-30 grids can easily be handled this way. They are collected by putting a slip of paper, about the size of the whole film, on top of the film, then peeling the adherent film and grids off the water (Fig. 2.3B). These grids are allowed to air-dry in a dust-free area. Alternatively, grids may be placed dull side up, over the openings of a submerged perforated metal plate (Fig. 2.3C), which may be made by any shop with a drill press. In this method, films are floated and positioned over the plate containing the grids (Fig. 2.3D), while the water is siphoned off and the film is lowered onto the grids. Air-dry the holder in a dust-free area. We prefer the second of these methods because the yield of unmarred films is greater.

Carbon evaporation, now routinely used to stabilize sections mounted on support film, and thus prevent drifting of the sections due to thermal effects on the Formvar, is also useful in increasing the strength of the support film. This is important when 1 X 2 mm slot grids are used, because the film area is substantial. For this reason we use 4-5 evaporated coats of carbon, which is in excess of the routine coating for a wire mesh grid. The precise technique of evaporating the carbon will depend on the evaporator used; however, the amount of carbon deposited is monitored by the use of white porcelain chips upon which a small

Fig. 2.3 (A) Formvar film supporting grids. Sixteen 1 X 2 mm slot grids are seen dull side down on this Formvar support film (*F*) as it floats in the water bath. Additional grids may be placed on this film until there is no further useful space available.

Fig. 2.3 (B) Paper pick-up technique. A sheet of paper is placed on the Formvar film containing the grids. The film adheres to the sheet, holding the grids in place, which may then be picked up and dried.

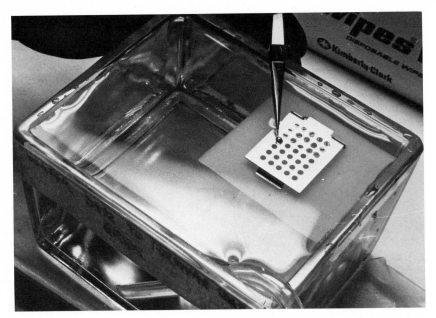

Fig. 2.3 (C) Grid orientation. Grids are oriented over openings in a metal plate, dull side up.

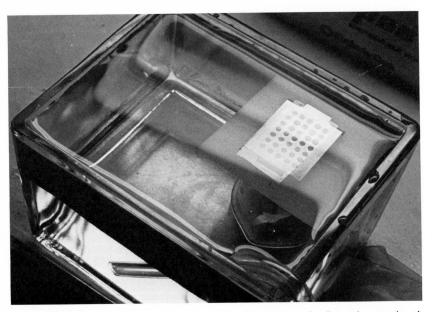

Fig. 2.3 (D) Film orientation. The Formvar film has been gently directed to overlay the grids on the metal plate. As the water level drops by siphoning, the film contacts the grids and coats them.

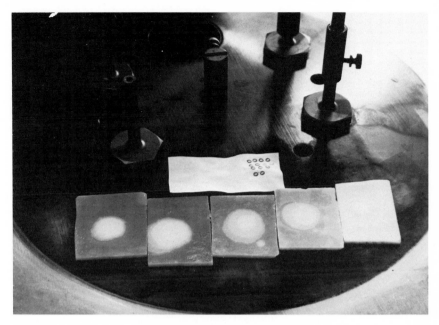

Fig. 2.4 Carbon coating. Porcelain chips containing vacuum pump oil are demonstrated left to right after 5, 4, 3, 2, and 1 evaporations. A paper containing grids is in the background as it would be during evaporation. The light clear area on the porcelain chips reflects the zone not carboned because of the coating of oil.

drop of diffusion pump oil is placed. A clean chip is placed in the chamber before each evaporation. In this way the chip turns tan to brown when the carbon is deposited, while the oil spot retains its white color for comparison. A very light tan color is sought in each successive evaporation, which after several runs builds a very strong carbon coat (Fig. 2.4). This stepwise buildup is cleaner, stronger, and more uniform than it would be if it were deposited in one prolonged evaporation from a very long carbon rod tip. Long evaporation times yield sparking with subsequent grid breakage more frequently than short evaporation times.

SECTIONING

For the successful cutting of serial thin sections it is important to have a good embedment, an appropriately trimmed block face, a sharp knife, and practical experience with the operation of the microtome to be used, including the appropriate knife angles, cutting speeds, and specimen advance mechanisms. All considerations routinely arising in thin sectioning of biological materials apply with respect to serial thin sectioning as well. Sjöstrand (1967) and Hayat (1970) deal with various aspects of sectioning and trouble-shooting in much greater

Fig. 2.5 Sectioning accessories. Note the can of Tackiwax (Cenco Scientific Co.) used for aiding ribbon adhesiveness. The white tape in the forceps on the right is Teftape [Teflon (Du Pont) Thread Seal Tape, Tape Industries, Inc., Farmingdale, N.Y.], which is useful in manipulating sections in the knife trough. The self-locking forceps are used for holding grids during section collection. The syringe is used for adjusting the water level in the knife boat. The Allen wrench is used for adjusting the jaws of the chuck for the LKB III ultratome.

detail than can be dealt with in the present discussion. However, because this aspect of the technique is of surpassing importance, we will present a very detailed description of the sequence as practiced in our laboratory. The few tools necessary for serial sectioning in our laboratory are seen in Fig. 2.5.

A 1-2-μm section of the entire block face is obtained after rough-trimming of the block with a razor blade to remove excess plastic. In the section one may observe depth of penetration of the fixative (if immersion-fixed), estimate the quality of fixation, and discern the presence of sites of interest in tissues where focal distribution of such sites occur. Even if the cells or structures under analysis form the bulk of the tissue being examined, as would the hepatic parenchymal cell in the liver, a thick section of the entire block face would allow an investigator to determine landmarks in the tissue, such as portal areas and central veins, and localize particular cells with respect to the landmarks. This is particularly important in the liver, since hepatic parenchymal cell cytology varies with placement in the lobule (Loud, 1968). Such thick sections also may be useful for light microscopic serial section analysis (Sétáló, 1969; Holländer, 1970; Merzel, 1971) or, once fine-trimmed, for alternating thick–thin studies combining light and electron microscopy (Webster, 1971; Favard and Carasso, 1972; Martin and Webster, 1973; Webster *et al.*, 1973).

The block may be hand-trimmed to the area of interest with a single edge razor blade, especially if a landmark observed in the thick section can be discerned on the block face. If one cannot be observed, precise location of the area of interest in the thick section can be made with an ocular reticle, and the location of the area on the block face may be derived from these measurements. The end result should be a block face which is a rectangle or a trapezoid; if the latter, the inferior edge must be slightly longer than the superior edge. If the block has been trimmed to a peak, with the intention of creating a face during sectioning, we find that it is best to trim the block so that the face which forms during sectioning is a rectangle. The included angle formed by an extension of the superior and inferior surfaces should be from 90 to 120°. We customarily leave a thin edge of plastic on the left lateral surface. Another thick section should be made at this time to verify the position of the area of interest.

Parallelism of the inferior edge of the block face with the knife edge can be checked by verifying that the reflection of the knife edge appears simultaneously across the whole inferior edge of the block face during its pass behind the knife. If the block face shows horizontal rotation with respect to the knife edge, the reflection of the edge would appear on one side of the block face before it appears on the other, and would travel across the face angled with respect to the presumably parallel superior and inferior surfaces. Correct any deviation from parallel by appropriate rotation of the block on its horizontal axis. To verify that the face of the block is parallel to an imaginary vertical line dropped from the edge of the knife, estimate the distance between the knife edge and its reflection on the block face (the reflection of the back of the knife). This distance is a function of the proximity of the knife edge to the block face (the closer the knife, the narrower the visible reflection). If the reflection on the block face remains a constant height across the block face as it passes behind the knife edge, the face is parallel to the aforementioned vertical line (Fig. 2.6 A-D). If the height of the band of reflection increases on passing the face, the top of the block face is farther from the knife edge than the bottom (Fig. 2.6 E-H); if it decreases, the reverse is true (Fig. 2.6 I-L). If the face of the block is not parallel to the vertical line, tilt the block in the proper direction.

With practice one may become sufficiently able to align a block by reflections so that the whole face is obtained on the first section. If the whole face is not obtained, appropriate manipulation of the block should be accomplished, taking care to retract the knife slightly before shifting the position of the face.

Once the area of interest is approximately centered in the block face, final trimming, which is essentially smoothing the superior and inferior surfaces near the block face, is accomplished. A glass knife with a relatively straight edge should be selected for the final trimming. The superior and inferior edges of the block face are set perpendicular to the knife edge, and the face of the block

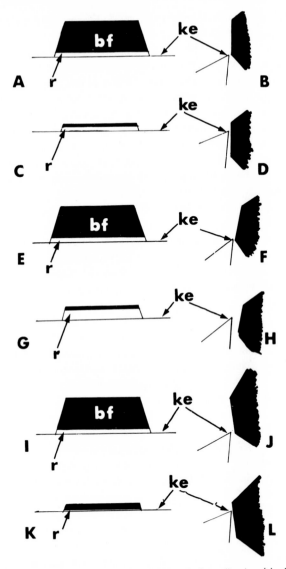

Fig. 2.6 Schematic illustration of the relationship of the reflection (*r*) of the knife edge (*ke*) on the block face (*bf*) as affected by block face angle. (A) The reflection (*r*) of the knife edge on the block face with the block face just beginning to pass behind the knife edge. (B) Side view of the position in (A). (C) The reflection (*r*) is the same height as it was in (A); much of the block face has passed behind the knife edge. (D) Side view of the position in (C). (E) The reflection (*r*) is the same height as in (A). (F) Side view of the position in (E). (G) The reflection (*r*) has increased in height on the block face as it passes behind the knife edge. (H) Side view of the position in (G). Note the increase in the distance between the knife edge and the block face. (I) The reflection of the knife (*r*) is the same height as it was in (A) and (E). (J) A side view of the position in (I). (K) The reflection of the knife (*r*) is extremely short. (L) A side view of the position in (K). Note that the gap between block face and knife has decreased because of the angle of the block.

must be checked by reflection again to verify that rotation of the specimen did not yield an angulation of the block face with respect to a vertical dropped from the edge of the knife. If this occurred, correct it, and make a test section to verify position of the face. With practice at orienting the specimen by reflection, it will not be necessary to make a test section at this time. Rotate the knife to parallel either the superior or inferior surface (if the knife is fixed in position, the block must be turned), advance, and make a section. From the first contact one should adjust the orientation of the knife to minimize the number of sections required to obtain a smooth surface. When the sections encroach on the block face, be sure that the vertical orientation of the edge is not destroyed. If the edge of the face is being sectioned at an angle, rotate the block a few degrees in the appropriate direction to restore the vertical edge.

After smoothing the first surface, retract the knife, rotate it and approach the opposite surface. The same approach is made to this surface as the previous surface. Great care should be taken at this time to be certain that the superior and inferior edges of the block face are parallel. Even though one does not rotate the specimen, the very slight curve present even in relatively straight glass knives may yield nonparallel edges; if this is observed to occur, one must rotate the specimen an appropriate amount (generally only a few degrees), to yield parallel edges. Nonparallel superior and inferior edges is one of the reasons for curved ribbons.

Trimming the lateral surfaces of the truncated pyramid with the glass knife is optional. It may facilitate sectioning, and it will minimize edge irregularities which can yield folds, but we have found that irregularities on the lateral surface facilitate section examination by serving as reference points, especially where the block face is large.

Remove the block from the microtome, and mount it in a trimming stand to apply an adhesive substance to the superior and inferior surfaces. We use a commercially available preparation, Tackiwax (Cenco Scientific Co.), although a glue made by dissolving scotch tape in chloroform has been used for this purpose by Macagno et al. (1973). Daub an excess amount on both superior and inferior surfaces with a razor blade, then reduce thickness and make the film uniform by dragging a clean single-edge razor blade over the surfaces, taking care not to damage the block face.

The block is now ready for thin sectioning; insert it in the microtome. Replace the knife with another, preferably a diamond knife, but if the ribbon desired is short (under 50 sections) a glass knife may be used. Add water to the trough and obtain appropriate reflection from the water surface.

Approach the face of the block, and when very close, align the knife edge and the inferior edge of the block as nearly parallel as one may by visual inspection. The contrast between the edge and block may be heightened by inserting a white index card behind the knife and the block face, which will assist the operator in determining any deviation from parallel. Verify the position of the block face with respect to the knife edge by reflection of the knife edge; correct it if necessary, then advance and begin sectioning.

The number of sections is counted, and when a ribbon of sufficient length is obtained (enough to fill several grids), dislodge it from the knife edge by applying the tip of a piece of Teflon tape (Teftape) to a section of the ribbon removed from the knife edge and pull the ribbon free (Fig. 2.7A). Initial efforts will probably destroy substantial portions of the ribbon being dislodged, because of the tendency of the sections to adhere to the Teftape if contacted too firmly. With

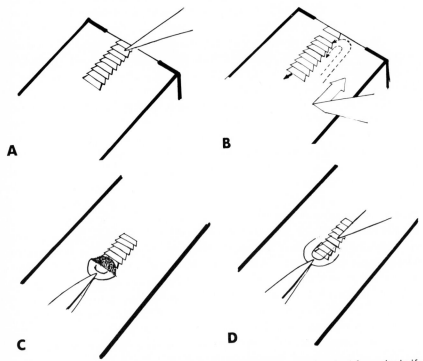

Fig. 2.7 (A) A ribbon of sections of appropriate length may be detached from the knife edge by contacting a section other than the one on the knife edge with the tip of a piece of Teftape or a hair and pulling the ribbon free from the section or sections preceding it. (B) The ribbon is directed toward the grid by generating currents in the trough water with the tape. By immersing the tip of the Teftape in the water and stroking toward the knife edge (*broad white arrow*), currents are generated (*dashed lines*) which will direct the ribbon away from the knife edge and toward the previously mounted grid. The direction of the ribbon will be at an angle to the currents generated (*short black arrow*) and must be corrected by stroking on the opposite side of the floating ribbon. (C) The ribbon approaches the immersed grid and may be attached to it by currents alone. If attachment cannot be made by currents, the rear of the ribbon may be tapped lightly to affix it to the grid. Note that the shadow area (*stippled region*) caused by the meniscus conceals the site of contact of the sections with the grid. (D) Retract the grid until at least two-thirds of the ribbon is on the grid. Decide which section would be the last to appear safely on the grid, then detach the remainder of the ribbon from that section by touching the following section with a hair or a piece of Teftape, and push it away.

practice, the tip of the tape will be applied with sufficient delicacy to allow removal of the ribbon from the section adhering to the knife edge without sticking to the Teftape. The ribbon may be allowed to float free on the trough while a grid is mounted on the LKB section collector or the "third hand" (Behnke and Rostgaard, 1964), immersed in the trough liquid completely, and then retracted until about $\frac{1}{2}$ mm of grid is exposed above the surface of the fluid.

The hydrophobic nature of the carbon film on the grid will yield a substantial meniscus whose depths will remain dark no matter what angle the microtome illumination is set at. One must be able to guide the ribbon into the dark zone and adhere it to the carbon surface without seeing the contact. It is a trying period, and quite difficult, but the difficulty imposed by the dark zone of the meniscus is surmounted with only a small amount of practice. We do it routinely, and the method is described below.

Once the ribbon is floating in the trough and the grid is in place, the ribbon is guided toward the grid by generating currents in the trough fluid with the Teftape (Fig. 2.7B). The Teftape is swept toward the knife edge and lateral to the ribbon, which generates a current in the opposite direction, central to the tape. The ribbon will move away from the knife edge and at an angle to it. The angle is corrected and the direction maintained by stroking the fluid surface on the opposite side of the ribbon. By this means the ribbon is moved near the grid, and strokes are modified to direct the ribbon precisely, maintaining the long axis of the ribbon parallel with the sides of the trough and the length of a 1 × 2 grid opening. When the ribbon approaches the grid, strokes are shortened and modified, but one should avoid contacting the ribbon with the tape if possible. When the conditions are good, the ribbon may be directed and affixed to the grid by means of the currents generated in the trough without any contact of the ribbon by the tape, although it may be necessary to tap the rear of the ribbon with the Teftape to ensure adherence. When the first sections of the ribbon enter the dark area of the meniscus and are hidden from view, one may estimate whether or not the ribbon has contacted the grid by the length of ribbon that has disappeared into the dark area of the meniscus (Fig. 2.7C).

With practice, it will be seen that this method is quite satisfactory, although initially one may wish to have visible proof of contact before retracting the grid. This may be accomplished by exerting an upward finger pressure upon the forceps, which will alter the angle of the meniscus at the grid, providing momentary views of the ribbon at or near the site of attachment. When the microtomist feels that the ribbon may be successfully obtained, the grid is retracted smoothly. After the grid is retracted far enough to include the number of sections desired, the Teftape is lightly applied to the section following, and the remainder of the ribbon is detached from the portion on the grid by pushing it off the section attached to the grid and away from it (Fig. 2.7D). Once the detached ribbon is away from the grid, remove the grid from the water, dry it carefully, count the number of sections on the grid, and store it in a dust-free site temporarily.

Compare the number of sections originally present in the ribbon with the sum of those present on the grid and floating on the trough to verify that none have been lost. Continue the process until the entire ribbon has been collected.

Generally collection continues in an uneventful manner, but sometimes the ribbon on the trough floats too close to the side of the trough to be affected by currents the microtomist generates when attempting to move it. Should this occur, wipe the tip of the Teftape between the thumb and index finger, and insert it between the trough wall and the ribbon. The Teftape should now repel the ribbon from the wall without contacting the ribbon. After collecting all sections, verify the number per grid and see if the sum is equal to the number in the original ribbon. Any loss should be explained by locating the missing sections, whether on the trough or on the sides of a grid. If loss of sections occurred, a judgment should be made about retaining the ribbon for inspection or discarding it and beginning anew. The grid obtained may be stored in any commercial grid holder for indefinite periods of time. If one is unable to direct the ribbon and affix it to the grid because of the meniscus, one may render the carbon film hydrophilic for a period of time by treating it with a glow discharge unit, available for most evaporators. One may also simply avoid use of precarboned grids (see the next section).

There are, however, certain considerations that deserve special mention. While glass knives may be used to cut serial thin sections, experience over the years has shown that for providing ribbons of greater than about 30 sections, with section thickness in the 60–100 nm range, diamond knives are a necessity. Furthermore, it is helpful to retain an individual diamond knife for specific tissues and embedments rather than use a single knife for many different types of blocks. Although more than a simple nick or two in a 4-mm diamond knife may render it unusable for some purposes, we have occasionally found such a defect of use as a marker in the study of the sections, as long as its location is known and does not pass through the area being studied. It is important to maintain the sharpness and cleanliness of the knife edge. Diamond knives must be cleaned after and sometimes during use to avoid buildup of debris.

SECTION COLLECTION

Four methods of section collection exist and are listed in Table 2.3. In all methods of collection, sections are brought to a portion of the trough where collection can safely be done without disturbing the knife edge. Such maneuvers are facilitated by maintaining a slightly convex liquid surface in the knife trough, by adjusting the water level with a syringe.

In the simplest method of section collection, a support film coated grid is placed either (film side down) onto the surface of the floating sections, or brought underneath them (film side up) (Dowell, 1959, Karlsson, 1966). When the grid is lifted from the surface of water, the sections adhere to the film. Two faults in

Table 2.3   Section Collection Methods

I. Simple collection onto filmed grids.
   a) Grid onto sections from above
   b) Grid lifted from beneath sections
      1) Dowell (1959)
      2) Karlsson (1966a)
II. Filmed loop collection, transfer to grids.
   a) Gay and Anderson technique
      1) Williams & Kallman (1954)
      2) Sjöstrand (1958, 1967, 1974)
      3) *Thin sectioning*, Sorvall Inc. (1967)
      4) Barnes & Chambers (1961)
   b) Modifications of the Gay and Anderson technique
      1) Thaemert (1966)
      2) Poritsky (1969)
      3) Mazziotta et al. (1973)
III. Filmed grid collection methods
   a) Lowering water level below grid
      1) Westfall (1961)
      2) Westfall and Healy (1962)
      3) Karlsson (1966)
   b) Mechanical grid retrieval from water
      1) Behnke and Rostgaard (1964)
      2) Ward (1972)
IV. Unfilmed grid collection-transfer to film methods
   a) Transfer to filmed grid–Galey and Nilsson (1966)
   b) Transfer to sol-gel then to grid–Anderson and Brenner (1971)
   c) Mount grid onto film–McCarthy and Peters (unpublished), as cited in:
      1) Hinds and Hinds (1972)
      2) Vaughn and Peters (1973)

---

this freehand method are the occurrence of wrinkles (although they are less frequent when collection is from below) and difficulty in orienting the sections on the grid.

The first published report of a technique specifically for the collection of serial thin sections came from Gay and Anderson (1954). It is a two-step method: sections are picked up on a filmed loop, then transferred to a grid. The technique involves the use of a Formvar-covered wire loop of ~4mm diameter, held in the jaws of a self-locking forceps. The film is attached to the loops by placing them on a Formvar film floating on a water bath, as described earlier in the section on grid preparation. Each loop is placed in the trough and used to pick up sections, freehand, from below. These sections are mounted on a serial section grid in the following way (Fig. 2.8A). A fresh unfilmed grid (dull side up) is placed on the end of a short plastic rod supported on top of a microscope substage condenser, which is racked up so that the grid is almost in level with the microscope stage

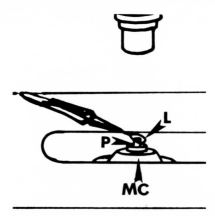

Fig. 2.8A. A Formvar film covered wire loop (L) containing sections is held in the jaws of a self-locking forceps. A slot grid is placed upon a plastic peg (P) which is mounted on the microscope condenser (MC). The sections are oriented to overlay the slot by moving the forceps, then the condenser is racked up to contact the film and pass through it. More elaborate devices which afford more consistency to the operation are described and illustrated by Sjöstrand (1967). (After Gay and Anderson, 1954).

surface. The wire loop is then oriented on the micrscope stage in such a way as to perfectly overlay the grid opening, and simultaneously be in focus through the microscope using the lowest-power objective. "The condenser mount is raised to bring the grid and the Formvar film into contact; further elevation causes the rod with the grid to pass through the wire loop" (Gay and Anderson, 1954). The Formvar film including the sections, is firmly attached to the grid in this manner. This technique with only slight modification has been used or advocated by Williams and Kallman (1955) and Sjöstrand (1958, 1967, 1974), and was suggested in the Sorvall manual on thin sectioning (1967).

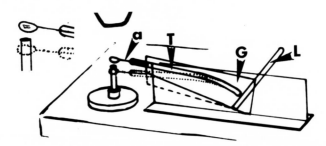

Fig. 2.8B. The micromanipulator is made from two 1 × 3 microscope slides glued together with epoxy glue. A third piece of glass (G) is affixed to the vertical with grease. To the third piece of glass a bent tube (T) and a lever (L) are glued. A glass rod (a) with a loop inserted are inserted in the tube, and after orienting the sections on the loop over the slot grid on the peg, the lever is pushed forward and the loop passes down over the peg (dotted line) and affixes the film to the grid in passing.

Barnes and Chambers (1961) mounted their sections utilizing a makeshift micromanipulator, instead of a microscope substage condenser (Fig. 2.8B). It consisted of a sliding glass rod supported with petroleum jelly on the vertical portion of an inverted T-shaped frame, composed of two glass slides. The wire loop containing the sections is a permanent part of the glass rod. After sections have been collected freehand, the glass rod is positioned on the frame and is used to lower the sections onto a grid which is supported on a plastic rod. This procedure is carried on while viewing the sections through a dissecting microscope.

Thaemert (1966) collected individual sections from the knife trough using a sable hair brush, and transferred them to support film coated loops. These loops were lowered onto 3-mm grids with a 0.5-mm circular opening, using a commercial micromanipulator while viewing the sections through a dissecting microscope. Poritsky (1969) collected each thin section with a toothpick and placed the sections into sequential compartments on a support film coated wax plate (Fig. 2.8C, D). The plate in this method contains three rows, with five compartments per row. The compartments are made by drilling 5-mm-diameter holes in the wax plate. A small sheet of X-ray film may also be used in this fashion (Wettstein and Grauer, 1973). The wax plate or X-ray film is coated with support film by placing it into a water bath, aligning the support film over it, and then siphoning off water to lower the film onto the sheet. Each thin section is then mounted on a 3-mm grid with an 0.8-mm circular hole, using the rod on the microscope condenser to support the grid and pass it through the film in the compartment. In addition, the rod has two 1 × 2 mm slot grids permanently mounted as a platform, with the same end of each cut out to provide easy access for removal to the grid containing the sections. Mazziotta *et al.* (1973) suggested the use of a

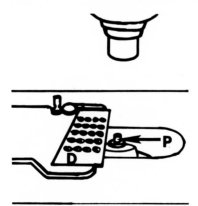

Fig. 2.8C. A plate of dental wax (D) containing 20 holes is covered with a Formvar film. Sections are floated on water drops (see below) and allowed to dry on the film. A slot grid is placed on the top of a peg (P) which is placed on a microscope condenser. The stage controls are used to orient the sections over the slot on the grid, then the microscope condenser is racked up to pass through the film.

Fig. 2.8D. A detailed view of two of the Formvar film covered holes of the holder in Fig. 2.8C. On the left, the drop of water has a section (S) floating on its surface. On the right, a toothpick (T) is immersed in the drop and the section (S) which was picked up from the trough with the toothpick is floating off the toothpick.

stereotaxic instrument which can be moved in a vertical and horizontal plane, for making contact between the film containing the section and the grid.

The major drawback in the use of these methods is their two-step nature. Sections or ribbons must first be collected and then be transferred to grids. Although the method ensures accurate section orientation, freehand collection of sections from the knife trough does leave room for introduction of wrinkles. An advantage to this method is that Formvar film is not as hydrophobic as carbon, which makes collection of the sections relatively simple.

Other techniques of section collection utilize a single-step pickup method, directly onto a filmed grid. They depend on lowering the water level in the knife trough after bringing the sections into alignment with the grid (Westfall and Healy, 1962), or drawing sections out of the water freehand (Dowell, 1959; Karlsson, 1966) or by mechanical means (Behnke and Rostgaard, 1964; Ward, 1972). The LKB IV ultramicrotome comes equipped with a syringe system for raising and lowering the water level in the knife trough. We have been using the "third hand" of Behnke and Rostgaard (1964) with the LKB III ultramicrotome, and the modification by Ward (1972) with the Reichert microtome in our laboratory (Knobler and Stempak, 1973; Knobler et al. 1974, 1976; Stempak and Laurencin, 1976). The procedure involves the use of a rack and pinion mechanism containing a clamp, all of which is mounted onto a ringstand. A self-locking forceps is held in the clamp, which is used to mechanically lower a grid into the knife trough; once the sections have been brought into alignment, they are withdrawn smoothly and precisely from the water bath, wrinkle-free. This method avoids the additional handling of methods requiring transfer of sections following pickup from the knife trough.

The first three basic methods of section collection, already described, have in common collection of sections onto support films or support film coated grids. This poses no handling problems if the tissue has already been stained *en bloc*. However, further handling is required if the sections are to be stained, and the fourth collection method utilizes an unfilmed 1 X 2 mm slot grid or small unfilmed loop for the initial collection of sections, which minimizes the possibility of film breakage while staining.

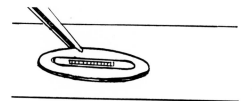

Fig. 2.8E. An unfilmed slot grid is placed on the surface of the water of the trough around the ribbon of sections. The grid may be removed and the sections will float on the surface of the droplet which will remain in the slot of the grid.

The method of Galey and Nilsson (1966) utilizes a 1 X 2 mm slot grid; one places the opening of the grid over the ribbon as it floats on a flat or slightly convex water bath in the knife trough and contacts the water surface with the grid (Fig. 2.8E). On its removal, a small drop remains within the slot of the grid, and the sections float on the drop. The grid is transferred to a drop of water on a hydrophobic surface, such as a wax plate, and stored until the remaining sections are collected (Fig. 2.8F). To prevent contamination, a plate containing such sections is kept in an enclosed dish. Sections contained in a fluid drop adherent to a slot grid may be stained by transfer to a drop of stain, and rinsed by transfer to appropriate drops of rinsing solutions. These sections can then be transferred to a grid containing a support film and carbon coating by aligning the grid containing the sections over the filmed grid, as the latter (film side up) is supported by forceps. This procedure may be viewed on the stage of a dissecting microscope. Excess water from the grid containing the sections is removed by carefully inserting a piece of pointed filter paper between the two grids (Fig. 2-8G). On removal of the water, the sections adhere to the film beneath. When the grids are dry, the upper grid may be removed, and the lower grid is ready for viewing in the electron microscope.

Anderson and Brenner (1971) suggest a method for the collection of individual thin sections from block faces greater than 1 mm square. The sections are col-

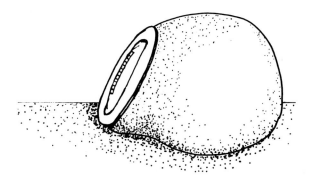

Fig. 2.8F. The grid may be transferred to a drop of water on a wax base for storage until sectioning has been completed or to a drop of aqueous stain.

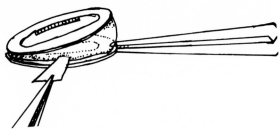

Fig. 2.8G. The grid is transferred to a filmed grid by overlaying slots accurately, then draining the fluid from between the grids with filter paper. A second pair of forceps assures that the two grids will not be completely apposed.

lected with an unfilmed wire loop and then floated on the surface of liquid gelatin, which immobilizes the sections when it gels. The gel is prepared by dissolving 3 gm of gelatin in 100 ml of boiling water; the solution is then poured into a small Petri dish, where it is allowed to cool until it becomes viscous but not set. Sections transferred to the liquid gel surface may be spread by exposure to trichloroethylene vapors. Individual sections or ribbons are arranged in their sequential order, and may be oriented on the surface with a hair. The liquid gel is set by lowering its temperature in a refrigerator. The sections may be viewed in a dissecting microscope; tilting the dish toward the light source aids visualization. A Formvar and carbon coated grid is oriented over the sections (film side down), and then released, but is not picked up or moved while the gel is set, since it might break the film. "The gelatin is liquefied in a 60°C oven, the grid picked up and drained, and then floated section side down for 30 min in a 2% solution of acetic acid at 60°C. The acid is neutralized by a 3–5 min flotation on a bath of tris buffer, pH 7.1, and a final washing of 30 min in distilled water is given; all treatments being at 60°C" (Anderson and Brenner, 1971). Grids so prepared may then be stained if they previously were not. Although it is not described by these authors, sections can be stained prior to floating on the liquid gel, in the fashion described by Galey and Nilsson (1966). Because some structural details are seen on the surface of the set gel, this technique can be most useful for accurate placement if unfilmed mesh grids are used; it is also useful for arranging placement of sections from different blocks on the same grid to facilitate comparison.

Finally, McCarthy and Peters (unpublished; cited by Hinds and Hinds, 1972; Vaughn and Peters, 1973) utilized 0.4 × 2.0 mm single slot grids to collect sections. The surface area of these grids is less than half that of the 1 × 2 mm grids, and they hold up better to handling. In this method sections are picked up with an unfilmed grid (dull side down), as in the technique of Galey and Nilsson (1966). However, McCarthy and Peters then place the grid onto a freshly made Formvar film floating on a water bath, pick up the film containing the grid using a wire loop (Ladd Research Industries, Burlington, Vt.), and allow it to air-dry. After drying, the grid is separated from the surrounding film for staining and viewing. Once again these sections can be stained, as described by Galey and Nilsson (1966), prior to mounting on a film, thus reducing the handling of the support film to a minimum. Grids should be stored sequentially in numbered grid boxes, which are commercially available, until ready for use.

MICROSCOPY

Early electron microscopes were not designed to facilitate serial section analysis. Consequently, some laboratories involved in this work attempted special modifications of existing microscopes to increase output. Most others utilized existing microscopes. Improvements or optional accessories devised for electron microscopes over the years, although incorporated to facilitate examination of single sections, assisted examination of serial sections.

It is apparent that a microscope that could accomodate several grids protected from contamination by a cold finger and containing a large film load are what a microscopist examining serial sections would wish. Recently these wishes have been realized on some microscopes.

Before examining the situation in microscopes ideally suited to serial section analysis, we can point out that virtually any electron microscope can be used to examine them. Our first efforts (Stempak and Laurencin, 1966) were made on an RCA-2E electron microscope in spite of five faults which limit its value in serial section work, namely, (1) a single grid specimen holder, (2) small film capacity (only five shots per cassette), (3) no specimen airlock, (4) no film airlock, and (5) no cold finger.

Design defects (2) and (4) require that the entire microscope column be brought to atmospheric pressure after each five micrographs. The resultant flow of air causes considerable wastage in ribbons, because the grid films are not capable of withstanding vectorial pressure changes. To reduce film breakage, air-flow restriction is accomplished by inserting 23-gauge needles through rubber stoppers, and then placing them in the air inlet and outlet ports on the dessicant container. This device retains more grids, but the result is a recycle time of 7–10 min for each five micrographs. Another older microscope (the Philips 100), which has the advantage of a 35-mm roll film camera and a specimen airlock, is an obvious choice for microscopists doing serial section studies.

Ribbon length is restricted by the size of the grid available—in virtually every case 3 mm is the standard size; in it the longest slot opening is 2 mm. Only Siemens utilized 2.4-mm grids, a definite detriment to serial section work, but their current microscopes now utilize the 3-mm grid size. Since then some microscopes have modified specimen holders to allow loading more than one grid at a time. AEI Ltd. accomplished this on the EM6B. Examining three specimens at a loading allows investigators of intracellular organelle relationships to be certain that the structure of interest is totally included in the ribbon before any micrographs are made, a great savings in time and material. At the present time several manufacturers offer microscopes with multiple specimen loading; they are AEI, JEOL, and Zeiss. Sjöstrand (1974) further modified the specimen holder of the AEI-EM6B to allow a long (1 cm) single slot grid to be utilized.

Virtually as important as specimen loading facilities is recording capacity of the microscope. In most early microscopes, glass plates were considered most desirable for electron microscopy because of their dimensional stability, although it was apparent to many even then that the advantage of glass over acetate-based film was slight compared to dimensional errors introduced by variations in microscope magnification and photographic printing. Since then, estar base film has reduced the small problem to negligible proportions. Although large-volume plate cameras exist, use of them in serial section work is difficult because glass plates are bulky and require individual handling of each plate, thus slowing the process. Philips microscopes utilized 35-mm film as a record in their first "100" series electron micrscopes, and have retained it as an option in all microscopes except the most recent. Seventy-millimeter film cameras are offered on several microscopes; the AEI, Hitachi, JEOL, Philips, and Siemens, and Zeiss, all offer microscopes with this recording option. Sjöstrand (1974) in the AEI microscope modified to his design, installed a 70-mm camera capable of holding a 100-foot roll of film.

All modern microscopes afford the investigator the capacity to restrict contamination by liquid nitrogen–cooled cold traps. They are essential to the first stages of an investigation, when several hours can be spent scrutinizing the central grid or those adjacent for a specific area; but once the cell or region has been decided upon, they are not as critical, since location of the area, focusing on it, and recording the image can take place in less than a minute in optimal circumstances, and not much more in less than ideal conditions.

Large-area examinations, those studying intercellular relationships, present another problem which complicates the contamination problem. An increase in illumination to allow objective focusing causes plastic sublimation from the area of interest in our material. Thus, the area subjected to the higher-intensity beam shows more contrast than adjacent areas owing to removal of substrate. When examining large areas, it has been our custom to "heat" the area of interest and its surrounding with a uniform beam of the intensity normally used for focusing the area, in order to equalize the contrast of the area of interest and the surrounding. This maneuver is rather time-consuming, but quite necessary.

A factor rarely considered of consequence in microscopy is the slope of the specimen pyramid, provided only that it is large enough to support the specimen. However, in serial section microscopy, it is important that both slopes of the pyramid near the peak be identical to prevent possible loss of the area of interest as one examines sections near the beginning of the ribbon. If the investigator selects for analysis a cell that is in the center of the section and begins analysis on the central section of the ribbon, any deviation from equal slopes would shift the area of interest nearer the edge with the smaller included angle (as measured from a horizontal line bisecting the block face). If the centrally located cell deviates from center, the cell or part of it could be lost to examination, perhaps rendering previous examination valueless. It is at this point that the value of multiple specimen holders becomes obvious; since one has the opportunity to verify that the cell or area of interest does not shift position to a site which removes it from previous sections, the problem need never occur in short ribbons (under 100 sections). In long ribbons the problem may still occur, but multiple grid holders would allow one to assess the degree of slope and select another area of interest more accurately than one might when examining an individual grid. In single specimen microscopes, ribbons of 30 to 100 sections can be examined by the expedients of utilizing small block faces and folding the ribbon. This technique requires considerable skill on the part of the technician, and is subject to disappointments, primarily during the period of manipulation of the specimen on the trough and retrieval of it from the trough. However, it can be accomplished, and the benefits of verifying presence of the area of interest and its artifact-free condition might make it worthwhile in specific cases.

A first consideration in examining ribbons obtained is where to begin. It is our custom to begin examination in the central section and then to examine the sections on both sides of it. If a multiple specimen loading microscope is available to the investigator, the grids immediately surrounding the central grid should also be examined. This examination could provide clues to the direction to take in examining other grids. If it does not, or if the project requires examination in both directions, as our recent work does, we take a fairly substantial number of low-magnification micrographs of the section in the ribbon next to the central section and opposite to the direction of immediate examination. This is done to prevent loss of orientation when we finally return to the unexamined section next to it, which may take weeks or months.

Size of the block face must be dictated by the nature of the work in progress. The rule in our laboratory is "the smaller the better" because the surplus tissue is simply a distraction once the area is chosen. Large block faces soon restrict the number of sections that can be accommodated on a grid, and increase the difficulty in locating the area of interest. In tissues where relatively uniform distribution of the cell type or material under investigation suggests that the cell or organelle of choice will be easily located, very small block faces can be obtained by trimming the block to a peak and sectioning the peak. After a few thick (1 μm) sections the block face will probably be of a size adequate to begin thin sectioning ($\sim$ 50 μm vertical distance).

Reference structures assist the investigator in locating the site of interest in subsequent sections. It is essential that the reference structure be present over many sections; in some cases it may be present for the extent of the organelle being examined. Where long ribbons are being examined, new reference points will have to be established as the old ones are lost.

The most commonly used reference in locating a cell or an organelle in immersion-fixed tissue is a red blood cell, owing to its density and unusual form in thin sections. Red blood cells are easily observed at low magnifications and can be located very near the cell of interest. During our studies of mitochondrial form it was not uncommon for us to select a red blood cell of unusual form and then decide to examine a hepatic parenchymal cell within a short distance of it.

Perfusion-fixed materials lack blood cells, and one must rely on other methods to locate the site of interest in subsequent sections. One method involves tracking an approximately measured distance from an identified corner of the specimen, along the outside edge until one reaches an external tissue marker or the approximate site where one begins travel into the section from the exterior. Travel into the section towards the site of interest may be by a measured course (using calibrations on specimen travel controls) or by noting landmarks such as blood vessels or unusual nuclei until one reaches the site of interest. We usually demarcate the site of interest with several nuclei so that a pattern of nuclei may be recognized. This allows recognition of the site even if one of the nuclei is lost because of sectioning. When such a loss occurs, another nucleus or two should be incorporated into the pattern.

ANALYSIS AND RECONSTRUCTION

The goal of analyzing serial section electron micrographs is the accurate reconstruction of three-dimensional organization and interrelationships of components of interest. The reconstructive process can be accomplished by one of several methods suggested in Table 2.4. Despite their differences these methods yield either a two-dimensional graphic representation, emphasizing a particular aspect of the reconstruction, or an actual model that can be viewed from different angles. Recent computer techniques provide advantage of a model, in that computers can be programmed to give views from different angles without time-consuming model-building efforts.

Irrespective of the method of reconstruction utilized, several considerations ought be kept in mind for the accurate interpretation of the individual electron micrographs. These include: (1) section thickness, (2) section size, (3) compression, (4) other technical imperfections such as chatter or knife scratches, (5) membrane haziness due to obliquity, (6) fixation artifacts, and (7) picture sharpness. These factors are most important in evaluating the image quality and hence the usefulness of individual sections in the determination of the presence or absence of membrane or cytoplasmic continuity.

It is important to assess thickness because of the impact it will have on the final dimension of the reconstructed image. If in reconstruction each section is assumed to be of uniform thickness, but unaccounted variability exists, the final reconstruction will be proportionately short or long compared to the true object size. This variation may lead to errors in dimensions calculated for various intracellular or extracellular components. Section thickness should be

Table 2.4   Reconstructive Techniques

I. Graphic techniques
    Dunn (1972)
    His (1880)
    Krieg (1967)
    Mitchell and Thaemert (1965)
    Yamada and Yoshida (1972)
II. Alternate thick–thin reconstruction techniques
    Webster (1971)
III. Transillumination techniques
    Bang and Bang (1957)
    Fuscaldo and Jones (1959)
IV. Model-building techniques
    Born (1883)
    Sjöstrand (1958, 1967, 1974)
V. Cinematographic techniques
    Senior (1970) (16 mm)
    Buchanan and Chmielewski (1973) (35 mm)
VI. Computerized techniques
    Hillman *et al.* (1977)
    Levinthal and Ware (1972)
    Llinás and Hillman (1975)

---

thinner than the smallest component to be reconstructed, to make it visible as a profile that can be identified and followed. There should be no missing sections in most cases, since alterations in configuration and continuity can occur in the thickness of even one section in the 50–60-nm range. Section thickness may be estimated by noting the interference colors generated by sections floating in the trough (Peachy, 1958). This estimate may be corroborated by counting the number of sections required to pass through spherical structures in sections.

Consideration of section size is important when difficulties in alignment of adjacent sections arise. Larger-size sections are more subject to compression, which may be variable in extent throughout the section, depending on both tissue and embedment homogeneity.

For best results in alignment of sections, the component of interest should be followed with several adjoining reference points. Each adjacent section should be aligned with the previous for the best fit. Small allowances for section distortion are usually handled with no difficulty in this manner, but some occasionally require careful interpretation because of their importance in determining orientation and continuity. Similar adjustments may be necessary if other technical imperfections such as chatter, scratches, etc., are encountered. Such distortions, however, are infrequent with present technical developments, and the limitation of section size to include only the most essential structures.

The least-controllable parameter of electron micrograph interpretation is the appearance of membrane haziness due to membrane obliquity. As pointed out by Williams and Kallman (1955), "the contour of an object in the plane of the 'top' surface of one section should match that in the plane of the bottom surface of the section immediately preceding it, but one frequently sees no matching of this nature." This statement in part reflected the state of the art with respect to problems of embedding media, susceptibility to cutting artifacts, and rapid sublimation in the electron beam. It also points to the effect of membrane orientation with respect to the direction of the electron beam in providing sharp images. When a membrane is oriented parallel to the impinging electron beam, a sharp line appears. If the membrane is sloped to any degree, a haze appears. Such haziness has been interpreted by some as membrane discontinuity analogous to fixation-induced membrane disruption; however, the basis of this haziness is membrane slope. If recognized during initial section examination, it may be corrected by tilt analysis. However, it frequently goes unrecognized until a leisurely analysis of a series of electron micrographs can be made. Furthermore, since electron micrography of serial sections can be a tedious effort when one is simply following and photographing a component of interest from section to section, it would add an unreasonable burden to require analysis of each section; thus, haziness due to obliquity must be treated by careful interpretation, and may in some instances be uninterpretable.

It should be obvious at this point that virtually every aspect of serial sectioning electron microscopy has an impact on the image interpretation. This is especially true of the quality of the photographs, which in turn depend on factors such as section thickness, uniformity of staining, focus, magnification, and graininess. Accurate focusing is essential, and although virtually guaranteed by a through focal series, it is difficult to obtain for the same reasons that a tilt analysis is not carried out on every image. Thus, we are forced to cut corners in some areas for practicality. In the analysis of isolated portions of short ribbons (up to 40 sections) these techniques may be feasible.

Several graphic techniques of reconstruction have been in use since late in the nineteenth century. His (1880) developed a means of projected reconstruction by producing a drawing of the object in cross section and then projecting the outline of the object perpendicular to the plane of the section. With foreshortening in the cross-sectional drawing and projection lines added, the illusion of depth is created. There is no difficulty in this method if the object outline is uncomplicated and the transition from section to section is unambiguous. In essence, this method is used in alternate thick section–thin section reconstruction techniques.

Mitchell and Thaemert (1965) suggested the utilization of a perspective chart (commercially available at art supply stores) to form a reconstruction by tracing pertinent structures, overlaying them, and connecting them in the most logical manner to form an illustration in perspective. Krieg (1967) "projects on a sheet of paper the outline of any structure in a series of sections and draws its shifting trace in proper relation, so that one obtains a contour map of the object. . . . By use of the slice reconstruction method one can make a fresh start wherever too much would get hidden and the reference to the data source is kept by the relations on the front surface of each slice." Dunn (1972) suggested the use of an inscribed plastic sheet as a guideline for tracing successive sections. Yamada and Yoshida (1972) suggested the use of a mechanical "stereo-deviator" to guide the tracing of successive sections. Simply stated, each of these techniques leads to a graphic reconstruction by tracing adjacent sections and connecting them to form an illusion of three-dimensional form.

Webster (1971) has utilized alternate thick (1 μm) and thin (70–100 nm) sections to formulate a graphic reconstruction. This modification is most useful in studies of long ribbons of tissue where the center of interest is the arrangement of external cell membranes rather than the more intricate arrangement of intracellular organelles.

Bang and Bang (1957) suggested tracing components of interest in adjacent sections onto rigid plastic sheets and stacking them equal distances apart. They would be transilluminated and viewed from the opposite end of the reconstruction, providing an illusion of the third dimension. Fuscaldo and Jones (1959) utilized photographic enlargements of adjacent sections on lantern slides, which were then stacked and transilluminated.

Model building has often produced an elegant representation of the external features of the reconstructed tissue, although it is done infrequently because of the time and work involved, as well as its yielding an inadequate display of internal structures. Born (1883) was the first to build models, using plates made from wax, which could readily be cut and molded into the necessary form. Sjöstrand (1958) applied the graphical and model building techniques to studies of serial sections with the electron microscope.

We have also used both graphical and model-building techniques in our laboratory (Knobler et al., 1974; Stempak and Laurencin, 1976). Components of interest and several reference points were traced onto acetate sheets of 1.5 mm or 2 mm thickness, to allow for adjustment of section thickness. Adjacent sheets were then aligned for best fit, and three holes were drilled for post guides of the completed model. The material to be reconstructed was cut out, oriented, and glued together after aligning the reference points. Drawings or photographs could be made from such models.

Cinematographic reconstructive techniques essentially use the principles of animation. They were originally explored at the light microscopic level. The effect is that of traveling through the specimen, with a three-dimensional illusion being created by the after images in the eye of the observer. One such technique using 16-mm film is described by Senior (1970). A modification for teaching purposes involves showing adjacent sections with two 35-mm projectors linked by a dissolve control (Buchanan and Chmielewski, 1973). It is best used for short series of sections.

Application of serial section cinematography to electron microscopic studies was first described by Levinthal and Ware (1972). Their method involved photography not of the sections as in light microscopy, but of negative trans-parencies of them.   This was to allow alignment of adjacent sections before inclusion in the film.   This technique has been used to provide information on the outline of a section that can be fed into a computer to derive a three-dimensional reconstruction (Macagno *et al.*, 1973; LoPresti *et al.*, 1973, 1974).   Also, quantitative data may be extracted.

Llinás and Hillman (1975) and Hillman *et al.* (1977) have modified the procedure in computerizing the tracings of adjacent sections to obviate the need for the cine film.   They outline the structure to be reconstructed and feed it directly to their computer via a Quantimet 720 computer (IMANCO, Metals Research Instrument Corp.), aligning the adjacent section by use of a computer-generated image of the previous section.   The computer-reconstructed image can be rotated on the screen and allow viewing from many different angles.

CONCLUDING REMARKS

The present chapter recognizes the increasingly important role to be played by serial section electron microscopical analysis in morphological studies of developing and mature tissues.   Various technical aspects such as tissue preparation, staining techniques, grid preparation, dehydration and embedding, sectioning, section collection methods, microscopy and photography, analysis, and reconstruction are discussed.   The emphasis is on the most practical approach to achieving high-quality serial sections for analysis.   Section collection and analytic reconstructive methods are especially emphasized, since these procedures are frequently troublesome, and the most time-consuming areas in serial section electron microscopy.   The methods described are primarily those as practiced in the laboratory of the authors.   Alternative methods where known are also described.   It is hoped that investigators will find the present report a useful starting point and impetus for future applications of the technique of serial section electron microscopical analysis and reconstruction.   An apology is extended to those authors whose work was not included in the present report for reasons of oversight.

References

Andersson, R. G. W., and Brenner, R. M. (1971). Accurate placement of ultrathin sections on grids; control by sol-gel phases of a gelatin flotation fluid. *Stain Technol.* 46, 1.

Andersson-Cedegren, E. (1959).   Ultrastructure of motor end plate and sarcoplasmic components of mouse muscle fiber as revealed by three-dimensional reconstruction from serial sections. *J. Ultrastruct. Res. Suppl* 1.

Anderson-Cedegren, E., and Karlsson, U. (1966). Demyelination regions of nerve fibers in frog muscle spindle as studied by serial sections for electron microscopy. *J. Ultrastruct Res.* 14, 212.

Bang, B. G., and Bang, F. B. (1957). Graphic reconstruction of the third dimension from serial electron micrographs. *J. Ultrastruct. Res.* 1, 138.

Barajas, L. (1970). The ultrastructure of the juxtaglomerular apparatus as disclosed by three-dimensional reconstructions from serial sections. The anatomical relationship between the tubular and vascular components. *J. Ultrastruct. Res.* **33**, 116.

Barajas, L., and Muller, J. (1973). The innervation of the juxtaglomerular apparatus and surrounding tubules–quantitative analysis by serial section electron microscopy. *J. Ultrastruct. Res.* **43**, 107.

Barbolini, U., Zitelli, A., Baroni, A., and Moretti, G. F. (1974). A technique for staining of ultrathin sections prior to their collection on a grid. *Sci. Tools* **21**, 4.

Barnes, B. G., and Chambers, T. C. (1961). A simple and rapid method for mounting serial sections for electron microscopy. *J. Biophys. Biochem. Cytol.* **9**, 724.

Behnke, O., and Rostgaard, J. (1964). Your "Third Hand" in mounting serial sections on grids for electron microscopy. *Stain Technol.* **39**, 205.

Berger, E. R. (1973). Two morphologically different mitochondrial populations in the rat hepatocyte as determined by quantitative three-dimensional electron microscopy. *J. Ultrastruct. Res.* **45**, 303.

Born, G. (1883). Die Plattenmodelliermethode. *Arch. Mikr. Anat.* **22**, 584.

Buchanan, J. W., and Chmielewski, S. (1973). Three-dimensional histology; a 35 mm photographic method. *J. Microscopy,* **99**, 353.

Campbell, F. R. (1968). Nuclear elimination from the normoblast of fetal guinea pig liver as studied with electron microscopy and serial sectioning techniques. *Anat. Rec.* **160**, 539.

Conradi, S. (1969a). Ultrastructure and distribution of neuronal and glial elements on the proximal part of a motoneuron dendrite, as analyzed by serial sections. *Acta Physiol. Scand. Suppl.* **332**, 49.

Conradi, S. (1969b). Observations on the ultrastructure of the axon hillock and initial axon segment of lumbosacrol motoneurons in the cat. *Acta Physiol. Scand. Suppl.* **332**, 65.

Conradi, S. (1969c). Ultrastructure of dorsal root boutons on lumbosacral motoneurons of the adult cat, as revealed by dorsal root section. *Acta. Physiol. Scand. Suppl.* **332**, 85.

Conradi, S., and Skoglund, S. (1969a). Observations on the ultrastructure and distribution of neuronal and glial elements on the motoneuron surface in the lumbosacral spinal cord of the cat during postnatal development. *Acta Physiol. Scand. Suppl.* **333**, 5.

Conradi, S., and Skoglund, S. (1969b). Observations on the ultrastructure of the initial motoraxon segment and dorsal root boutons on the motoneurons in the lumbosacral spinal cord of the cat during postnatal development. *Acta Physiol. Scand. Suppl.* **333**, 53.

De Kock, L. L., and Dunn, A. E. (1964). Ultra-structure of carotid body tissue as seen in serial sections. *Nature* **202**, 821.

Donelli, G., Guglielmi, F., Rosati-Valente, F., and Tangucci, F. (1970). Determination of the shape of viral particles by means of serial sections. *Ann. Ist. Super. Sanita.* **6**, 88.

Dowell, W. C. T. (1959). Unobstructed mounting of serial sections. *J. Ultrastruct. Res.* **2**, 388.

Dunn, R. F. (1972). Graphic three-dimensional representations from serial sections. *J. Microscopy,* **96**, 301.

Elfvin, L. G. (1963a). The ultrastructure of the superior cervical sympathetic ganglion of the cat. I. The structure of the ganglion cell processes as studied by serial sections. *J. Ultrastruct. Res.* **8**, 403.

Elfvin, L. G. (1963b). The ultrastructure of the superior cervical sympathetic ganglion of the cat. II. The structure of the preganglionic end fibers and the synapses as studied by serial sections. *J. Ultrastruct. Res.* **8**, 441.

Elfvin, L. G. (1971a). Ultrastructural studies on the synaptology of the inferior mesenteric ganglion of the cat. I. Observations on the cell surface of the postganglionic perikarya. *J. Ultrastruct. Res.* **37**, 411.

Elfvin, L. G. (1971b). Ultrastructural studies on the synaptology of the inferior mesenteric ganglion of the cat. II. Specialized serial neuronal contacts between preganglionic end fibers. *J. Ultrastruct. Res.* 37, 426.

Elfvin, L. G. (1971c). Ultrastructural studies on the synaptology of the inferior mesenteric ganglion of the cat. III. The structure and distribution of the axodendritic and dendrodentric contacts. *J. Ultrastruct. Res..* 37, 432.

Elias, H., Hennig, A., and Schwartz, D. E. (1971). Stereology: Applications to biomedical research. *Physiol. Rev.* 51, 158.

Famiglietti, E. V., and Peters, A. (1972). The synaptic glomerulus and the intrinsic neuron in the dorsal lateral geniculate nucleus of the cat. *J. Comp. Neur.* 144, 285.

Favard, P., and Carasso, N. (1972). The preparation and observation of thick biological sections in the high voltage electron microscope. *J. Microscopy.* 97, 59.

Fuscaldo, K. E., and Jones, H. H. (1959). A method for the reconstruction of three-dimensional models from electron micrographs of serial sections. *J. Ultrastruct. Res.* 3, 1.

Galey, F. R., and Nilsson, S. E. G. (1966). A new method for transferring sections from the liquid surface of the trough through staining solutions to the supporting film of a grid. *J. Ultrastruct Res.* 14, 405.

Gay, H., and Anderson, T. F. (1954). Serial sections for electron microscopy. *Science* 120, 1071.

Gillies, C. B. (1972). Reconstruction of the *Neurosspora crassa* pachytene karyotype from serial sections of synaptonemal complexes. *Chromosoma* 36, 119.

Gray, E. G., and Willis, R. A. (1968). Problems of electron stereoscopy of biological tissues. *J. Cell Sci.* 3, 309.

Hayat, M. A. (1970). *Principles and Techniques of Electron Microscopy: Biological Applications*, Vol. 1. Van Nostrand Reinhold Company, New York and London.

Hayat, M. A. (1975). *Positive Staining for Electron Microscopy.* Van Nostrand Reinhold Company, New York and London.

Helmcke, J. G. (1965). Determination of the third dimension of objects by stereoscopy. *Lab. Invest.* 14, 195.

Hildebrand, C. (1971a). Ultrastructural and light-microscopic studies of the nodal region in large myelinated fibers of the adult feline spinal cord white matter. *Acta Physiol. Scand. Suppl* 364, 43.

Hildebrand, C. (1971b). Ultrastructural and light microscopic studies of the developing feline spinal cord white matter. I. The nodes of Ranvier. *Acta. Physiol. Scand. Suppl.* 364, 81.

Hildebrand, C. (1971c). Ultrastructural and light-microscopic studies of the developing feline spinal cord white matter. II. Cell death and myelin sheath disintegration in the early postnatal period. *Acta Physiol. Scand. Suppl.* 364, 109.

Hillman, D. E., Llinás, R., Chujo, M. (1977). Automatic and semi-automatic analysis of nervous system structure. In: *Computer Analysis of Neuronal Structures* (Lindsay, R. D., ed.), p. 73. Plenum Press, New York and London.

Hinds, J. W., and Hinds, P. L. (1972). Reconstruction of dendritic growth cones neonatal mouse olfactory bulb. *J. Neurocytol.* 1, 169.

Hinds, J. W., and Hinds, P. L. (1974). Early ganglion cell differentiation in the mouse retina: an electron microscopic analysis utilizing serial sections. *Develop. Biol.* 37, 381.

Hinds, J. W., and Hinds, P. L. (1976a). Synapse formation in the mouse olfactory bulb. I. Quantitative studies. *J. Comp. Neur.* 169, 15.

Hinds, J. W., and Hinds, P. L. (1976b). Synapse formation in the mouse olfactory bulb. II. Morphogenesis. *J. Comp. Neur.* 169, 41.

Hinds, J. W., and Ruffett, T. L. (1971). Cell proliferation in the neural tube: An electron microscopic and Golgi analysis in the mouse cerebral vesicle. *Z. Zellforsch.* 115, 226.

Hinds, J. W., and Ruffett, T. L. (1973). Mitral cell development in the mouse olfactory bulb: reorientation of the perikaryon and maturation of the axon initial segment. *J. Comp. Neur.* **151** 281.

His, W. (1880). *Anatomic menschlicher Embryonen.* Vogel, Leipzig.

Hoffman, H. P., and Avers, C. J. (1973). Mitochondrion of yeast: Ultrastructural evidence for one giant, branched organelle per cell. *Science* **181**, 749.

Holländer, H. (1970). The section embedding (SE) technique. A new method for the combined light microscopic and electron microscopic examination of central nervous tissue. *Brain Res.* **20**, 39.

Jones, D. G., and Brearley, R. F. (1973). An analysis of some aspects of synaptosomal ultrastructure by the use of serial sections. *Z. Zellforsch. Mikrosk. Anat.* **140**, 481.

Karlsson, U. (1966a). Three-dimensional studies of neurons in the lateral geniculate nucleus of the rat. I. Organelle organization on the perikaryon and its proximal branches. *J. Ultrastruct. Res.* **16**, 429.

Karlsson, U. (1966b). Three-dimensional studies of neurons in the lateral geniculate nucleus of the rat. II. Environment of perikarya and proximal parts of their branches. *J. Ultrastruct. Res.* **16**, 482.

Karlsson, U., Andersson-Cedegren, E., and Ottoson, D. (1966). Cellular organization of the frog muscle spindle as revealed by serial sections for electron microscopy. *J. Ultrastruct. Res.* **14**, 1.

Karnovsky, M. J. (1971). Use of ferrocyanide-reduced osmium tetroxide in electron microscopy. *J. Cell Biol. Suppl.* **51**, 146.

Keddie, F. M., and Barajas, L. (1969). Three-dimensional reconstruction of *Pityrosporum* yeast cells based on serial section electron microscopy. *J. Ultrastruct. Res.* **29**, 260.

Keddie, F. M., and Barajas, L. (1972). Quantitative ultrastructural variations between *Pityrosporum ovale* and *P. orbiculare* based on serial section electron microscopy. *Int. J. Dermatol.* **11**, 40.

Kishi, K. (1972). Two types of cytoplasmic nucleolus-like bodies. Observation of the serial sections of rat neurons. *Arch. Histol. Jap.* **35**, 83.

Knobler, R. L., and Stempak, J. G. (1973). Serial section analysis of myelin development in the central nervous system of the albino rat: an electron microscopical study of early axenal ensheathment. In: *Progress in Brain Research,* Vol. 40, p. 407, (Ford, D. H., ed.), Elsevier, Amsterdam, London and New York.

Knobler, R. L., Stempak, J. G., and Laurencin, M. (1974). Oligodendroglial ensheathment of axons during myelination in the developing rat central nervous system. A serial section electron microscopical study. *J. Ultrastruct. Res.* **49**, 34.

Knobler, R. L., and Stempak, J. G. (1976). Report on efforts to fix five-day old rat central nervous system. *Anat. Res.* **184**, 450.

Knobler, R. L., Stempak, J. G., and Laurencin, M. (1976). Nonuniformity of the oligodendroglial ensheathment of axons during myelination in the developing rat central nervous system. A serial section electron microscopical study. *J. Ultrastruct. Res.* **55**, 417.

Krieg W. J. S. (1967). Reconstruction from serial sections. In: *Stereology* (Elias, H., ed.) p. 293. Springer Verlag, New York.

Levinthal, C., and Ware, R. (1972). Three-dimensional reconstruction from serial sections. *Nature* **236**, 207.

Llinás, R., and Hillman, D. E. (1975). A multipurpose tridimensional reconstruction computer system for neuroanatomy. In: *Golgi Centennial Symposium Proceedings* (Santini, M., ed.), p. 71. Raven Press, New York.

Locke, M., and Krishnan, N. (1971). Hot alcoholic phosphotungstic acid and uranyl acetate as routine stains for thick and thin sections. *J. Cell Biol.* **50**, 550.

Locke, M., Krishnan, N., and McMahon, J. T. (1971). A routine method for obtaining high contrast without staining sections. *J. Cell Biol.* **50**, 540.

LoPresti, V., Macagno, E. R., and Levinthal, C. (1973). Structure and development of neuronal connections in isogeneic organisms: Cellular interactions in the development of the optic lamina of *Daphnia. Proc. Nat. Acad. Sci.* 70, 433.

LoPresti, V., Macagno, E. R., and Levinthal, C. (1974). Structure and development of neuronal connections in isogeneic organisms: Transient gap junctions between growing optic axons and lamina neuroblasts. *Proc. Nat. Acad. Sci.* 71, 1098.

Loud, A. V. (1968). A quantitative stereological description of the ultrastructure of normal rat liver parenchymal cells. *J. Cell Biol.* 37, 27.

Luft, J. H. (1961). Improvements in epoxy resin embedding methods. *J. Biophys. Biochem. Cytol.* 9, 409.

Macagno, E. R., LoPresti, V., and Levinthal, C. (1973). Structure and development of neuronal connections in isogeneic organisms: variations and similarities in the optic system of *Daphnia magna. Proc. Nat. Acad. Sci.* 70, 57.

Martin, J. R., and Webster. H. deF. (1973). Mitotic Schwann cells in developing nerve: Their changes in shape, fine structure and axon relationships. *Develop. Biol.* 32, 417.

Mazziotta, J. C., Hamilton, B. L., and Fenner-Crisp, P. A. (1973). A device for the precise transfer of serial sections for electron microscopy. *Stain Technol.* 48, 153.

Merzel, J. (1971). Preparation of semithin serial sections of Epon embedded material. *Experientia* 27, 611.

Mitchell, H. C., and Thaemert, J. C. (1965). Three dimensions in fine structure. *Science* 148, 1480.

Mugnaini, E. (1965). "Dark cells" in electron micrographs from the central nervous system of vertebrates. *J. Ultrastruct. Res.* 12, 235.

Nadelschaft, I., and Rose, L. M. (1974). Serial sectioning of Epon-embedded material for autoradiography. *Stain Technol.* 49, 178.

O'Leary, T. P., Bemrick, W. J. and Johson, K. H. (1973). Serial section analysis of cells in the microfilariae of *Dirofilaria immitis* containing neurosecretory-like granules. *J. Parasitol.* 59, 701.

Pachter, B. R., Davidowitz, J., and Breinin, G. M. (1973). A light and electron microscopic study in serial sections of dystrophic extraocular muscle fibers. *Invest. Ophthalmol.* 12, 917.

Pachter, B. R., Davidowitz, J., Eberstein, A., and Breinin, G. M. (1974). Myotonic muscle in mouse: A light and electron microscopic study in serial sections. *Exp. Neurol.* 45, 462.

Palay, S. L., McGee-Russell, S. M., Gordon, S., and Grillo, M. A. (1962). Fixation of neural tissues for electron microscopy by perfusion with solutions of osmium tetroxide. *J. Cell Biol.* 12, 385.

Peachey, L. D. (1958). Thin sections. A study of section thickness and physical distortion produced during microtomy. *J. Biophysic. and Biochem. Cytol.* 4, 233.

Peachey, L. D. (1965). Electron microscopy of tilted biological sections. *RCA Scient. Instrum. News* 10, 7.

Peters, A. (1970). The fixation of central nervous tissue and the analysis of electron micrographs of the neuropil, with special reference to the cerebral cortex. In: *Contemporary Research Methods in Neuroanatomy* (Nauta, W. J., and Ebbesson, S. O. E., eds.), p. 56. Springer-Verlag, New York.

Pinching, A. J., and Powell, T. P. S. (1971a). The neuron types of the glomerular layer of the olfactory bulb. *J. Cell Sci.* 9, 305.

Pinching, A. J., and Powell, T. P. S. (1971b). The neuropil of the glomeruli of the olfactory bulb. *J. Cell Sci.* 9, 347.

Pinching, A. J., and Powell, T. P. S. (1971c). The neuropil of the periglomerular region of the olfactory bulb. *J. Cell Sci.* 9, 379.

Poritsky, R. (1969). Two and three dimensional ultrastructure of boutons and glial cells on the motoneuronal surface in the cat spinal cord. *J. Comp. Neur.* 135, 423.

Price, J. L., and Powell, T. P. S. (1970a). The morphology of the granule cells of the olfactory bulb. *J. Cell Sci.* 7, 91.

Price, J. L., and Powell, T. P. S. (1970b). The synaptology of the granule cells of the olfactory bulb. *J. Cell Sci.* 7, 125.

Price, J. L., and Powell, T. P. S. (1970c). An electron-microscopic study of the termination of the afferent fibers to the olfactory bulb from the cerebral hemisphere. *J. Cell Sci.* 7, 157.

Price, J. L., and Powell, T. P. S. (1970d). The mitral and short-axon cells of the olfactory bulb. *J. Cell Sci.* 7, 631.

Rakic, P. (1972). Extrinsic cytological determinants of basket and stellate cell dendritic pattern in the cerebellar molecular layer. *J. Comp. Neur.* 146, 335.

Reiter, W. (1966). On the 3-dimensional system of the endoplasmic reticulum of skin-nerve fibers. Studies on serial sections. *Z. Zellforsch.* 72, 446.

Reynolds, E. S. (1963). The use of lead citrate at high pH as an electron-opaque stain in electron microscopy. *J. Cell Biol.* 17, 208.

Sabatini, D. D., Bensch, K., and Barnett, R. J. (1963). Cytochemistry and electron microscopy. The preservation of cellular ultrastructure and enzymatic activity by aldehyde fixation. *J. Cell Biol.* 17, 19.

Schultz, R. L. and Willey, T. J. (1973). Extracellular space and membrane changes in brain owing to different alkali metal buffers. *J. Neurocytol.* 2, 289.

Senior, W. (1970). Reconstruction of microscopic objects by photographing serial sections onto cine film for projection. *J. Microscopy.* 92, 223.

Sétáló, G. (1969). A simple method for cutting semithin serial sections. *Acta Morphol. Acad. Sci. Hung.* 17, 315.

Shimamura, T., and Sorenson, G. D. (1965). Electron microscopy of serial sections of the murine glomerular mesangium. *Anat. Rec.* 152, 141.

Sjöstrand, F. S. (1958). Ultrastructure of retinal rod synapses of the guinea pig eye as revealed by three-dimensional reconstructions from serial sections. *J. Ultrastruct. Res.* 2, 122.

Sjöstrand, F. S. (1967). *Electron Microscopy of Cells and Tissues,* Vol. 1, *Instrumentation and Techniques.* Academic Press, New York, and London.

Sjöstrand, F. S. (1974). A search for the circuitry of directional selectivity and neural adaptation through three-dimensional analysis of the outer plexiform layer of the rabbit retina. *J. Ultrastruct. Res.* 49, 60.

Sorvall, I. (1967). *Thin Sectioning and Associated Technics for Electron Microscopy.* Sorvall, Inc., Norwalk, Connecticut.

Sotelo, J. R., Garcia, R. B., and Wettstein, R. (1973). Serial sectioning study of some meiotic stages in *Scaptericus borrelli* (Grylloidea). *Chromosoma* 42, 307.

Špaček, J. (1971). Three-dimensional reconstructions of astroglia and oligodendroglia cells. *Z. Zellforsch.* 112, 430.

Špaček, J., and Lieberman, A. R. (1974). Ultrastructure and three-dimensional organization of synaptic glomeruli in rat somatosensory thalamus. *J. Anat.* 117, 487.

Špaček, J., Pařizek, J., and Lieberman, A. R. (1973). Golgi cells, granule cells and synaptic glomeruli in the molecular layer of the rabbit cerebral cortex. *J. Neurocytol.* 2, 407.

Stempak, J. (1967). Serial section analysis of mitochondrial form and membrane relationships in the neonatal rat liver cell. *J. Ultrastruct. Res.* 18, 619.

Stempak, J. G., and Laurencin, M. L. (1966). Serial section analysis of mitochondrial form and membrane relationships in the neonatal rat liver cell. *J. Ultrastruct. Res.* 18, 619-637.

Stempak, J. G., and Knobler, R. L. (1972). Bidirectionality in the tongue processes of the oligodendroglial cell investment of axons in the albino rat. *Am. J. Anat.* 135, 287.

Stempak, J., and Laurencin, M. (1976). Mitochondrial form in hepatic parenchymal cells in rats of several ages. *Am. J. Anat.* 145, 261.

Stempak, J., and Ward, R. T. (1964). An improved staining method for electron microscopy. *J. Cell Biol.* **22**, 697.

Stensaas, L. J., and Stensaas, S. S. (1968). Astrocytic neuroglial cells, oligodendrocytes and microgliacytes in the spinal cord of the toad. II. Electron microscopy. *Z. Zellforsch.* **86**, 184.

Suganuma, A. (1968). Electron microscopic studies of *Staphylococci* by serial sections. *J. Electron Microscopy.* (Tokyo) **17**, 315.

Sumi, S. M. (1969). The extracellular space in the developing rat brain: Its variation with changes in osmolarity of the fixative, method of fixation and maturation. *J. Ultrastruct Res.* **29**, 398.

Swash, M., and Fox, K. P. (1975). Abnormal intrafusal muscle fibers in myotonic dystrophy: a study using serial sections. *J. Neurol. Neurosurg. Psychiatry* **38**, 91.

Tennyson, V. M. (1973). Personal communication.

Thaemert, J. C. (1966). Ultrastructural interrelationships of nerve processes and smooth muscle cells in three dimensions. *J. Cell Biol.* **28**, 37.

Thaemert, J. C. (1969). Fine structure of neuromuscular relationships in mouse heart. *Anat. Rec.* **163**, 575.

Thaemert, J. C. (1970). Atrioventricular node innervation in ultrastructural three dimensions. *Am. J. Anat.* **128**, 239.

Thaemert, J. C. (1973). Fine structure of the atrioventricular node as viewed in serial sections. *Am. J. Anat.* **136**, 43.

Vaughn, D. W., and Peters, A. (1973). A three dimensional study of layer I of the rat parietal cortex. *J. Comp. Neur.* **149**, 355.

Vivier, E., and Petitprez, A. (1972). Serial sections and three-dimensional reconstitution of the vacuolar system of the hematozoan *Anthemosoma garnhami. J. Ultrastruct. Res.* **41**, 219.

Ward, R. T. (1972). A section lifter designed for attachment to an ultramicrotome. *Stain Technol.* **47**, 257.

Ward, S., Thomson, N., White, J. G., and Brenner, S. (1975). Electron microscopical reconstruction of the anterior sensory anatomy of the nematode *Caenorhabditis elegans. J. Comp. Neur.* **160**, 313.

Ware, R. W., and LoPresti, V. (1975). Three-dimensional reconstructions from serial sections. *Int. Rev. Cytol.* **40**, 325.

Webster, H. deF. (1971). The geometry of peripheral myelin sheaths during their formation and growth in rat sciatic nerves. *J. Cell Biol.* **48**, 348.

Webster, H. deF., Martin, J. R., and O'Connell, M. F. (1973). The relationship between interphase Schwann cells and axons before myelination: A quantitative electron microscopic study. *Develop. Biol.* **32**, 401.

Weibel, E. R. (1969). Stereological principles for morphometry in electron microscopic cytology. *Int. Rev. Cytol.* **26**, 235.

Weibel, E. R., and Bolender, R. P. (1973). Stereological techniques for electron microscopic morphometry. In: *Principles and Techniques of Electron Microscopy: Biological Applications* Vol. 3 p. 237. (Hayat, M. A., ed.), Van Nostrand Reinhold Company, New York and London.

Westfall, J. A. (1961). Obtaining flat serial sections for electron microscopy. *Stain Technol.* **36**, 36.

Westfall, J. A., and Healy, D. L. (1962). A water control device for mounting serial ultrathin sections. *Stain Technol.* **37**, 118.

Wettstein, R., and Grauer, A. (1973). Film supporting frames for mounting serial section grids. *J. Ultrastruct. Res.* **43**, 476.

White, E. L. (1972). Synaptic organization in the olfactory glomerulus of the mouse. *Brain Res.* **37**, 69.

White, E. L. (1973). Synaptic organization of the mammalian olfactory glomerulus: New findings including an intraspecific variation. *Brain Res.* **60,** 299.

White, E. L. (1976). Ultrastructure and synaptic contacts in barrels of mouse SI cortex. *Brain Res.* **105,** 229.

Williams, R. C., and Kallman, F. (1955). Interpretations of electron micrographs of single and serial sections. *J. Biophys. Biochem. Cytol.* **1,** 301.

Yamada, M., and Yoshida, S. (1972). Graphic stereo-reconstruction of serial sections. *J. Microscopy.* **95,** 249.

Zotikov, L., and Bernhard, W. (1970). Electron microscope localization of the activity of certain nucleases in ultrathin serial sections. *J. Ultrasruct. Res.* **30,** 642.

3. CALIBRATION OF MAGNIFICATION IN TRANSMISSION ELECTRON MICROSCOPY

Robert F. Dunn

Division of Vestibular Disorders, Department of Otolaryngology,
University of Pittsburgh School of Medicine, Pittsburgh, Pennsylvania

---

INTRODUCTION

An accurate determination of magnification of the electron microscope image has become more important with the increasing emphasis on quantitative studies. During the early days of electron microscopy the entire realm of biological fine structure was suddenly available for scrutiny, resulting in a wealth of qualitative information describing the various cellular membrane systems and organelles. For these purposes a 10 to 15% error in magnification was, and remains, an acceptable error rate, particularly when dimensions are expressed in relative terms as ratios. The increased interest in quantitative relationships demands not only that magnification be rigorously calibrated, but also that attention be directed to the reproducibility of a given magnification over long periods of time, that is, the ability to accurately duplicate a given magnification during successive operational sessions on the electron microscope.

FACTORS AFFECTING MAGNIFICATION

Magnification, very simply stated, is the ratio of the image size to the object size and may be expressed as:

$$M = S_i/S_o \tag{3.1}$$

Whereas this expression is useful for determining the object size when the magnification and image size are known, it does not immediately reveal the relationships between magnification and the many parameters upon which the image magnification depends. The final image size is dependent upon both physical and electrical properties of the image-forming lenses, the objective lens and the projector lens complex.

Dosse (1941), experimenting with an early Siemens electron microscope, was able to determine the relationship between magnification, M, of the image, the focal length, f, of the individual lens, and the distance, b, from the center of the lens being tested to the image plane (Fig. 3.1). The magnification was determined by measuring the image size of a test object solving Eq. (3.1). The magnification

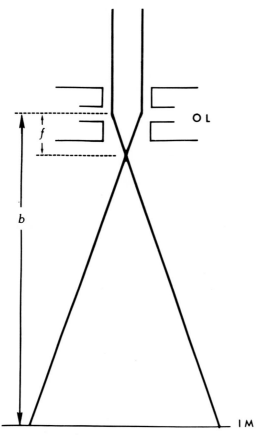

Fig. 3.1 This schematic diagram illustrates the relationships given as equation (3.2), where f is the focal length, and b is the distance of the image screen (*IM*) to the center of the objective lens (*OL*). (*Modified from Dosse, 1941, and Zworykin et al., 1945.*)

may then be expressed as:

$$M = (b - f)/f \qquad (3.2)$$

Hence, the magnification is related to the focal length as:

$$f = b/(M + 1) \qquad (3.3)$$

It should be noted that there is an inverse relationship between the focal length and the magnification such that a decrease in focal length will result in an increase in magnification. With this in mind, additional factors that cause changes in the focal length of the lens may also be easily related to magnification.

The basic electromagnetic lens consists of a wire having a number of coils, n, through which a current, I, flows, thus creating an electromagnetic field in the center of the coil. The pole piece and lens casing, being constructed of highly permeable iron, serve not only to restrict the gap diameter but, more important, to intensify the magnetic field, concentrating it into a restricted portion along the lens axis. These manipulations alter the focal length of the lens. In addition to the permeability properties of the casing and pole piece, the magnetic field strength, H, can be altered by a change in either current magnitude or the number of turns in the coil. This relationship may be expressed as:

$$\int_{-\infty}^{\infty} H \, dz = \frac{4\pi n I}{10} \qquad (3.4)$$

where the field intensity, H, is directly proportional to the ampere turns, nI (Zworykin et al., 1945). The field intensity is also inversely proportional to the focal length of the lens, which, with a constant configuration and number of ampere turns, is expressed as:

$$\frac{1}{f} = e(8 \, mc^2 \Phi)^{-1} \int_{-\infty}^{\infty} H^2 \, dz \qquad (3.5)$$

in which the expression $e(8 \, mc^2 \Phi)^{-1}$ relates the effects of the electron charge, e, the electron's mass, m, and the speed of light, c. The final expression, Φ, is the electrical potential along the axis and it is measured in volts. By solving for the constants, Eq. (3.5) becomes:

$$\frac{1}{f} = \frac{0.022}{V} \int_{-\infty}^{\infty} H^2 \, dz \qquad (3.6)$$

If a magnetic lens is considered as an equivalent single loop having the same magnetic properties as the lens, a loop whose axial radius is a, then the field intensity along the z-axis may be expressed as:

$$H_z = \frac{2\pi n I a^2}{10 \, (z^2 + a^2)^{3/2}} \qquad (3.7)$$

Provided that the axial radius, a, is much less than the focal length of the lens, then by substituting Eq. (3.6) with Eq. (3.7), the expression for the focal length becomes:

$$\frac{1}{f} = \frac{0.022}{V} \int_{-\infty}^{\infty} \left[\frac{2\pi n I a^2}{10 (z^2 + a^2)^{3/2}} \right]^2 dz \qquad (3.8)$$

Integrating this expression and solving for the constants (see Appendix), the equation becomes:

$$\frac{1}{f} = \frac{0.0206 \, n^2 I^2}{dV} \qquad (3.9)$$

where $d = 2a$. One final manipulation is required to arrive at an expression for magnification. By combining Eqs. (3.3) and (3.9):

$$M = \frac{0.0206 \, n^2 I^2 b}{dV} - 1 \qquad (3.10)$$

Equation (3.10) is a useful expression that summarizes the functional relationships of lens current, I, accelerating voltage, V, the effective lens diameter, d, and the height factor, b, upon the magnification, M.

CALIBRATION SPECIMENS

Instrument magnification ranges of modern electron microscopes, from less than 1,000 to 200,000X, preclude the opportunity of having a single calibration specimen adequate for this wide operational range. Thus, it becomes necessary to utilize different calibration specimens at various magnification ranges, thereby necessitating primary and secondary standards.

Early in the development of the electron microscope, it was recognized that one of the more important features of this instrument was its high resolving power and that magnification represented a method to translate the resolving power into image distances required by the human eye. Strict calibration then became an integral and necessarily important part of determining the magnification. Initially, comparisons between photomicrographs and electron micrographs of the same well defined specimen were used for calibration purposes. Although the light microscope could be calibrated against a stage micrometer, there are several problems associated with this method. First, any measurement errors gained in the light microscope are transferred directly to the electron microscope calibration, and in some cases amplified. Burton et al. (1942) also cautioned that the larger the particle the more precise is the light microscope measurement. Hence, with a specimen larger than 2.5 μm, a two-stage calibration becomes necessary with all its attendant problems and error sources.

Diffraction Grating Replicas

Mahl (1940) introduced techniques by which metal oxides could be used to produce thin film replicas from the surface of opaque materials, replicas suitable then for electron microscopic observation. These techniques were used to replicate diffraction gratings whose spacings could be accurately determined by a spectrometer. Zworykin and Romberg (1941) devised a double system to produce diffraction grating replicas. First, silver was evaporated upon the grating surface, and then stripped off. Next, a thin film of collodion was deposited by allowing a 1% solution of collodion in amyl acetate to flow over the surface of the silver replica, which was then dissolved by nitric acid, leaving only the collodion replica. The advantage of this method, though time-consuming, was thought to be that the silver film would be less subject to mechanical distortions during the stripping steps. Burton *et al.* (1942) introduced diffraction grating replicas made with Formvar 15/95. A 0.5% solution of Formvar 15/95 in ethylene dichloride was placed directly onto the diffraction grating surface and allowed to dry thoroughly. The Formvar replica was subsequently floated off by carefully immersing the grating surface under a water surface which had been "well swept." Formvar replicas appeared to be more sturdy than those made with collodion, and they were subject to much less, but constant, shrinkage in the electron beam. Burton *et al.* (1942), preparing their replicas from a 30,000-parallel-line reflection grating, ascertained the shrinkage of these replicas by comparing the spacing of the original grating to that of the replica, and determined a periodicity of 0.844 μm for the former and 0.832 μm for the latter. They further noted, however, that at magnifications of 20,000 to 25,000X, the usefullness of these replicas decreases because of the increased diffusion of the image, thereby making accurate measurements difficult.

Hass and McFarland (1950) suggested another two-step method of forming replicas. A thick film (>1 μm) of aluminum was evaporated onto the grating in a vacuum chamber, followed by a thin layer of magnesium (<0.1 μm). The replica was then removed with Scotch Tape. The specimen was anodized to form a thin layer of aluminum oxide on the aluminum surface. The entire film was next immersed in hydrochloric acid to dissolve the magnesium, whereupon the aluminum oxide–aluminum layer surfaced. The hydrochloric acid also eliminated the aluminum film, with only the aluminum oxide replica, a positive replica of the original surface, remaining. This replica was then washed and mounted on a grid, at which point it could be shadow-cast to enhance the contrast. Hass and McFarland (1950) used germanium for their shadowing. The advantages of replicas produced by this method include: minimal specimen damage, uniform thickness, and stability in an electron beam of high intensity. Sjöstrand (1967) also suggested the use of an aluminum replica. First, a collodion replica, 0.5 to 1 mm thick, was made. Aluminum was then evaporated onto the collodion replica, which was then dissolved away. Sjöstrand indicated that these aluminum replicas were very stable and did not exhibit dimensional changes in the electron beam.

Rouze and Watson (1953) suggested that a "parting layer" of Victawet 35B be used between the original surface and the replica to facilitate the stripping process. The Victawet was applied by vacuum evaporation onto a diffraction grating, after which a layer of chromium was shadow-cast (Williams and Wyckoff, 1946) at a 27° angle, followed by a normal angle evaporation of aluminum-beryllium. Next, the replicas were floated off onto a water surface and picked up on a 720-mesh grid. Victawet apparently proved useful in minimizing mechanical distortions to the replica during the stripping process. In addition, this method of replica production was not deleterious to the original sample and resulted in replicas of superior optical qualities that facilitated the calibration of these specimens spectroscopically.

The diffraction grating replicas enjoyed immediate and widespread usage as an internal standard by which the magnification could be determined. Their greatest advantage appears to be that they can be calibrated by independent methods. Their popularity has continued, and they remain readily available from many electron microscope accessory suppliers. Initially the diffraction gratings were parallel lines spaced at 15,000 to 30,000 lines/inch, but the usefulness of the replicas as internal standards was limited to magnifications of ~25,000X. Grating replicas are currently available either as parallel-line or crossed-line grating replicas, generally carbon replicas that have been metal-shadowed and mounted on suitably sized grids. The spacings have been variously listed as 15,240, 28,800, 54,800, and 54,864 lines/inch. Silicon monoxide replicas are also available and have the advantage of being quite stable in the electron beam; however, damage to the replica does become apparent after prolonged exposure. The crossed-line or "waffle type" replica has several advantages over the parallel-line type in that the former allows an assessment of: magnification in two directions; image distortion due to electron lens effects and magnification changes imposed by tilting the specimen. Aqueous suspensions of replica fragments are also available currently to serve as an internal calibration standard directly on the specimen.

Glass Spheres

Fullam (1943) suggested using glass spheres of predetermined size as suitable calibration specimens for magnifications up to 30,000X. The glass spheres were prepared by pulverizing Pyrex glass to a predetermined size, followed by heating the glass particles to form the final spheres. These spheres could then be dispersed in distilled water so that they could be mounted either directly on the specimen support film to provide an internal standard during normal operation, or on to a diffraction grating replica for purposes of calibrating the spheres. Hence, they are secondary standards, requiring measurement against another standard. Several other uses were suggested by Fullam; for example, the spheres could be utilized to evaluate the image distortion throughout the field of view. He indicated that since the glass beads were nearly perfect spheres, any ellipticity of the spheres, when located on the optical axis of a properly aligned microscope would be due to faulty microscope alignment. The accuracy to be expected with these spheres was estimated at 2–3%, provided that proper care was exercised and that a sufficient number of measurements were made during calibration.

Latex Spheres

Latex spheres did serve and continue to serve as an internal calibration standard. Backus and Williams (1948, 1949) described a particular batch of latex spheres from Dow Chemical Company (Latex 580-G, lot 3584) having a mean diameter of 259 ± 2.5 nm. These authors suggested that the spheres would be suitable as a standard to determine the magnification of the instrument. Unique clusters of the latex particles were identified with a light microscope equipped with a calibrated eyepiece reticule, and the distances between clusters were measured. These same clusters were then located in the electron microscope, and the distances between these unique clusters were measured on overlapping exposures montaged to cover the entire distance. The various fields of view also allowed the diameter of several individual latex spheres to be determined. Gerould (1950) tabulated the reports of Latex 580-G, lot 3584 dimensions from seventeen laboratories, and remarked that "thirteen of the seventeen determinations show an average size of 2588Å with a range from 2520Å to 2630Å." Assuming that the original figure quoted by Backus and Williams was ± one standard deviation, then ± 2 standard deviations should include 95% of the dimension range, that is 254 to 264 nm, or a variation of 3.9%. Utilizing this latter range, and with the further constraint that the instruments used were either the electron microscope or light microscope–electron microscope, the mean diameter from Gerould's tabulation is 259.8 ± 2 nm, still remarkably good agreement. The problem, of course, rests with the acceptability of the 3.9% variation. Should such a level be acceptable, then an attempt must be made to assess its effect on any quantitative manipulations of data.

The latex spheres had several advantages as standards aside from the uniformity of their dimensions. Among these was their capability to be dispersed directly on grids containing sections, thereby creating a convenient internal standard for use during specimen observation. Gerould (1950) did, however, comment that the latex spheres could be altered by dessication, bacterial contamination, exposure to high beam intensities, and metal shadowing. These points were later substantiated by Watson and Grube (1952) and Bradford and Vanderhoff (1955), who cautioned that while single specimens may contain latex spheres of uniform size, they need to be calibrated from specimen to specimen. Unfortunately, the supply of these particular latex spheres has long since been exhausted, but their consideration is predictive of many problems associated with the choice of calibration standards. However, latex particles, in a variety of sizes (0.091–100 μm), continue to be available and remain useful calibration specimens in many applications.

Crystal Lattices

The size characteristics of both diffraction grating replicas and latex spheres effectively preclude their use as calibration specimens at magnifications greater than approximately 30,000X. Several organic crystals have proved suitable for use as calibration specimens over a wide range of magnifications, but their particular advantage is in the scale of 20,000X and greater. These organic crystals include: indanthrene olive (Labaw, 1964), phtholocyanine crystals, faujasite, tremolite, and especially the crystalline form of the enzyme catalase (Beeston et al., 1973; Horne, 1965). For the calibration specimens, the dimensions of the crystalline lattice can be determined by means of X-ray diffraction, thereby establishing the sample spacing independently of the electron microscope to be calibrated.

In solids having a crystalline array, the atoms are arranged in a regular configuration formed by repeating a basic pattern in three dimensions, a pattern that is a fundamental feature of crystals (Dekker, 1957). The basic pattern, the unit cell, is defined as the smallest unit which, when repeatedly translated in two dimensions, will produce a single planar lattice of the crystal. The unit cells consist of parallelograms with either one atom cluster at each corner or along two faces. Whereas several parallelograms may be defined in a crystalline array, it is customary to choose that parallelogram having the shortest sides. The final structure of the lattice may be arrived at by repeating the planar lattice along the z-axis in a similar manner, resulting in an ordered parallelepiped arrangement. The three vectors, defined by the three-dimensional repetition, are referred to as a, b, and c, which correspond to the x, y, and z axes, respectively (Fig. 3.2). These vectors are used to define the directional distance between adjacent atom clusters, a notation that may be encountered in the literature.

The Miller index is another notation frequently used in the more recent literature. Since the lattice points in a crystalline lattice may be considered a system of parallel planes, the ordered lattice may be denoted numerically as a set of parallel planes, with the notation denoting both the direction and offset. Figure 3.3 gives an example of various ways to divide a planar lattice. The integers may be either positive or negative to signify the direction of the intercepts. Specific sets of planes are denoted in the form (100), (010), (001), corresponding to the x-axis, y-axis, and z-axis planes, respectively, of those cube faces (Dekker, 1957; Heidenreich, 1964).

Menter (1956) studied the lattice spacings of thin crystals of platinum phtholocyanine and compared the results from X-ray diffraction analysis, 1.194 nm in

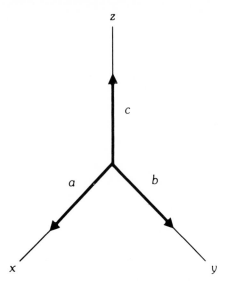

Fig. 3.2 The crystalline lattice directions *a*, *b*, and *c* are shown in their relations to the normal axes, *x*, *y*, and *z*.

the $(20\bar{1})$ planes, and direct electron microscopic examination, 1.2 ± 0.02 nm. While the primary purpose of the study was related to the structure of the crystals, it did indicate the potential use of Pt phtholocyanine crystals as an internal calibration specimen.   Luftig (1967) used copper phtholocyanine β crystals to calibrate the instrument magnification, and indicated that X-ray diffraction data revealed a line spacing 1.25 nm in the (100) planes, and 0.95 nm in the $(10\bar{2})$ planes (Fig. 3.3). The phtholocyanin crystals formed, as long flat structures, had a ribbon-like form, a geometry that possessed many advantages for electron microscopic observation. The most serious handicap was that the crystals were contaminated rapidly at high electron beam intensities, particularly in instruments having only a single condenser lens. Violent motions of the crystals upon irradiation by the beam were also noted and were attributed to deformation problems.

Crystals of the enzyme catalase, isolated from beef livers, have proved to be a reliable calibration specimen and remain readily available from a number of electron microscopic accessory suppliers. Catalase is thermally stable in the electron beam, and is easily prepared for observation. Hall (1950) determined a spacing of $6.4 \times 6.4 \times 15.9$ nm in the a_{01}, b_0, and c_0 directions, for the unit cell of catalase, which was unfixed and metal-shadowed. He further indicated that the initial instrument magnification calibration was $\pm 2\%$. Somewhat different dimensions were reported by Labaw (1967), who used shadowed

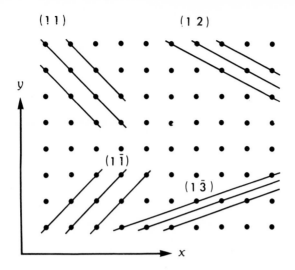

Fig. 3.3 This schematic diagram illustrates the methods by which the square lattice pattern may be divided in the xy plane with the Miller indices indicated for each set shown. Projecting each of these lines along the z-axis, the lattice planes would be (110), (120), ($1\bar{1}0$), and ($1\bar{3}0$), respectively.

replicas of catalase and reported the dimensions to be 7.3 X 14.1 X 8.4 nm for the catalase unit cell. Luftig (1967) found an 8.8 ± 0.3 nm spacing on negatively stained preparations of catalase crystals, which was in good experimental agreement with an 8.7 ± 0.3 nm periodicity he learned from R. W. Horne as a personal communication. In both cases, extensive care was taken in calibrating the electron microscope. It should be noted here that Luftig cautions that measurements should be made directly on the electron micrograph negatives, since a 2-3% decrease of the catalase spacing was found when measurements were made on prints. The figures of 8.8 ± 0.3 nm and 8.7 ± 0.3 nm determined by Luftig and Horne can be compared to those of Hall and Labaw when it is realized that the latter figures represent the value $\frac{1}{2}c_0$, a value which is normally used during calibration. For comparison purposes, these figures are: 6.95 nm (Hall, 1950); 7.95 nm (Labaw, 1967); 8.8 ± 0.3 nm (Luftig, 1967); 8.7 ± 0.3 nm (Horne, cited by Luftig); 8.7 ± 0.3 nm (Wrigley, 1968).

Wrigley (1968) reported a_0 spacings of 6.85 ± 0.075 nm, and c_0 spacings of 17.5 ± 0.2 nm for bovine liver catalase crystals which had been recrystallized, fixed with glutaraldehyde, and negatively stained with 4% sodium silicotungstate. These spacings were determined by low angle electron diffraction. Comparison of dimensions to unfixed catalase indicated that the glutaraldehyde treatment slightly altered the spacings; however, the change was not statistically significant.

Viruses

Sjöstrand (1967) mentioned that virus particles may be used as internal calibration specimens for high magnification, high resolution work. Klug and Caspar (1960) reviewed the X-ray analysis and structure of the tobacco mosaic virus (TMV), in which there are long rod-like structures consisting of a nucleic acid core surrounded by a large number of protein units arranged in a helical fashion. The pitch of the helix is 2.3 nm, as determined by X-ray analysis. This same dimension has been confirmed by electron microscopy on specimens of TMV that had been negatively stained with either uranyl formate, sodium phosphotungstate, or potassium phosphotungstate (Finch, 1964, 1969). Because the periodicity of negatively stained TMV preparations may not be clear enough to distinguish the 2.3-nm spacing, simple optical diffraction methods developed by Klug and Berger (1964) nicely delineate the average protein unit spacings.

In addition to TMV, T2 bacteriophage particles may also prove to be useful calibration specimens. X-ray diffraction studies on the tail sheath of T2 phage particles have shown the protein units have a periodicity of 4.1 nm (Moody, 1971), dimensions which have been confirmed on negatively stained preparations by electron microscopy (Kellenberger, unpublished results). F. Eiserling, of the Department of Biology at UCLA (personal communication), has suggested that the tail length of the T2 phage particles, approximately 95 nm, provides yet another useful dimension for spot-checking the magnification at lower ranges.

LENS DISTORTIONS

Any consideration of magnification generally assumes measurement at the optical axis, an assumption that negates the effects of lens distortion. For example, pincushion distortion will be detected by an increased spacing as the measurements are taken progressively away from the optical axis. Conversely, a decreased spacing is encountered when barrel distortion is present. Reisner (1965) points out that determining the linear distance separating two reference image points demands that, if one point is on the optical axis and the other off, different magnifications exist at these two points and the measured distance becomes an average magnification. When both points are off the axis, the error, the change in magnification compared to that of the axis, becomes greater the farther from the axis the measurements are made. This error is in effect due to the presence of either pincushion or barrel-type distortions, and the error magnitude increases as the distortion increases.

Several situations must be considered to attain the most accurate measurements. If the specimen permits, the measurements must be confined within a short radial distance from the optical axis, as determined from alignment criteria; then the change in magnification, M, will be within acceptable limits. However, should the measurements need to be made either on an extended specimen or within a limited area at the image periphery, then it becomes necessary to calculate a distortion factor. Reisner (1965) considers these as two cases. The first case dealing with extended specimen objects requires that the change in displacement of the image points, Δr, be expressed as a function of the radial distance, r', of those points from the axis. The correction factor for axial magnification, M_a, is calculated as a function of r' when all the structures are confined to a limited area off the optical axis. To calculate these factors Reisner (1965) suggests that a diffraction grating replica be used. The radial distance r', measured in millimeters, is plotted as a function of the number of lines, n, within that radial distance. The object position, r, corresponding to the image position, r', is calculated by:

$$r = ndM_a \qquad (3.11)$$

where n is the number of spacings along the image radial distance r', d is the spacing distance of the diffraction replica, and M_a is the axial magnification. When r' is plotted as a function of r, the slope of the curve yields the local correction factor by which the axial magnification value must be adjusted. Reisner (1965) further suggests that magnifications of off-axis structures can readily and conveniently be estimated by plotting the $r'r$ slope as a function of r'. The correction factors thus calculated may be considered as the product of the total lens distortion when multiple lenses form the image. It may be possible to adjust the current to the imaging lenses so that the pincushion and barrel distortions offset each other, or at least the total distortion becomes minimized. This is of particular advantage when one is working at the lower magnification ranges.

Reisner (1970) has described a much simpler method of measuring the radial distortion of the image. Using a specimen consisting of approximately circular holes, a circle of about 1 cm diameter is centered on the optical axis. The image is slightly underfocused, and a photographic exposure is made. The hole is then translated to a corner of the image area, and without refocusing another photographic exposure is taken. The process is facilitated if both images can be taken on a single negative, i.e., a double exposure.

Presupposing that the electron microscope has been properly aligned, the next step is to scribe a mark at the center of the negative, then mark the estimated center of the hole in the corner, and then join these two scribes with a fine line. The distance, r, along the radial line is measured between the two centers. The diameter of the center hole, d_0, and that of the corner hole, d_1, are both measured in a direction perpendicular to the radial line. The radial distortion at some radial distance, r, from the image center is:

$$r_0 = (d_1 - d_0)/d_0 \qquad (3.12)$$

Since this ratio is proportional to the square of the radius, Reisner (1970) indicates that a more useful expression is

$$Kr^2 = (d_1 - d_0)/d_0 \qquad (3.13)$$

since the constant K may be computed and used to compare distortions at different operating conditions. He further indicates that with distortions less than 10% the value r need not be corrected for distortion, nor do the center of the plate and the center of the center hole need to coincide. Two additional cautions are presented: first, all measurements should be directly on the negative; and second, the measuring device must be capable of measuring in the range of 0.02 mm.

Pincushion and barrel distortions contribute most to the change in radial magnification, M_r. These distortions are planar defects, and they do not affect the image sharpness. They vary as the cube of the radial distance from the optical axis in the following x and y relationships:

$$\Delta x = 0 \qquad \Delta y = Sy_0^3 \tag{3.14}$$

where S is the image shift and y_0 is the radial distance of the object. Equation (3.14) describes both pincushion and barrel distortions, and they differ only in the sign of the value S (Zworykin et al., 1945). Anisotropic distortion, also a planar defect, is peculiar to magnetic lenses, and it too can contribute at least as significantly to a change in radial magnification, since the image shift in this case is also proportional to the cube of the radial distance from the optical axis, as:

$$\Delta x = -Sy_0^3 \qquad \Delta y = 0 \tag{3.15}$$

These three distortions are considered as planar distortions in which the image sharpness is not altered. Conversely, curvature of field is not a planar distortion, and it varies not only proportionally to the square of the off-axis distance, but also directly proportionally to a curved surface tangent to the image plane. Hence, with this distortion alone, the image sharpness changes within the image plane. Owing to the resulting image quality, the presence of curvature of field is easily detected. Astigmatism is another type of distortion varying with the square of the off-axis radius, as well as with two curved image surfaces, such that there exist both tangential and sagittal image surfaces. This distortion is also readily recognizable and easily corrected, with the degree of correction often directly related to the patience of the microscopist. For a full consideration of these and other lens distortions affecting the quality of the image, the interested reader is referred to Zworykin et al. (1945) and Hall (1966).

Heckman and Roper (1962) described a technique by which image distortion can be evaluated. Beginning with the first measurement of replica spacing distances at the center of the negative, successive measurements were made in 5-mm increments on both sides of the center to the edge of the negative. The difference between the increment and center measurements was expressed as a percentage value, which was plotted as a function of the radial increment distance. They indicated that their resulting curves, asymmetric parabolic curves concave upward, showed their instrument had an anisotropic pincushion distortion.

METHODS OF CALIBRATION

Numerous methods exist by which the magnification of the electron microscope can be calibrated. All are characterized by an effort to attain accuracy, reliability, and ease of application, and an attempt to account for as many errors as possible. Since the final image magnification is a result not only of accelerating voltage, a combination of three lens currents, the magnetic state, and the history of the lenses, but also of the physical characteristics of the test specimen and the mechanical tolerances of the specimen stage and grid, it is necessary to standardize as many of these factors as possible so that only a few variables need be considered during calibration. In addition, electron microscopes whose magnifications are varied in a stepwise fashion as tap settings are only as accurate as the resistor circuit network involved. These resistor values can, and often do, change over the operational life of the microscope, and therefore they can alter the reproducibility of attaining an accurate magnification. In this author's opinion, the tap setting type of magnification change appears to have two basic fallacies: first, it precludes the opportunity of fine tuning; second, and more important, the magnification level of the study becomes instrument-determined instead of there being the optimum situation where the study fully determines the magnification.

The first step in any calibration is a decision made by the microscopist, who must determine the level of accuracy he is willing to tolerate. Most instrument manufacturers specify an accuracy in magnification from 3 to 10%; however, a rigorous calibration should be the first priority of operation, and one which should be repeated with regularity.

Accelerating voltage plays a significant role in variations of magnification. The accelerating voltage is generally changed in a stepwise fashion in most modern instruments. The operator control therefore becomes limited to assuring that small variations at any given accelerating voltage level are minimized. Fortunately, high-quality voltage regulators and stabilizers are available to accomplish this, and should be incorporated in the initial installation of any electron microscope. With proper regulation and stabilization of the accelerating voltage, this factor approximates a constant value, and is no longer a variable with which the microscopist must deal.

Pease (1950) developed a method of calibrating the magnification in which measured image movements were related to known object shifts by means of a Fabry-Perot interferometer mounted in the specimen stage of the microscope. Pease reported that an average error in magnification of 0.5% was possible. Interestingly, the error associated with repeatability was ~2.5% in successive calibrations, but the agreement between the magnifications as determined from interferometer and replica (test specimen) calibrations was within 0.25%. The error of 2.5% may be attributed partially to specimen error but mostly to non-reproducibility of lens electrical conditions. This latter problem is a handicap encountered where there is no mechanism available by which the current to the imaging lenses can be monitored or manipulated independently.

Elbers and Pieters (1964) reported a technique of calibration having an accuracy of 1 % in which they were able to account for the hysteresis effects, changes in accelerating voltage, distortion of the projector lens, and differences in the position of the object plane. The last of these factors is important, since they calculated a 10% variation alone due to the specimen position, which may be caused by a bent grid or use of different specimen holders. They further determined that an objective lens current change of 1 mA was sufficient to cause a 0.5% variation in magnification, a value that agrees fairly well with the figures of 0.2-0.36% calculated by Dunn and Preiss (1974). The effect of hysteresis can be minimized by switching off the objective and the intermediate lens currents, setting the current control to maximum, switching on the currents, and reducing the lens currents to focus (Elbers and Pieters, 1964). Three repetitions of this procedure were sufficient, since they noted that after the third time, the focus was the same as the second. It should be noted that they were using a Siemens Elmiskop I in which it is possible to couple the objective and intermediate lenses in such a way that the same lens current energizes both lenses. The accelerating voltage was determined by what appears to be a simple modification of the high tension circuit, and kept at a preset level through the high tension regulator. They offered the details of a compensation circuit that had a precision sufficiently accurate to measure the objective lens current, an important factor in establishing the specimen height within the objective lens. The optimum current to the projector lens, the final lens contributing to the image, was determined by plotting the projector lens current as a function of the percent magnification variation as measured in both the radial and tangential directions. When plotted in this fashion, the level of projector lens current producing the minimum distortion becomes obvious. Heckman and Roper (1962) had employed a similar technique.

Elbers and Pieters (1964) further suggested an alternative method to calibrate the carbon grating replica which was utilized as the calibration specimen. They determined the number of replica spacings between two recognizable points—in their case, opposite corners on a grid square. They indicated that their choice was facilitated, since the lines were approximately parallel to the diagonal between the opposing corners. The distance between the two identified points was measured in a light microscope equipped with an ocular micrometer that had been previously calibrated. The details of the latter calibration are given by the authors. The mean repeat distance of the replica spacing was then determined simply by dividing the identified point distance by the number of replica lines between those points. With the test specimen thus calibrated, the projector lens current standardized, and the lenses properly normalized, the image magnification was determined by the setting of their compensating potentiometer at the point of object focus, and numerically calculated by Eq. (3.1).

Many electron microscopes in use today rely upon a stepwise change in lens currents.   For example, with the Siemens Elmiskop I Series, these steps are unlabeled.   Without the capability to monitor directly the lens currents, the operator is forced into continually counting the number of steps from either extreme.   Weibel (1962) constructed a device consisting of two cold cathode counter tubes, and supporting circuitry, which allows a direct numerical display to facilitate identification of the tap setting.   The reader is directed to that paper for the details of this display device.

Eades and van Dun (1972) described a piezoelectric crystal device that served to facilitate the calibration of magnification on instruments equipped with a goniometer stage.   Their device produces a lateral specimen displacement of a measured distance.   They reported that an applied voltage from −500V to 500V produced a specimen shift of 0.31 μm, which was measured independently of the microscope operating conditions.   The image displacement is measured on a doubly exposed negative, and the object displacement is calculated using Eq. (3.1).   This, of course, presupposes that the magnification was accurately known. The advantage of this technique is that, working at a standard precalibrated magnification, the operating magnification could be easily verified on the working specimens under the conditions of observation.   The authors do, however, caution that their device shows electrical hysteresis, which can be eliminated by normalization.   A second, and major, limitation appears to be one of mechanical hysteresis, a fault that appears to be related to position of the specimen within the stage.   This device does, however, have many potential advantages should these pitfalls be resolved.

A method of magnification calibration has been suggested by Dunn and Preiss (1973, 1974) by which a 1.5% error is easily attained on a silicon monoxide grating replica.   In this case, the error was in the same range as the test specimen variation, which appeared to be the limiting factor.   Despite this specimen limitation, the reproducibility of the magnification has been calculated to be less than 0.2% over a 26-month time.   The operator is thus assured of attaining a magnification variability of 1.5 ± 0.2% throughout an extended time period. The method involved the use of simple statistics and a digital-display milliampere meter capable of monitoring the objective, intermediate, and projector lens currents.   In this case, a Model 2000-01 digital lens current monitor (M&W Systems) with a 3.5 place digital 7-segment display unit was installed so that the lens currents to the imaging lenses could be monitored independently.

Prior to any calibration, the microscope should be turned on and allowed to warm up for at least one hour before operation (Bahr and Zeitler, 1965).   The high voltage should be on and properly regulated to ensure stability.   Only the highest-quality replica should be used, and calibrated independently of the microscope—for example, spectrographically or by light microscopy (Elbers and Pieters, 1964).   Alternatively, crystals of the highest purity should be used, and calibrated, in this case, by means of diffraction analysis. The specimen grid must be free of bends, and where possible the same specimen holder should be used throughout. The latter points are worth considering, since mechanical tolerances may result in a magnification error of 5% or greater (Bahr and Zeitler, 1965; Elbers and Pieters, 1964; Reisner, 1965).

Determination of a standard specimen plane may be established by setting a standard magnification and recording the objective lens current when the image is in focus. Once this plane has been established, it is possible, with practice and patience, to return to this same plane in subsequent studies by repeating the objective lens current and standard magnification, after which the specimen height is adjusted along the optical axis until focus, ±1–2 mA objective lens current. The manipulation of the specimen height is facilitated if a z-axis movement is possible, as for example in the Siemens Elmiskop I's and the Zeiss electron microscopes. As an alternative, spacer rings may need to be used with bayonet-type specimen holders. A 4–5 mA difference in the objective lens current was calculated to cause a 1.7% error in the magnification. Therefore, a careful calibration of the objective lens is required after the intermediate-projector lens complex has been calibrated.

Calibration of the projector lens, in instruments where the projector is separated from the intermediate lens, requires that a subjective decision must be made to establish a standard projector lens current. This should be preferably at the current of least distortion (Elbers and Pieters, 1964). Once the standard projector lens current has been established, the operator simply returns to this same digital value after normalizing the lens. Variability due to operator decision variation is thereby eliminated. The contribution of the projector lens to image magnification becomes a constant.

Calibration of the intermediate lens is by far the most important single step in a rigorous calibration, and it is completed utilizing a test specimen of known spacing using standard methods of calibration (Agar, 1965; Beeston et al., 1973; Hall, 1953, 1966; Sjöstrand, 1967). With the object in the standard specimen plane, the projector lens set at the standard projector lens current, and all lenses normalized, electron micrographs are taken at various intermediate lens currents, and each digital value is noted. The image spacing distance is measured directly on the negative within a 3-cm-diameter circle of the optical axis (Elbers and Pieters, 1964). Prints of the electron micrographs should not be used for measurements because of the errors introduced during photographic processing. It may also be necessary to take the pictures close to a grid bar in order to minimize the effect of film sag, which can be detected by a change in objective lens current. The replica lines are quite sharp at lower magnification ranges, and little difficulty is encountered in measuring. However, at higher magnifications, it becomes more difficult to determine precisely the points of measurement, owing to the ambiguity of the lines (Fig. 3.4). Hence, it is necessary to measure a number of spacings, preferably 10 to 15, and to calculate a mean spacing. If a "waffle-type" replica is used, a predetermined number of measurements, say 13, are made, first in one direction and then at a 90° direction, for a total of 26 computation values for each set.

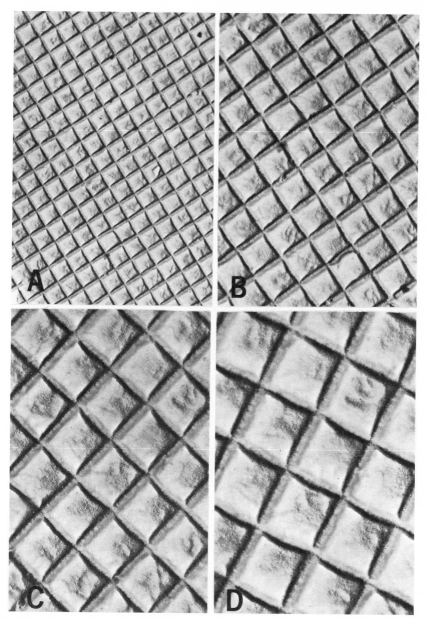

Fig. 3.4 Four electron micrographs of a "waffle-type" diffraction grating replica. At the lower magnifications (A, B) the spacings are relatively sharp and distinct, minimizing the ambiguity encountered when the spacings are more diffuse at higher magnification (C, D). At lower magnifications, more spacings are available, in which case acceptable averaging of the spacing dimensions can be accomplished on a single negative. This task is complicated at higher magnifications, where the number of spacings on the negative is greatly reduced. Print magnifications—A: 10,731×; B: 18,662×; C: 29,218×; D: 35,342×.

The measurements are then converted to magnification values using Eq. (3.1); and the mean, $\bar{x}$, the standard deviation, SD, and the standard error, SE, are calculated using standard statistical methods. The standard deviation may then be expressed as a percent of error ($\%E$) using the following relationships:

$$\%E = 100\,(SD/\bar{x}) \tag{3.16}$$

To test the precent of error, multiple sets must be made at each magnification, so that an F-ratio may be determined. Five or six sets are usually sufficient. When a significantly high F-ratio results, a one-way analysis of variance may be completed (Snedecor and Cochran, 1967), with the coefficient of variation, CV, expressed as a percent, as follows:

$$CV = \frac{(\text{Treatment } MS/n)^{1/2}}{\text{overall } \bar{x}} \times 100 \tag{3.17}$$

The results of one such series of calculations are shown in Table 3.1. The magnification is then plotted as a function of intermediate lens current, as shown in Fig. 3.5, for each accelerating voltage level. If the plots are completed accurately, the curves may be used to ascertain the intermediate lens current required for intermediate magnifications, which can readily be verified by the calibration methods just outlined.

The same test negatives may be further used to calculate both the radial and the tangential variations in magnification, to determine a correction factor for off-axis measurements (Reisner, 1965, 1970). To calculate the radial factor, measurements should be taken randomly within a region 1.5–2.5 cm from the optical axis, 2.5–3.5 cm, etc., and the mean magnification calculated as outlined. The radial factor may easily be calculated by dividing the radial mean magnification by the center mean magnification. The tangential correction factor may be determined by measuring across the image replica spacings on the negative in the x-direction at successive y values of 1-cm increments from the central 3-cm-diameter circle to the edge of the negative. The complementary calculation is determined in the y direction, at successive x values of 1-cm increments. These

Table 3.1   Magnification at 100 kV (n = 156)

Intermediate Lens Current	104 mA	115 mA	123 mA	164 mA	207 mA	251 mA
$\bar{x}$	4,061	5,343	6,336	11,768	17,480	22,231
SD	50.1	84.0	90.5	131.1	164.5	162.6
SE	9.8	16.5	17.0	25.7	32.3	32.9
$\%E$	1.2%	1.6%	1.4%	1.1%	0.9%	0.7%
CV	1.2%	1.6%	1.4%	1.1%	0.9%	0.7%

(Reproduced from Dunn and Preiss, 1974, with permission of the editor and publishers of *The Journal of Microscopy*.)

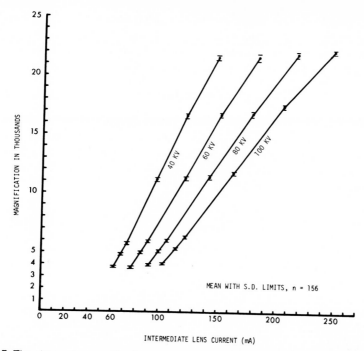

Fig. 3.5 The change in magnification as a function of increased intermediate lens current is shown for all four accelerating voltages on the ×20,000 projector pole piece of a Siemens Elmiskop IA. When accurately plotted, such a graph may be used during daily operation to determine intermediate magnifications. The points are plotted ± one standard deviation, $n = 156$. *(Reproduced from Dunn and Preiss, 1974, with permission of the editor and publishers of* The Journal of Microscopy.)

correction factors may then be applied later in determining the dimensions of small structures outside the center of the negative (radial correction), or large structures well off axis (tangential correction).

The final step is that of calibrating the objective lens to ascertain its contribution to image magnification, that is, a change in magnification as a function of the objective lens current. This can be accomplished by setting the currents to the objective, intermediate, and projector lenses to their standard digital values, moving the specimen along the optical axis, focusing, noting the objective lens current, and taking a negative. This procedure is repeated to obtain several additional values. The change in magnification is then proportional to the change in objective lens current, a value that can be utilized in future observations. Knowledge of this factor can be utilized operationally, and a return to the standard specimen height is necessary only when a full recalibration is required. Results of calculations for the objective lens have shown a 4–5-mA difference in objective lens current to be sufficient to introduce a magnification error of 1.7%, as mentioned previously.

Table 3.2 Magnification Error

Dial Setting	40 kV		60 kV		80 kV		100 kV	
	$\bar{x}$ Mag	%E	$\bar{x}$ Mag	%E	$\bar{x}$ Mag	%E	$\bar{x}$ Mag	%E
×3,000	3,732	24.4	3,742	24.7	3,987	32.9	4,061	35.4
×4,000	4,856	21.2	4,961	24.0	5,053	26.3	5,343	33.6
×5,000	5,728	14.6	5,911	18.2	5,973	19.5	6,336	26.7
×10,000	11,151	11.5	11,245	12.4	11,411	14.1	11,768	17.7
×15,000	16,556	10.4	16,719	11.5	16,842	12.3	17,480	16.5
×20,000	21,615	8.1	21,659	8.3	21,951	9.8	22,231	11.2

(Reproduced from Dunn and Preiss, 1974, with permission of the editor and publishers of *The Journal of Microscopy*.)

A method by which the individual lens currents can be monitored has long been recognized as an important prerequisite for high precision in calibrating the magnification of the electron microscope (Hall, 1953). The use of digital meters for these measurements eliminates much of the operator's subjectivity, and is an important means of ensuring consistency of operational conditions. It is possible to measure variations in lens current drifts to the sensitivity of the meter. The resolution of a 3.5 digit meter, as an example, has been calculated at 0.1% of the full range, and of a 4.5 digit meter as 0.01% of the full scale. Hence, it is practical to deal with small current variations and to determine their effect upon magnification.

The value of all the operator's efforts can be determined, once the magnification calibration has been completed, by comparing the results of the calibration to the magnification shown on the instrument. Table 3.2 is offered as a positive indication of the advisability of completing an accurate calibration of the magnification.

The author wishes to express appreciation to D. P. O'Leary for his assistance with the mathematics and for his critical appraisal, to H. Shinozuka and L. Estes for the use of the Central Electron Microscope Facility, and particularly to Beverly Robinson for preparing the manuscript.

APPENDIX: DERIVATION OF EQUATION (3.9)

The intermediate steps used in developing formulae that are important in an argument are often omitted for simplicity. This omission many times presumes too much on the reader's part, and makes it difficult to verify the arguments. It is for this reason that the following derivation of Eq. (3.9) has been included here.

The solution resulting in Eq. (3.9) begins with Eq. (3.6):

$$\frac{1}{f} = \frac{0.022}{V} \int_{-\infty}^{\infty} H^2 \, dz \tag{3.6}$$

where the field intensity, H, in units of gauss, is to be integrated along the optical axis as indicated by dz. Since

$$H_z = \frac{2\pi n I a^2}{10 (z^2 + a^2)^{3/2}} \tag{3.7}$$

we may, when the focal length is much larger than the axial radius of an equivalent simple loop, then substitute Eq. (3.6) with Eq. (3.7):

$$\frac{1}{f} = \frac{0.022}{V} \int_{-\infty}^{\infty} \left[\frac{2\pi n I a^2}{10 (z^2 + a^2)^{3/2}} \right]^2 dz \tag{3.8}$$

By defining

$$c = \frac{0.022}{V} \quad \text{and} \quad k = \frac{2\pi n I a^2}{10}$$

then we may express Eq. (3.8) as:

$$\frac{1}{f} = c \int_{-\infty}^{\infty} \left[\frac{k}{(z^2 + a^2)^{3/2}} \right]^2 dz \tag{3.8a}$$

$$\frac{1}{f} = ck^2 \int_{-\infty}^{\infty} \frac{dz}{(z^2 + a^2)^3} \tag{3.8b}$$

The integral may be solved by first using the general integral solution number 63 twice (Hodgman, 1951), with the partial solution expressed as:

$$\frac{1}{f} = ck^2 \left[\frac{1}{4a^2} \cdot \frac{z}{(z^2 + a^2)^2} + \frac{3z}{8a^4 (z^2 + a^2)} + \frac{3}{8a^4} \int_{-\infty}^{\infty} \frac{dz}{z^2 + a^2} \right] \tag{3.8c}$$

The final integration using the general integral form solution number 41 (Hodgman, 1951) yields:

$$\frac{1}{f} = ck^2 \left\{ \frac{z}{4a^2 (z^2 + a^2)^2} + \frac{3z}{8a^4 (z^2 + a^2)} + \frac{3}{8a^5} \tan^{-1} z/a \right\} \Bigg|_{-\infty}^{\infty} \tag{3.8d}$$

The first two expressions within the brackets become zero for $z = \pm\infty$. Hence, expression (3.8d) becomes:

$$\frac{1}{f} = ck^2 \left[\frac{3}{8a^5} \cdot (\tan^{-1} \infty/a - \tan^{-1} -\infty/a) \right] \tag{3.8e}$$

The arc tangent value ∞/a is $\pi/2$, while that of $-\infty/a$ is $3\pi/2$. Hence Eq. (3.8e) becomes:

$$\frac{1}{f} = ck^2 \left[\frac{3}{8a^5} (-\pi) \right] \tag{3.8f}$$

If we substitute the values previously defined as c and k^2, Eq. $(3.8f)$ becomes:

$$\frac{1}{f} = \frac{0.022}{V} \cdot \frac{4\pi^2 n^2 I^2 a^4}{100} \cdot \frac{3}{8a^5} \cdot -\pi \qquad (3.8g)$$

Since the sign indicates the direction of the field along the z-axis, the absolute value of this expression is more useful; and $d = 2a$. Then, after solving for the constants, the final expression is:

$$\frac{1}{f} = \frac{3\pi^3 n^2 I^2}{100 dV} = \frac{0.0206 n^2 I^2}{dV} \qquad (3.9)$$

References

Agar, A. W. (1965). The operation of the electron microscope. In: *Techniques for Electron Microscopy* 2nd ed. (Kay, D. H., ed.), 1–42, F. A. Davis, Philadelphia.

Backus, R. C., and Williams, R. C. (1948). Some uses of uniform sized spherical particles. *J. Appl. Phys.* **19**, 19.

Backus, R. C., and Williams, R. C. (1949). Small spherical particles of exceptionally uniform size. *J. Appl. Phys.* **20**, 224.

Bahr, G. F., and Zeitler, E. (1965). The determination of magnification in the electron microscope. II. Means for the determination of magnification. *Lab. Invest.* **14**, 880.

Beeston, B. E. P., Horne, R. W., and Markham, R. (1973). Electron diffraction and optical diffraction techniques. In: *Practical Methods in Electron Microscopy* (Glauert, A. M., ed.), pp. 393–401. North-Holland Pub. Co., Amsterdam.

Bradford, E. B., and Vanderhoff, J. W. (1955). Electron microscopy of monodisperse latexes. *J. Appl. Phys.* **27**, 864.

Burton, C. J., Bowling Barnes, R., and Rochow, T. G. (1942). The electron microscope calibration and use at low magnifications. *Ind. Eng. Chem. Ind. Ed.* **34**, 1429.

Dekker, A. J. (1957). *Solid State Physics.* Prentice-Hall Inc., Englewood Cliffs, N. J.

Dosse, J. (1941). Über optische Kenngrössen starker Elektronenlinsen. *Z. Phys.* **117**, 772.

Dunn, R. F., and Preiss, G. W. B. (1973). Digital display of lens current for accurate calibration of the electron microscope. In: *Electron Microscopy* (Arceneaux, C. J., ed.), pp. 98–99. Claitor's Pub. Div., Baton Rouge., La.

Dunn, R. F., and Preiss, G. W. B. (1974). Reproducibility of electron microscope magnification with digital display of individual lens currents. *J. Microscopy* **101**, 317.

Eades, J. A., and van Dun, A. (1972). A device for magnification measurement. In: *Electron Microscopy* (Arceneaux, C. J., ed.), pp. 622–623. Claitor's Pub. Div., Baton Rouge, La.

Elbers, P. F., and Pieters, J. (1964). Accurate determination of magnification in the electron microscope. *J. Ultrastruct. Res.* **11**, 25.

Finch, J. T. (1964). Resolution of the substructure of tobacco mosaic virus in the electron microscope. *J. Mol. Biol.* **8**, 872.

Finch, J. T. (1969). The pitch of tobacco mosaic virus. *Virology* **38**, 182.

Fullam, E. F. (1943). Magnification calibration of the electron microscope. *J. Appl. Phys.* **14**, 677.

Gerould, C. H. (1950). Comments on the use of latex spheres as size standards in electron microscopy. *J. Appl. Phys.* **21**, 183.

Hall, C. E. (1950). Electron microscopy of crystalline catalase. *J. Biol. Chem.* **185**, 749.

Hall, C. E. (1953). *Introduction to Electron Microscopy*. McGraw-Hill Book Co., New York.

Hall, C. E. (1966). *Introduction to Electron Microscopy*. McGraw-Hill Book Co., New York.

Hass, G., and McFarland, M. E. (1950). Aluminum oxide replicas for electron microscopy produced by a two-step process. *J. Appl. Phys.* **21**, 435.

Heckman, F. A., and Roper, S. G., Jr. (1962). The reduction of error in magnification determination in electron microscopes. In: *Electron Microscopy*, Vol. 1 (Breese, S. S., Jr. ed.), pp. EE2–EE3. Academic Press Inc., New York.

Heidenreich, R. D. (1964). *Fundamentals of Transmission Electron Micorscopy*. Interscience Pub., New York.

Hodgman, C. D. (1951). *Mathematical Tables from Handbook of Chemistry and Physics*, 9th ed., Chem. Rubber Pub. Co., Cleveland.

Horne, R. W. (1965). The application of negative staining methods to quantitative electron microscopy. *Lab. Invest.* **14**, 1054.

Klug, A., and Berger, J. E. (1964). An optical method for the analysis of periodicities in electron micrographs, and some observations on the mechanism of negative staining. *J. Mol. Biol.* **10**, 565.

Klug, A., and Caspar, D. L. D. (1960). Structure of small viruses. *Adv. Virus Res.* **7**, 225.

Labaw, L. W. (1964). Preparation of a 25-Å spacing crystal for magnification calibration above 18,000X. *J. Appl. Phys.* **35**, 3076.

Labaw, L. W. (1967). On liver catalase crystal structure by electron microscopy. *J. Ultrastruct. Res.* **17**, 327.

Luftig, R. (1967). An accurate measurement of the catalase crystal period and its use as an internal marker for electron microscopy. *J. Ultrastruct. Res.* **20**, 91.

Mahl, H. Z. (1940). Metallkundliche Untersuchungen mit dem elektrostatischen Übermikroskop. *Z. Tech. Phys.* **21**, 17.

Menter, J. W. (1956). The direct study by electron microscopy of crystal lattices and their imperfections. *Proc. Roy. Soc. Lond., Ser. A.* **236**, 119.

Moody, M. S. (1971). Structure of the T2 bacteriophage tail-core and its relation to the assembly and contraction of the sheath. In: *Proc. First Europ. Biophys. Congr.* (Broda, E., Locker, A., and Springer-Lederer, H., eds.), pp. 543–546. Verlag der Wiener Medizinischen Akademie, Vienna.

Pease, R. S. (1950). The determination of electron microscope magnification. *J. Sci. Instr.* **27**, 182.

Reisner, J. H. (1965). The determination of magnification in the electron microscope. I. Instrumental factors influencing the estimate of magnification. *Lab. Invest.* **14**, 875.

Reisner, J. H. (1970). Measurement of radial distortion. In: *Electron Microscopy* (Arceneaux, C. J., ed.), pp. 350–351. Claitor's Pub. Div., Baton Rouge, La.

Rouze, S. R., and Watson, J. H. L. (1953). Metallic grating replicas as internal standards for calibrating electron microscopes. *J. Appl. Phys.* **24**, 1106.

Sjöstrand, F. S. (1967). *Electron Microscopy of Cells and Tissues*, Vol. 1, *Instrumentation and Techniques*. Academic Press Inc., New York.

Snedecor, G. V., and Cochran, W. G. (1967). *Statistical Methods*, 6th ed., Iowa State University Press, Ames, Iowa.

Watson, J. H. L., and Grube, W. L. (1952). The reliability of internal standards for calibrating electron microscopes. *J. Appl. Phys.* **23**, 793.

Weibel, J. (1962). A magnification control for the Siemens Elmiskop I electron microscope. In: *Electron Microscopy*, Vol. 1 (Breese, S. S., Jr., ed.) pp. EE4-EE5. Academic Press Inc., New York.

Williams, R. C., and Wyckoff, R. W. G. (1946). Applications of metallic shadow-casting to microscopy. *J. Appl. Phys.* **17**, 23.

Wrigley, N. G. (1968). The lattice spacing of crystalline catalase as an internal standard of length in electron microscopy. *J. Ultrastruct. Res.* **24**, 454.

Zworykin, V. K., Morton, G. A., Romberg, E. G., Hillier, J., and Vance, A. W. (1945). *Electron Optics and the Electron Microscope.* John Wiley and Sons, Inc., New York.

Zworykin, V. K., and Romberg, E. G. (1941). Surface studies with the electron microscope. *J. Appl. Phys.* **12**, 692.

4. CONTRAST ENHANCEMENT BY USING TWO ELECTRON MICROGRAPHS

D. L. Misell

Biophysics Division, National Institute for Medical Research,
Mill Hill, London, England

INTRODUCTION

Detrimental Effects of Inelastic Scattering

The contribution of inelastic scattering to biological electron microscopy is a totally negative one. Not only is this component of the transmitted electron beam responsible for the destruction of the fine structure of biological materials by radiation damage, but the contribution of the inelastic scattering to the image is mainly as a low resolution background (Beer *et al.*, 1975). The normal transmission electron microscope image comprises an elastic image (in either bright-field or dark-field microscopy) with intrinsically high resolution detail superimposed on an inelastic background blurred by the chromatic aberration of the objective lens. This leads not only to a severe loss of image contrast, but also a loss in image resolution, since high resolution detail of low contrast may be lost in the inelastic background. These detrimental effects of inelastic scattering are more evident in unstained biological macromolecules than in heavy atom–labeled macromolecules because inelastic scattering decreases relative to elastic scattering as the atomic number of the specimen increases.

Negatively stained materials must be considered as intrinsically low resolution specimens because of the large crystallite sizes of the negative stain (Beer *et al.*, 1975), but image contrast will still be lost owing to the contribution of inelastic scattering from the biological material and its substrate. In the following situations inelastic scattering is relevant to a loss in image contrast and resolution:

1. In the resolution of single atoms used to label macromolecules such as DNA on a carbon substrate. Under normal irradiation conditions it is certain that all the DNA comprises ashes, which predominantly cause inelastic scattering. The heavy atoms, such as platinum ($Z = 78$), contribute to the image by high resolution elastic scattering; the elastic scattering cross-section increases as $Z^{3/2}$, whereas inelastic scattering is almost independent of atomic number. The final image comprises high resolution heavy atoms on a background of low resolution due to the ashes and substrate (Whiting and Ottensmeyer, 1972).

2. In unstained biological macromolecules the relative contributions of inelastic and elastic scattering to the image may be as high as $10:1$ (Burge, 1973). Assuming that the ashes of the biological molecule preserve some structure of the original molecule (Ottensmeyer *et al.*, 1975), then the loss in contrast caused by inelastic scattering in the ashes and substrate may completely hide the elastic contribution to the image. For crystalline biological specimens this situation may be improved because the basic fact of a repeating unit can be used to increase the signal (the elastic contribution) to noise (photographic noise + the inelastic contribution), by either optical filtering or digital image processing (Unwin and Henderson, 1975).

3. Negatively stained specimens give a strong elastic contribution from the heavy metal atoms of the supporting stain, but a large background contribution of low resolution arises from the inelastic scattering in the biological macromolecules and the substrate.

4. Sections of biological material, stained or unstained, have a typical thickness of 50 nm, and the contribution to the image from the low-density material, biological material, and embedding medium is predominantly inelastic (Ottensmeyer and Pear, 1975). The inability of the transmission electron microscope to focus inelastic electrons because of their large energy spread (~20 eV) must give in all section work an image resolution limited mainly by chromatic aberration to about 5 nm (Cosslett, 1956, 1969). In such thick specimens the contribution to the image from the relatively small proportion of elastic scattering is probably limited in resolution to 1 nm by beam spreading rather than by spherical aberration. A similar consideration applies to images of whole viruses, stained or unstained.

Situations (1)–(3) correspond to high/medium resolution images (0.5–1.5 nm) where the resolution besides being specimen-limited (for example, by radiation damage, dehydration) is affected by spherical aberration and defocus of the objective lens in the case of the elastic contribution, and by chromatic aberration and defocus for the inelastic component. These are situations where the specimen is quite thin (usually less than 10 nm) and the image formation can be described by wave optics. It is emphasized that only a wave-optical treatment of image formation in the transmission electron microscope will explain the well-known phase contrast effects observed in bright-field and to a lesser extent in dark-field microscopy (Cowley, 1976, Vol. 6 of this series). However, in case (4) given above the image is not high resolution, and the resolution will be limited by the angular spread of the beam and chromatic aberration to 2-5 nm; the image formation in this case can be adequately explained by geometrical optics. Defocus effects may still be important because of preferential focusing at a particular depth in the thick specimen. All cases discussed above will benefit in contrast from the removal of the inelastic scattering from the image, and this benefit may also show as an increase in image resolution for high resolution detail of low contrast previously masked by the inelastic scattering.

There are two instrumental developments that give a successful separation of the elastic and inelastic components. The first method is to use an electron spectrometer below the objective lens or first projector lens of a normal transmission electron microscope (Castaing, 1971; Henkelman and Ottensmeyer, 1974; Egerton et al., 1975) to allow only those electrons which have lost no energy to contribute to the image. These energy-selecting electron microscopes can also be used to form images from a particular energy loss of the inelastic component (Colliex and Jouffrey, 1970; Castaing, 1971). The energy-selecting microscope has not been tested at a high spatial resolution, because the electron spectrometer usually degrades the spatial performance of the microscope. However, experimental results on biological materials show a clear gain in contrast from the removal of most of the inelastic scattering.

Figure 4.1 shows (a) a normal and (b) an elastic image of a stained section of liver taken in bright-field microscopy (Henkelman and Ottensmeyer, 1974). The gain in contrast is evident, although because of the nature of the specimen no higher resolution or even additional detail is evident in the energy-filtered image (b). Figure 4.2 shows an even greater gain in contrast for a thick ($t = 60$ nm) stained section of visual cortex tissue taken in normal bright-field microscopy (Egerton, unpublished result). There is some additional detail in the filtered image (b) as compared with the unfiltered image (a). Again because of the intrinsic low resolution of the specimen, the main effect of removing the inelastic scattering background is a significant increase in image contrast. It is doubtful whether the gain in contrast would be as good in thick unstained sections, where there is very little elastic scattering to contribute to a bright-field image and the differential elastic scattering between the embedding medium and biological tissue is small (Ottensmeyer and Pear, 1975).

The images shown in Fig. 4.3 show (a) a normal and (b) an elastic image of polyoma virus, unstained and unshadowed, taken in dark-field microscopy (Henkelman and Ottensmeyer, 1974). In the normal image (a), the predominant inelastic scattering contributes to an overall blurring of the virus particles;

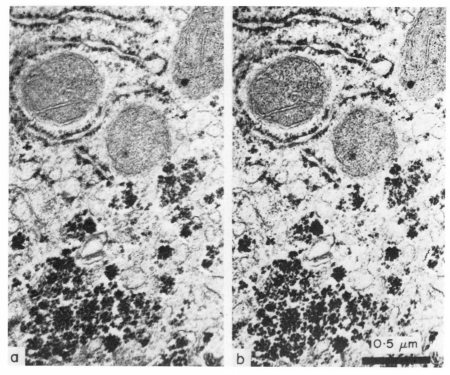

Fig. 4.1 Images of a stained section of liver in bright-field microscopy: (a) normal (elastic + inelastic) image, (b) filtered (elastic only) image. Incident electron energy, E_0 = 60 keV, magnification, M = 36,000. (*From Henkelman and Ottensmeyer, 1974.*)

whereas in the elastic image (b), surface detail of about 3.5 nm resolution can be resolved, although this detail is difficult to interpret because of the superposition of detail from the top and bottom of the virus.

The second instrumental development that gives a separation of the elastic and inelastic images is the scanning transmission electron microscope perfected by Crewe and his co-workers (Crewe, 1971). In this microscope all the lenses precede the specimen, and as a result of the absence of lenses after the specimen, the inelastic image does not suffer from chromatic aberration. The principle of separation of the elastic and inelastic signals is based on the differences in the angular distributions of the two scattering processes (Crewe, 1971). Essentially, the inelastic scattering is predominantly at small angles (less than 0.001 rad), whereas the elastic scattering dominates at larger angles of scattering (greater than 0.01 rad). An annular collector with a hole corresponding to a maximum angle of 0.01 rad allows the inelastic and unscattered electrons (which are undeviated in passing through the specimen) to be detected separately from the elastic signal, which is collected on the annulus. The unscattered and inelas-

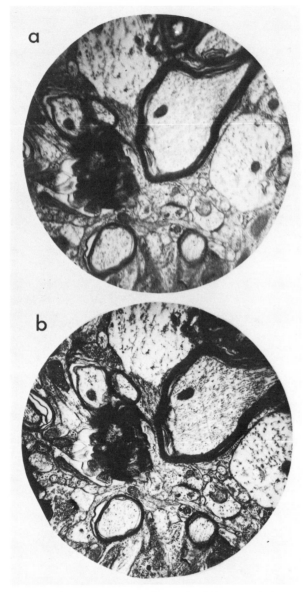

Fig. 4.2 Images of visual cortex tissue sections, about 60 nm in thickness, in bright-field microscopy: (a) normal (elastic + inelastic) image, (b) filtered (elastic only) image. E_0 = 80 keV, M = 10,000. (*Courtesy of Dr. R. F. Egerton.*)

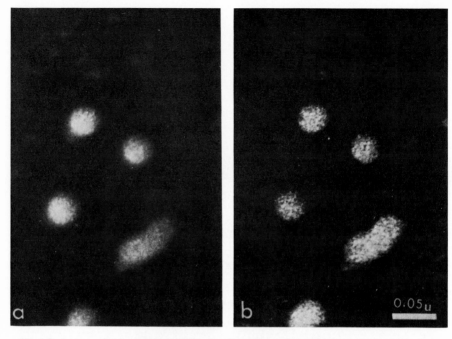

Fig. 4.3 Images of polyoma virus, critical point dried and unstained, unshadowed in dark-field microscopy: (a) normal image, (b) filtered image. E_0 = 60 keV, M = 240,000. (*From Henkelman and Ottensmeyer, 1974*)

tically scattered electrons can be separated by an electron spectrometer beneath the detector.

Figure 4.4 shows (a) the inelastic and (b) the dark-field elastic images of thallium (Z = 81)-labeled DNA (Crewe, 1971). The elastic image (b) shows a resolution of ~0.5 nm compared with the inelastic image (a) resolution of ~1 nm. The lower contrast of the inelastic image is evident and is caused by a large inelastic contribution from the ashes of the DNA and the substrate; the elastic image contrast will only be slightly affected by a small amount of elastic scattering from the DNA plus substrate. These images show clearly that even in the absence of chromatic aberration the inelastic image is inferior in both contrast and resolution to the elastic image. The addition of chromatic aberration to the inelastic image in a conventional transmission electron microscope will lead to a further blurring of the inelastic image and loss in contrast. It is evident that in the conventional transmission microscope the addition of images (a) and (b) will lead to a loss in contrast and probably to a loss in some of the high resolution detail of the thallium atoms. Crewe (1971) also shows (Fig. 4.4c) an image obtained from dividing the elastic by the inelastic component, and further contrast enhancement is evident.

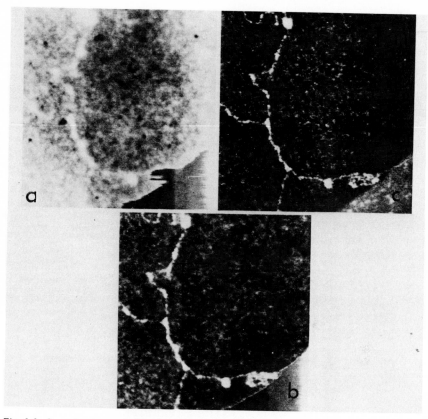

Fig. 4.4 Scanning transmission electron microscope images of thallium-stained DNA: (a) inelastic image, (b) elastic dark-field image, (c) elastic divided by inelastic image. $E_0 = 25$ keV, $M = 1,800,000$. (*From Crewe, 1971.*)

If one is using a high resolution energy-selecting electron microscope or a scanning transmission electron microscope, the method of contrast enhancement using two electron micrographs described in this review will be irrelevant. The review will describe a method removing most of the low resolution inelastic image by the subtraction of two micrographs taken at slightly different defocus values, ~50 nm difference, in a conventional transmission microscope (Misell and Burge, 1975). The contrast enhancement is based on theoretical calculations showing that whereas the elastic image resolution and contrast depends critically on defocus at high resolution (0.5 nm), the inelastic image varies slowly with defocus (Misell, 1975). Thus, in principle an image taken at the optimum defocus for the elastic image (underfocus by 60 to 100 nm) will contain a high resolution elastic image superimposed on a low resolution (2 nm) inelastic background. A second image taken ±50 nm from this optimum defocus will contain a low resolution, low contrast elastic image superimposed on an almost unchanged inelastic image. A subtraction of the aligned micrographs will leave essentially the high resolution elastic image and a reduced inelastic background.

The main part of this review will describe how inelastically scattered electrons form images, the defects in these images, a comparison between elastic and inelastic images in terms of contrast and resolution, and a theoretical assessment of contrast enhancement by image subtraction. The practical problems in implementing the image subtraction, such as defocus measurements and image alignment, will be discussed. Because dark-field microscopy of biological specimens is more seriously affected by the contribution from inelastic scattering than bright-field microscopy, it seems relevant to go into some detail in a comparison of dark-field and bright-field microscopy in terms of contrast and signal to noise. The review by Dubochet (1973) in Volume 3 of this series gives a detailed assessment of the advantages and problems in dark-field microscopy but does not include a detailed account of the effects of inelastic scattering on image resolution and contrast.

Other methods of contrast enhancement, such as the use of phase plates (Unwin, 1972; Johnson and Parsons, 1973), will be discussed. At lower resolution (2 nm) an assessment will be given of the viability of image subtraction for contrast enhancement of biological sections and whole viruses, unstained and stained. Photographic methods of contrast enhancement of electron micrographs, which can produce improvements in contrast similar to those shown in Figs. 4.1 and 4.2, will be omitted from this review (see, for example, Farnell and Flint, 1970 and in vol. 5 of this series).

IMAGE FORMATION BY INELASTICALLY SCATTERED ELECTRONS

In order to distinguish between elastic and inelastic electron scattering, we consider not the spatial distribution but the energy distribution of electrons transmitted by the specimen. Unscattered and elastically scattered electrons are transmitted through the specimen with their original energy E_0, whereas inelastic electrons lose energy E and have a final energy $E_0 - E$. Typically E is only 10–30 eV, which may not seem a large energy loss in 100,000 eV. But in practice this energy loss has two detrimental effects on the inelastic image. First, because the objective lens suffers from chromatic aberration, electrons of different energy are not focused at a single point in the image but give a broadening of each image point. Geometrically, for electrons scattered through an angle θ with energy loss E a measure of this blurring r_c is given by (Marton et al., 1955, Cosslett, 1956)

$$r_c = \frac{C_c E \theta}{E_0} \tag{4.1}$$

for unit magnification; C_c is the chromatic aberration constant of the objective lens, usually about 2 mm. For biological materials E is about 25 eV, so that $r_c = 0.5$ nm for an angle of scattering $\theta = 0.001$ rad and 100 keV incident electron energy. This is an oversimplification of how to calculate r_c geometri-cally, since account should be taken of the angular and energy distributions of the inelastically scattered electrons (Crick and Misell, 1971; Nagata and Hama, 1971). Another way of looking at this effect of chromatic aberration on the inelastic image is to calculate the out of focus distance, Δf, for an energy loss E, $\Delta f = C_c E/E_0$ (Misell, 1971), which is 500 nm for $E = 25$ eV. Thus the inelastic image can be focused approximately by an objective lens defocus of about -500 nm ($-$ indicates underfocus), but the elastic image with $E = 0$ is then 500 nm out of focus. However, since the elastic image displays higher contrast and resolution than the inelastic image, which cannot in any case be represented by a single energy loss E, such large underfocus values cannot be recommended in biological electron microscopy.

The second reason for not focusing on the inelastic image arises from the nature of the energy loss processes. Whereas elastically scattered electrons are coherently scattered, that is, the elastically scattered waves exhibit definite phase relationships with each other and the unscattered beam, inelastically scattered electrons of different energy have unrelated phases and certainly can-not interfere with each other in the wave-optical sense, or with the elastic/unscattered waves. This is important because most biological electron micros-copy uses the interference between the elastically scattered and unscattered electrons to give enhanced image contrast, as in bright-field microscopy. In fact the analysis of inelastic image formation given in this review considers inelastic electrons to form an image completely independent of the elastic/unscattered image. The final image is then a superposition of the intensities arising from elastic scattering and inelastic scattering in the specimen, taking account of the lens defects such as spherical aberration and chromatic aberra-tion, and including the effects of objective lens defocus.

For thin specimens (thickness t less than $\sim$10 nm) wave optics are used to calculate both elastic and inelastic images. For such thin specimens the geo-metrical theory of image formation, which neglects all interference or phase contrast effects (Crick and Misell, 1971), is inadequate. The geometrical theory may be a valid treatment of inelastic scattering, which is essentially incoherent, but it cannot be used to predict the resolution and contrast of elastic images, which may be dominated by phase contrast effects. In thick specimens ($t = 50$ nm), particularly when unstained, inelastic scattering dominates, and so qualita-tively the neglect of elastic scattering, or its treatment as an incoherent phenom-enon, may give a satisfactory interpretation of image contrast. Even for thick specimens this approach may be inadequate if even a small amount of high resolution elastic scattering occurs in the specimen. This situation is clearly shown in the results of Nagata and Hama (1971) of platinum particles (sizes $\sim$0.6–1 nm) and a thin crystal of pyrophillite (lattice spacing = 0.46 nm) on a thick (30–300 nm) biological specimen. Although the contrast is very low, the high resolution information can still be observed, enhanced by phase contrast effects.

Scattering contrast, which evaluates only scattered intensities, would certainly not predict a resolution or image contrast as high as that observed. So in this review a clear distinction is made between the scattering from thin specimens, where potential high resolution detail is imaged by a phase contrast mechanism, described by wave optics, and scattering from thick specimens, where the differential scattering within the objective aperture from different specimen areas (with geometrical optics) can be used to predict image contrast and resolution. The images of thin specimens must be treated by wave optics, whereas geometrical optics may be adequate in describing the images of thick, low resolution specimens (van Dorsten, 1971; Burge, 1973; Misell, 1975).

The wave-optical treatment of elastic images from thin specimens (Lenz, 1971; Cowley, 1976) has been extended to inelastic images by considering each energy loss as forming an "elastic image" (Misell, 1971; Misell and Burge, 1973). The sum of the intensities from all these energy loss images, taking into account the actual distribution of energy loss and chromatic aberration effects in addition to spherical aberration and defocus, gives the final inelastic image. The wave-optical calculation of the inelastic image then allows a direct comparison with the corresponding elastic image, in both bright-field and dark-field microscopy; this is important if an assessment is to be made of the relative contributions of elastic and inelastic scattering to image resolution and image contrast.

Besides the loss in resolution of the inelastic image due to chromatic aberration, there is the low resolution nature of the inelastic image arising from the delocalization of inelastic scattering. This effect is clearly shown in images taken from the scanning transmission electron microscope where chromatic aberration is absent (Crewe, 1971; Isaacson *et al.*, 1974).

The following sections will detail experimental results on the nature of inelastic images, the theoretical evaluation of image resolution, the effect of chromatic aberration on inelastic images, and a comparison between elastic and inelastic images in different imaging conditions.

Inelastic Images: Experimental Results

The use of energy-selecting electron microscopes (El Hili, 1967; Colliex and Jouffrey, 1970; Castaing, 1971; Egerton *et al*, 1975) enables a particular energy loss band (width 1–2 eV) to be selected for image formation. Thus it is possible to compare experimentally the differences in the images formed by elastic and inelastic scattering (for a fixed energy loss). The energy loss spectrum depends on the nature of the material; for example in Fig. 4.5, (a) and (b) show, respectively, spectra of thin aluminum and carbon films. The zero loss at 0 eV corresponds to those electrons that have lost no energy, that is, elastically scattered and unscattered electrons. For the aluminum spectrum the principal energy loss at 15 eV has an energy half-width of only 2 eV, comparable with the energy spread of the elastic peak (due to the thermal energy spread of the electron source).

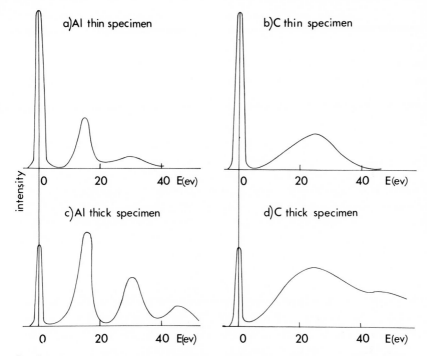

Fig. 4.5 The energy loss spectra of aluminum and carbon for thin and thick specimens.

Because of the narrow energy loss peak it seems reasonable to expect similar elastic and inelastic images to be obtained by forming an image from respectively, the zero loss and 15 eV loss electrons in aluminum. This is in contrast to the inelastic images that would have been obtained from carbon or biological materials, which have a broad energy distribution, ~20 eV half-width, centered at 20-25 eV. Clearly it is not possible to form a single image from all the inelastically scattered electrons as in the case of aluminum. Images formed from a selected energy loss band, say 25 ± 1 eV, will give images similar to the elastic image but generally of lower contrast. As the specimen thickness increases, inelastic scattering leading to two or more energy loss events is evident. In the case of aluminum (Fig. 4.5c) this leads to peaks at integral multiples of 15 eV, and it is now not possible to form a single inelastic image; but it is possible to focus on either the 15 or 30 eV loss electrons, both of which give similar images, with the latter energy loss of reduced contrast because of the larger energy spread of the 30 eV loss electrons.

For thicker specimens of biological materials (Fig. 4.5d) the distinction between elastically and inelastically scattered electrons is even more evident; the double loss at ~50 eV overlaps with the 25 eV loss so that even choosing a particular energy loss no longer ensures that the inelastic image is formed from only those electrons that have been inelastically scattered once. These observations are consistent with experimental results obtained using energy-selecting electron microscopes.

For materials that exhibit discrete energy loss spectra, typically non-transition metals, the contrast and resolution of the elastic image and first energy loss inelastic image are very similar. Figure 4.6 shows (a) a dark-field elastic and (b) the 15 eV inelastic image from an aluminum foil (El Hili, 1967). The es-

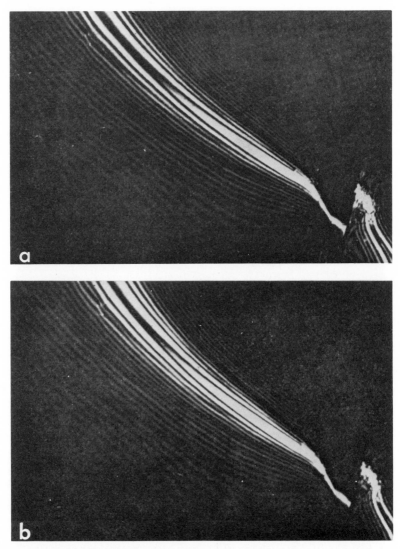

Fig. 4.6 Dark-field bend contour images of an aluminum foil: (a) elastic image, (b) inelastic image for 14.6 eV loss electrons. E_0 = 80 keV, M = 53,000. (*From El Hili, 1967.*)

sential similarity between the bend contour images in (a) and (b) indicates that the inelastic scattering within a narrow energy loss band can exhibit interference effects in the same way as the elastic scattering. There is, however, no experimental evidence that electrons of different energy loss can interfere with each other or with the elastically scattered electrons (Castaing, 1971). These results and others in metallurgical specimens support the model used here for calculating inelastic images; that is, inelastically scattered electrons of a given energy are treated as "elastic" within a narrow energy loss band, but the contributions from scattered electrons in different energy loss bands are summed incoherently.

For biological or amorphous materials the experimental results for elastic and inelastic images are sparse.   Colliex and Jouffrey (1970) show results for amorphous silicon, and in the scanning transmission electron microscope Stroud *et al.* (1969) show elastic and inelastic images for a thick stained section. Both these papers show the similar, although weaker contrast, images due to inelastic scattering; the resolution of these images is, however, rather low (>2 nm).

The scanning transmission electron microscope used at high resolution shows clearly the inferior contrast and resolution of the inelastic image (Fig. 4.4). Quantitative measurements of the intrinsic resolution of inelastic images have been made in the scanning microscope by Isaacson *et al.* (1974) by scanning the electron beam across the edge of a carbon film. The elastic image profile shows a resolution of $\sim$0.4 nm limited by spherical aberration, whereas the inelastic image profile is limited to a resolution of $\sim$0.8 nm. Thus in the absence of chromatic aberration it is clear that the intrinsic resolution of an inelastic image is a factor of 2 worse than the elastic image resolution. In normal transmission microscopy chromatic aberration will lead to a further blurring of the inelastic image.   At high resolution (0.5 nm) these results indicate the importance of eliminating the contribution from inelastic scattering to the image. However, at medium resolution (1–2 nm) in thin negatively stained specimens it is not clear that the inelastic image is grossly inferior in resolution to the elastic image (Beer *et al.*, 1975); in this situation the elastic image resolution is limited by the inherent size of the negative stain particles or crystallites.

A final example will show clearly that inelastic scattering in the conventional electron microscope mainly lowers contrast, and resolution may be only a secondary factor.   Figure 4.7 shows images of the lattice of pyrophillite (spacing $d = 0.46$ nm) when the crystal is mounted on (a) 30 nm, or (b) 300 nm of biological material (Nagata and Hama, 1971). Although the contrast of the lattice is very low superimposed on an inelastic image from 300 nm of biological material, the lattice is clearly resolved. The explanation for this high resolution arises from the nature of lattice imaging where the contrast of a single spacing can be enhanced by a single defocus value with consequent blurring of the background structure. It is doubtful whether such a high resolution could be obtained if the lattice were distributed at several different depths in the specimen; the difference in focus between the top and bottom of the specimen, $\sim$300 nm,

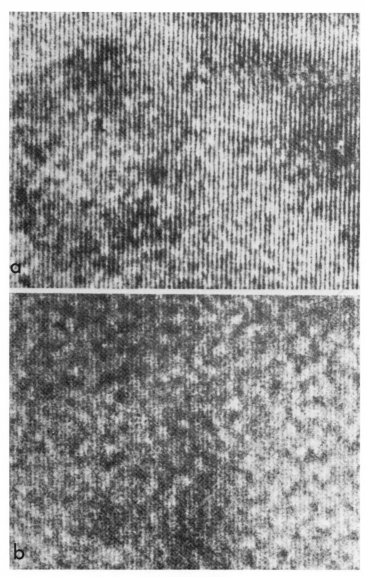

Fig. 4.7 Bright-field lattice images of a thin crystal of pyrophillite ($d = 0.46$ nm) on a biological specimen: (a) 30-nm-thick specimen, (b) 300-nm-thick specimen. $E_0 = 75$ keV, $M = 2,800,000$. (*From Nagata and Hama, 1971*.)

would be sufficiently large to lose any high resolution detail. But there remains the interesting possibility in unstained biological sections of imaging high resolution detail even in the presence of gross inelastic scattering. In the Nagata and Hama (1971) results (Fig. 4.7b), approximately 98% of the incident electron beam has been scattered inelastically in the biological material; phase or inter-ference effects from the remaining 2%, or less, of elastically scattered electrons seem to be responsible for the contrast observed from the high resolution detail.

Inelastic Images: Theoretical Results

In order to calculate the inelastic image, it is necessary to consider the inelastic scattering with particular energy loss E; electrons within ΔE of this energy loss are considered to be coherent and are treated in the same way as a monochro-matic elastic image. The choice of ΔE is not a critical factor in these calcula-tions, and ΔE may vary between 0.1 and 2 eV. The transmitted inelastic wave immediately after the specimen is $\psi_0(r_0, E)$, where $r_0 \equiv (x_0, y_0)$ is the two-dimensional coordinate of a point in the specimen from which the inelastic scat-tering occurs. We assume that the object is thin enough to neglect the effect of thickness on the inelastic image. The scattered wave, $S_0(\nu, E)$, in the back focal plane of the objective lens is calculated from the Fourier transform of $\psi_0(r_0, E)$ (Lenz, 1971):

$$S_0(\nu, E) = \int \psi_0(r_0, E) \exp{(2\pi i \nu \cdot r_0)} \, dr_0 \qquad (4.2)$$

where ν represents the spatial frequency corresponding to a scattering angle $\theta = \lambda \nu$ for electrons of wavelength λ. We prefer to use spatial frequencies ν (nm^{-1}) to θ because ν can be directly related to resolution $\sim 1/\nu$ (nm). The objective aperture diameter d_{obj} can be related to the maximum spatial fre-quency ν_{max} by $d_{obj} = 2f\lambda\nu_{max}$ for an objective lens of focal length f. Experi-mentally it is not possible to measure S_0 but we can measure the scattered intensity $|S_0|^2$ as a function of angle of scattering. In practice S_0 is modified in the electron microscope by lens aberrations and an objective aperture, which stops all electrons scattered at angles greater than θ_{max}. The wave aberration function $W(\nu, E)$ measures the deviation of the electron wave front from the ideal spherical wave front in a perfect lens (Lenz, 1971), and for a lens subject to spherical aberration (coefficient C_s), defocus (Δf), and chromatic aberra-tion (coefficient C_c):

$$W(\nu, E) = \frac{C_s \nu^4 \lambda^4}{4} + \frac{\Delta f \nu^2 \lambda^2}{2} + \frac{C_c E \nu^2 \lambda^2}{2E_0} \qquad (4.3)$$

corresponding to a phase shift $\gamma(\nu, E) = 2\pi W(\nu, E)/\lambda$, for the scattered wave in the direction ν. The scattered wave for energy loss E is then modified to

$$S_i(\nu, E) = S_0(\nu, E) \exp{[-i\gamma(\nu, E)]} B(\nu) \qquad (4.4)$$

where $B(\nu)$ is the objective aperture function, which is unity for the hole in the aperture and zero for the opaque part of the aperture.

One interesting result from Eq. (4.3) is the similar dependence of the defocus and chromatic aberration terms on ν; this implies that for a given energy loss E, an effective partial cancellation of the chromatic aberration term can be made by appropriate underfocusing of value $\Delta f \simeq -C_c E/E_0$ (Misell, 1971).

From Eq. (4.4) the contribution to the image wave function from electrons of energy loss E is calculated by taking the inverse Fourier transform of S_i, $\psi_i(\mathbf{r}_i, E)$, where $\mathbf{r}_i$ refers to the image coordinate. The intensity at an image point $\mathbf{r}_i$ is calculated by weighting $|\psi_i(\mathbf{r}_i, E)|^2$ with the number of electrons, $f(E)$, which actually lose energy E:

$$j(\mathbf{r}_i, E) = |\psi_i(\mathbf{r}_i, E)|^2 \, f(E) \qquad (4.5)$$

The total inelastic image for a thin specimen, where only one inelastic scattering event occurs, is then obtained by summing all the intensities for each energy loss E_n:

$$j(\mathbf{r}_i) = \frac{\displaystyle\sum_{E_n} |\psi_i(\mathbf{r}_i, E_n)|^2 \, f(E_n)}{\displaystyle\sum_{E_n} f(E_n)} \qquad (4.6)$$

The division by the total number of electrons which have lost energy normalizes $j(\mathbf{r}_i)$ so that it represents the fraction $I(\alpha)$ of the incident beam that is inelastically scattered within the objective aperture $\alpha = \theta_{max}$, that is,

$$\int j(\mathbf{r}_i) \, d\mathbf{r}_i = I(\alpha) \qquad (4.7)$$

Thus $j(\mathbf{r}_i)$ represents the inelastic image of a point in the object as affected by lens defects and defocus, and will give a direct measure of the effect of chromatic aberration on the inelastic image resolution.

Figure 4.8 shows the radial profiles of inelastic images for scattering from 10 nm of carbon, with an energy loss distribution $f(E)$ based on an experimental spectrum for carbon (Misell and Atkins, 1973). Similar profiles would be expected for biological materials. The in-focus inelastic image ($\Delta f = 0$) displays a resolution of ~1.8 nm (based on the half-width of the profile), decreasing to 0.4 nm for an underfocus value of 500 nm. Further underfocus overcompensates for chromatic aberration, and the resolution in the image deteriorates to 1 nm for $\Delta f \simeq -1,000$ nm. These calculations are consistent with an estimate of the optimum defocus for the inelastic image of

$$\Delta f_I \simeq -C_c E_{mp}/E_0 \qquad (4.8)$$

for focusing on the most probable energy loss E_{mp}. For carbon and biological materials $E_{mp} = 20$–25 eV, giving $\Delta f_I = -400$ to -500 nm with $C_c = 2$ mm and $E_0 = 100$ keV. At the lower incident electron energies sometimes used in biological electron microscopy, the defocus Δf_I is appropriately increased, say to -670 to -830 nm at 60 keV, with a corresponding increase in the half-widths of the radial profiles shown in Fig. 4.8.

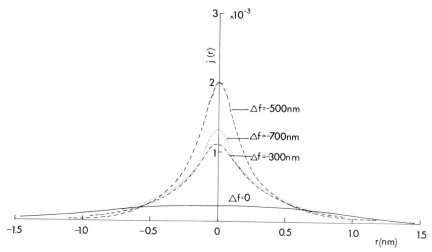

Fig. 4.8 Image profiles for inelastic scattering in carbon showing the blurring of a point in the object by chromatic aberration, and the effect of defocus Δf. E_0 = 100 keV, C_s = 2 mm, C_c = 2 mm, α = 0.01 rad.

The resolution and contrast of inelastic images are virtually unaffected by spherical aberration or objective aperture size (α rad), because of the small-angle nature of inelastic scattering for thin specimens (Misell and Burge, 1973); this is in contrast to the elastic image resolution and contrast, which vary with C_s and α.

Inelastic Images: Effect of Chromatic Aberration

Resolution is a very difficult quantity to define, and in order to give some figures for the effects of chromatic aberration on inelastic image resolution, we adopt a rather arbitrary definition in terms of the radial half-width of the image profile of a point in the object. This definition of resolution measures approximately how close two "object" points can be before only one peak is resolved. However, it is noticeable that the inelastic image profiles have very shallow tails at large radial distances; this implies that a significant number of electrons are imaged well outside the half-width limits. In fact, defining resolution in terms of the diameter that contains 50% of the inelastically scattered electrons gives a resolution figure two to three times as high as the resolution based on the radial half-width. With this limitation on the exact definition of resolution in mind, Fig. 4.9 shows the variation in resolution r, based on radial half-width, of an inelastic image with objective lens defocus Δf. Only underfocus values are considered, since this is the region for normal biological electron microscopy; overfocus values give lower resolution (>2 nm) inelastic images and are certainly unsuitable for high resolution elastic images.

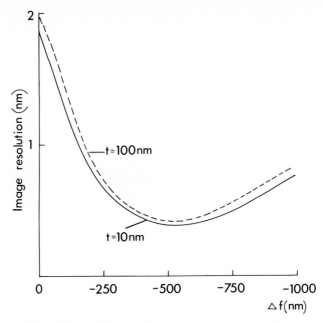

Fig. 4.9 The variation of the resolution in the inelastic image with defocus Δf. (Microscope parameters are the same as for Fig. 4.8.)

Figure 4.9 shows that in-focus inelastic images have a resolution of 1.8 nm decreasing to 0.4 nm at the optimum inelastic defocus, ~ -500 nm for a specimen with a most probable energy loss of $E_{mp} = 25$ eV at 100 keV ($C_c = 2$ mm). The increase in chromatic aberration with a reduction in incident electron energy E_0 can be calculated approximately from the 100-keV result in Fig. 4.9, using

$$r_{E_0} \simeq r_{100} \frac{100}{E_0(\text{keV})},$$

giving a resolution limit due to chromatic aberration of 3 to 0.7 nm at 60 keV for a defocus range 0 to -830 nm (Δf_I at 60 keV). Conversely, high-voltage electron microscopy was strongly recommended (Cosslett, 1969) because of the reduction in chromatic aberration in the inelastic image despite the increased value of C_c (Misell, 1973b).

Figure 4.9 also shows the corresponding resolution curve for a specimen of thickness $t = 100$ nm; there is only a slight deterioration in the image resolution, and the optimum defocus Δf_I shifts to ~ -600 nm, corresponding to an increase in E_{mp} to 30 eV for a thicker specimen. However, a detailed examination of the inelastic image profiles for $t = 100$ nm shows a very long flat tail at large r values, and the realistic estimate of image resolution may be a factor of 4 above that shown in Fig. 4.9.

Experimental results on the effect of chromatic aberration on the inelastic image resolution are inconclusive. The results of Nagata and Hama (1971) imply that the inelastic image contributes mainly to a loss in image contrast, and high resolution detail down to 0.5 nm can still be observed in the presence of gross inelastic scattering. But in this case it is very difficult to separate the elastic image of the platinum particles or of the lattice from the inelastic image arising from the biological material.

An additional factor in evaluating the resolution of inelastic images is the effect of the nonlocalized nature of inelastic scattering. The difference in resolution between the elastic and inelastic images of a carbon film edge, 0.5–1.0 nm, measured in the scanning transmission electron microscope reflects this delocalization (Isaacson et al., 1974). The results in Fig. 4.9 should be modified to take this additional effect into account. Assuming that both chromatic aberration and delocalization have a gaussian fall-off with increasing radial distance from the point of scattering in the object, the resolution r is given by the quadratic addition of r_c, due to chromatic aberration, and r_d, due to delocalization; that is, $r = (r_c^2 + r_d^2)^{1/2}$. Taking an upper limit of $r_d = 1$ nm for biological specimens, r will vary from 2.1 nm for $\Delta f = 0$ to 1.1 nm for $\Delta f = -500$ nm (cf., 1.8 to 0.4 nm in Fig. 4.9). Clearly this is low resolution information in the context of resolving single heavy metal atoms, but comparable in resolution with images taken of negatively stained macromolecules.

Dark-field images of biological specimens taken at the optimum inelastic defocus, 500 nm underfocus, appear to show a resolution of ~1.5 nm (Brakenhoff, cited in Beer et al., 1975), which confirms the theoretical results in this section for the resolution of inelastic images. However, in the context of recording image detail down to 0.5 nm or better in biological specimens, the inelastic scattering contribution to the total (elastic + inelastic) can only lead to a loss in image contrast of the high resolution information. The attainment of 0.5-nm resolution in biological materials, positively stained or even unstained, is not at all unrealistic (Henkelman and Ottensmeyer, 1971; Whiting and Ottensmeyer, 1972; Massover and Cowley, 1973; Ottensmeyer et al., 1975).

Comparison of Elastic and Inelastic Images

The main hope of separating by image processing the contributions to the image from elastic and inelastic scattering lies in the different behavior of the component images with changes in defocus. In this section we will show that the best elastic image displays a higher resolution and contrast than the best inelastic image, in both dark-field and bright-field microscopy. First, however, an outline will be given of the calculation of elastic image profiles for a thin object. This analysis is important because it gives the relationship between the image intensity and the object structure and is essential to the interpretation of electron micrographs. The analysis includes the effects of lens aberrations and defocus, and the results can be used in practice to determine the optimum electron-optical conditions for recording the image.

The object wavefunction, $\psi_0(\mathbf{r}_0)$, representing the transmitted electron beam, is written as (Lenz, 1971; Cowley, 1976):

$$\psi_0(\mathbf{r}_0) = 1 + i\phi(\mathbf{r}_0), \tag{4.9}$$

corresponding to the expansion of exp $[i\phi(\mathbf{r}_0)]$ to first order where unity represents the amplitude of the unscattered beam and $\phi(\mathbf{r}_0)$ is the projection of the potential field (due to the atoms in the specimen) onto the (x_0, y_0) plane. For a thin object of thickness t,

$$\phi(\mathbf{r}_0) = \frac{-2\pi m e \lambda}{h^2} \int_0^t V(\mathbf{r}_0, z) \, dz \tag{4.10}$$

The assumption that ϕ is much less than unity, so that higher order terms in exp $(i\phi)$ can be neglected, limits specimen thickness t to 10 nm for an unstained specimen and to 2.5 nm for a negatively stained specimen (Cowley, 1976). Cowley (1976) uses Eq. (4.9) in a slightly modified form, subtracting from ϕ the mean value of the potential field in the specimen ϕ_m; this then gives the local variations in the specimen structure:

$$\psi_0(\mathbf{r}_0) = 1 + i\left[\phi(\mathbf{r}_0) - \phi_m\right] \tag{4.11}$$

The calculation of the image intensity is made in the same way as a "monochromatic" inelastic image, only in this case the energy loss of the incident electron beam is zero; a slight chromatic defect does arise from the thermal energy spread (1–2 eV) of the incident electron beam, but it has only a small effect on the elastic image resolution.

The scattered wave is obtained from the Fourier transform of Eq. (4.11):

$$S_0(\boldsymbol{\nu}) = \delta(\boldsymbol{\nu}) + i\Phi(\boldsymbol{\nu}) \tag{4.12}$$

where the delta function $\delta(\boldsymbol{\nu})$ corresponds to the intense center spot observed in the diffraction plane of the electron microscope. In practice, in scattering experiments $|\Phi|^2$ is measured and Φ cannot be determined. The lens modifications to S_0 are introduced as discussed above, omitting now the chromatic aberration term from Eq. (4.3):

$$S_i(\boldsymbol{\nu}) = \left[\delta(\boldsymbol{\nu}) + i\Phi(\boldsymbol{\nu})\right] \exp\left[-i\gamma(\boldsymbol{\nu})\right] B(\boldsymbol{\nu}) \tag{4.13}$$

There are two distinct imaging conditions corresponding to dark-field and bright-field microscopy. In the ideal dark-field case the unscattered beam is intercepted by an axial beam stop at the center of the objective aperture, so that $B(\boldsymbol{\nu}) = 0$ for $\boldsymbol{\nu} = 0$. In practice, dark-field imaging may be achieved by using tilted illumination or conical illumination (Dubochet, 1973), but both techniques lead to a more complicated analysis than given here.

Setting $B(\nu) = 0$ for $\nu = 0$ for the dark-field case, with $\exp(-i\gamma) = \cos\gamma - i\sin\gamma$:

$$S_i(\nu) = \Phi(\nu)\sin[\gamma(\nu)]\,B(\nu) + i\Phi(\nu)\cos[\gamma(\nu)]\,B(\nu) \qquad (4.14)$$

Thus the Fourier transform of the object Φ is modified by the transfer functions $\sin\gamma$ and $\cos\gamma$, which are characteristic of the electron microscope and not of the specimen. Denoting the inverse Fourier transforms of $\sin\gamma$ and $\cos\gamma$ by $q(\mathbf{r})$ and $q'(\mathbf{r})$, respectively, the image wave function $\psi_i(\mathbf{r}_i)$ is given by (Cowley, 1976):

$$\psi_i(\mathbf{r}_i) = \int [\phi(\mathbf{r}) - \phi_m]\,q(\mathbf{r}_i - \mathbf{r})\,d\mathbf{r} + i\int [\phi(\mathbf{r}) - \phi_m]\,q'(\mathbf{r}_i - \mathbf{r})\,d\mathbf{r}$$

$$(4.15)$$

This equation represents the convolution of the object structure with a resolution function, q or q', which blurs out the object detail. By abbreviating the convolution integral by an asterisk ($*$) Eq. (4.15) becomes:

$$\psi_i(\mathbf{r}_i) = [\phi(\mathbf{r}_i) - \phi_m] * q(\mathbf{r}_i) + i[\phi(\mathbf{r}_i) - \phi_m] * q'(\mathbf{r}_i). \qquad (4.16)$$

The image intensity which is recorded in the electron microscope is calculated from $|\psi_i(\mathbf{r}_i)|^2$; that is,

$$j(\mathbf{r}_i) = |\psi_i(\mathbf{r}_i)|^2 = \{[\phi(\mathbf{r}_i) - \phi_m] * q(\mathbf{r}_i)\}^2 + \{[\phi(\mathbf{r}_i) - \phi_m] * q'(\mathbf{r}_i)\}^2$$

$$(4.17)$$

Or, if we neglect lens aberrations, the image intensity is proportional to $[\phi(\mathbf{r}_i) - \phi_m]^2$ and is not linearly related to the object structure.

In bright-field microscopy the analysis is simpler provided it is assumed that ϕ is much less than unity; the aperture function in bright-field is now unity up to a maximum spatial frequency ν_{max} determined by the objective aperture size. The image wave function is then:

$$\psi_i(\mathbf{r}_i) = 1 + [\phi(\mathbf{r}_i) - \phi_m] * q(\mathbf{r}_i) + i[\phi(\mathbf{r}_i) - \phi_m] * q'(\mathbf{r}_i) \qquad (4.18)$$

However, when $\psi_i(\mathbf{r}_i)$ is squared (ψ_i multiplied by its complex conjugate) to give the image intensity, only linear terms are retained:

$$j(\mathbf{r}_i) = 1 + 2[\phi(\mathbf{r}_i) - \phi_m] * q(\mathbf{r}_i) \qquad (4.19)$$

since squared terms such as those occurring in Eq. (4.17) are much smaller than the linear term (ϕ^2 is much less than ϕ, since ϕ is ~ 0.1). The important result in bright-field microscopy is that the image intensity is linearly related to the object structure $[\phi(\mathbf{r}_i) - \phi_m]$, although the image is modified by a resolution function $q(\mathbf{r}_i)$, which depends on the spherical aberration and the defocus of the objective lens.

For those who feel that the analysis given above is too difficult to follow, the essential result is that the dark-field image intensity depends on $[\phi(\mathbf{r}_i) - \phi_m]^2$, whereas in bright-field microscopy j depends on $[\phi(\mathbf{r}_i) - \phi_m]$; that is, j is linearly related to the variations in the potential field within the object. Since ϕ is about 0.1 for a thin specimen, the elastic image intensity in bright-field corresponds to a modulation of ~0.2 on a background of unity or 20% image contrast. However, the ideal dark-field elastic image intensity is only ~0.01 on a background of zero. These intensity figures will be reduced by the blurring effect of the objective lens aberration functions q and q'. This situation is shown in Fig. 4.10, which shows (a) dark-field and (b) bright-field elastic image profiles (E) of a point in the object in comparison with two inelastic image profiles (I).

The first inelastic image profile is calculated at the optimum elastic defocus $\Delta f_E \simeq -100$ nm, corresponding to a partial cancellation of the spherical aberration term by the defocus term in the aberration function $W(\nu) = (C_s \nu^4 \lambda^4 / 4 + \Delta f \nu^2 \lambda^2 / 2)$, Eq. (4.3); this defocus gives the highest resolution and contrast for the elastic image. The second inelastic image profile corresponds to $\Delta f_I = -500$ nm, the optimum inelastic defocus, and is to be compared with the best elastic images in Fig. 4.10. Although in dark-field (a) the maximum intensities of the optimum elastic (E) and optimum inelastic $(I, \Delta f = -500$ nm) images are com-

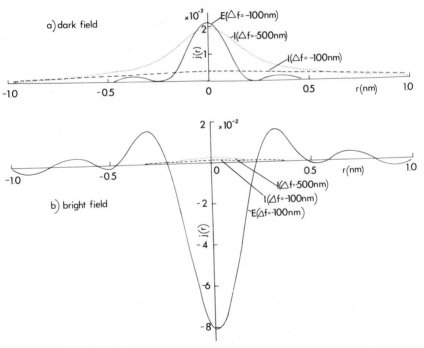

Fig. 4.10 A comparison of inelastic image profiles (I) with (a) dark-field, (b) bright-field elastic image profiles (E) for the optimum elastic defocus $\Delta f = -100$ nm. Note that the zero level of $j(r)$ for elastic profile (b) corresponds to the background level of the unscattered contribution.

parable at $r = 0$, the resolution of the elastic image is better by at least a factor of 2.   So in dark-field microscopy of biological materials at high resolution (0.5 nm) it is clearly advantageous to determine the optimum defocus for the elastic image (resolution $\sim$0.2 nm), when the inelastic contribution to the image will be as a relatively low resolution (1.5 nm) background.

The calculations in bright-field microscopy, Fig. 4.10b, clearly show the large difference in image contrast and resolution between the elastic image and an inelastic image at any defocus.   The contrast in the elastic image arises from the interference between the unscattered wave and the scattered wave, and this gives much larger variations in the image intensity than observed in dark-field microscopy (compare the intensity scales in Figs. 4.10a and b).   In bright-field microscopy the inelastic contribution to the image principally can be considered as an additional background to the already large contribution from the unscattered electron beam.   The elastic image profile is far from ideal; although the radial half-width of the main peak is less than 0.3 nm at optimum elastic defocus, the secondary peaks at $r =$0.4 and 0.7 nm indicate a resolution inferior to 0.3 nm, and it is also possible for secondary structures, not in any way related to the object structure, to be observed in the image (Beer *et al.*, 1975).

The critical dependence of the resolution functions q and q' on defocus is clear from these calculations of elastic image profiles.   In the bright-field case a variation of 50 nm from optimum elastic defocus is sufficient to give a low resolution elastic image with several undesirable peaks in the resolution function q (see section on "Determination of Defocus").   In contrast, for variations in defocus of 50 nm the inelastic image profile changes only slightly with the resolution changing from 1.3 nm for $\Delta f = -100$ nm to 1 nm for $\Delta f = -150$ nm. This differential behavior of the elastic and inelastic image profiles with variations in defocus forms the basis for image subtraction in order to reduce the contribution from inelastic scattering to the total image.

From these calculations it is also seen that inelastic scattering has less effect on the bright-field than on the dark-field image contrast.   Defining contrast C by:

$$C = \frac{\text{modulation of image intensity (signal)}}{\text{background intensity + signal}} \qquad (4.20)$$

for bright-field $C \simeq 2\phi/(1 + 2\phi)$ and for dark-field $C \simeq \phi^2/\phi^2$ in the absence of background inelastic scattering; the signal is included in the background term to avoid the embarrassment of infinite contrast in the dark-field case.   The addition of inelastic scattering $I(\alpha)$, to the bright-field background of $1 + 2\phi$ has only a small effect, but the addition of $I(\alpha)$ to the relatively small dark-field intensity, ϕ^2, may reduce contrast by a factor of about 10.   This assessment of the relative effects of inelastic scattering on dark-field and bright-field elastic images will be discussed in detail in the section on "Dark-field versus Bright-field Microscopy."

One danger in the electron microscopy of biological materials, particularly in dark-field microscopy, is focusing on the inelastic image rather than the elastic image. In dark-field microscopy the contrast is clearly comparable (Fig. 4.10a), although the resolution of the inelastic image is inferior to that of the elastic image. The problem is unlikely to occur in bright-field microscopy, unless the specimen is fairly thick, because the maximum elastic image contrast is a factor of 20 above the image contrast corresponding to the optimum inelastic defocus (Misell, 1973a). Also, in the analysis of focus series of electron micrographs, it is often assumed that only elastic scattering contributes to the intrinsic image resolution and image contrast. Image processing on this basis does not take into account the relative variations of the elastic and inelastic images over a range of focus as large as 1,000 nm. An exception where inelastic scattering may be partially eliminated by image processing occurs in stained or unstained crystalline specimens. Here the Fourier transform, determined numerically or by using an optical system, gives sharp diffraction spots for the elastic component of the image, whereas the inelastic scattering forms a relatively diffuse background (Unwin and Henderson, 1975). Computer or optical filtering, allowing only the diffraction spots to contribute to the reconstructed image, then eliminates a substantial proportion of the inelastic scattering from both the biological material and the substrate.

IMAGE SUBTRACTION

Theoretical Results

Image subtraction for reducing the effects of inelastic scattering depends on the different rates of change of the intensities of two images with changing defocus of the objective lens (Misell and Burge, 1975). The method uses the sensitivity of the elastic image close to the optimum defocus, Δf_E, for maximum contrast. For this region of defocus the inelastic image, far from the condition for optimum resolution ($\sim$500 nm underfocus), is relatively insensitive to changes in defocus. Subtraction of two images taken at Δf_E and $\Delta f_E \pm 50$ nm will substantially reduce the intensity of the inelastic component and differentially enhance the elastic image. In unstained specimens the ratio of inelastic to elastic scattering within the objective aperture, $I(\alpha)/E(\alpha)$, is $\sim$6, reducing to 1-2 for a stained specimen; heavy atom-labeled macromolecules give a small $I(\alpha)/E(\alpha)$ in the region of the heavy atoms and a larger ratio for the biological material. Thus, a differential enhancement of the elastic contribution with respect to the inelastic contribution to the image can lead to a substantial reduction in the low resolution background intensity, and a corresponding increase in image contrast. In this section are shown model calculations to give some indication of this gain under ideal conditions; practical problems, such as the accurate measurement of defocus and the alignment of the two images for subtraction, are discussed in the following section ("Practical Problems in Image Subtraction").

For thin specimens, less than 20 nm thick if unstained and less than 5 nm if stained, the wave-optical calculations of the previous section are used to calculate both elastic and inelastic image profiles. Although the results of these calculations are model-dependent, an indication is obtained of how the inelastic image behaves with defocus; both the elastic and inelastic images depend in detail on the scattering model used and other approximations, such as the use of a weak phase object ($\phi \ll 1$). The relative behavior of the elastic and inelastic images should be correct, provided that the specimen does retain some high resolution structure; in negatively stained specimens, where the resolution of the elastic image arising from the stain may be limited to 1–2 nm, the distinction between the elastic and inelastic images in terms of resolution is not well defined. The theoretical calculations presented below will indicate three important quantities as a result of image subtraction: (1) the change in resolution and contrast of the elastic image profiles, (2) the proportion of inelastic scattering left in the difference image, and (3) the overall gain in image contrast.

For dark-field microscopy it is assumed that axial illumination is used with a dark-field stop in the objective aperture; this is the ideal condition and ignores the problem of electrical charging of the center stop by the unscattered beam. If the angle of illumination is small, α_c less than about 0.001 rad, the imaging conditions correspond to coherent dark-field microscopy treated in the previous section. Figures 4.11 and 4.12 show two sets of dark-field image profiles near Δf_E (~ –70 nm for the electron-optical parameters used). Figure 4.11 shows (a) two elastic image profiles for $\Delta f = 0$ and –50 nm together with the difference curve and (b) the corresponding curves for the inelastic image.

Some indication of the differential gain in image contrast can be obtained from the ratio of the image intensities j_E/j_I at $r = 0$. In the original $\Delta f = $ –50 nm image j_E/j_I is ~10, and it increases to 24 in the difference profiles. The resolution of the elastic image is hardly affected by the subtraction, and the main result from the difference inelastic image is a marked reduction in the contribution from the low resolution (large r) tail; in fact, the half-width of the inelastic image profile is reduced from 1.2 nm to 0.6 nm, and even allowing for delocalization effects this is a significant improvement in the inelastic image resolution.

Quantitatively, the gain in image contrast can be determined from the change in the relative proportions of elastic and inelastic scattering. For an incident beam of unit intensity, the elastic contribution is proportional to ϕ^2 or approximately the fraction of electrons elastically scattered $E(\alpha)$; the intensity of the inelastic image is proportional to $I(\alpha)$. For a carbon specimen, thickness $t =$ 10 nm, $E(\alpha) = 0.016$ and $I(\alpha) = 0.100$, and the image contrast before image subtraction $C \simeq E(\alpha)/[E(\alpha) + I(\alpha)] = 14\%$. After image subtraction the elastic contribution in Fig. 4.11 is reduced by 24%, whereas the inelastic contribution is reduced by 72%, giving $C \simeq 31\%$, which is substantial gain in image contrast.

A similar gain in image contrast occurs for the two-dark-field images taken at $\Delta f = $ –100 nm and –150 nm (Fig. 4.12). Again the difference inelastic image shows improved resolution and a corresponding reduction in the low resolution region (large r). By image subtraction the image contrast is improved from 14% to 29%, and there is also a significant gain in the resolution of the difference

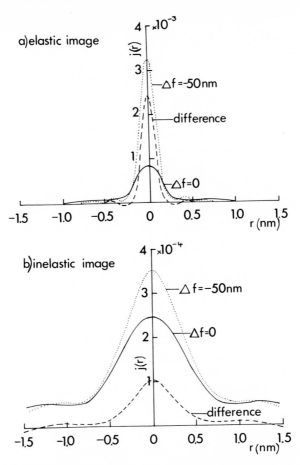

Fig. 4.11 Illustration of image subtraction for coherent dark-field microscopy. The image profiles for both (a) elastic and (b) inelastic scattering are shown for $\Delta f = 0$ and $\Delta f = -50$ nm. The respective difference curves are shown by dashed curves. $E_0 = 100$ keV, $\alpha = 0.01$ rad with a dark-field center stop of 0.001 rad, $C_s = 2$ mm, $C_c = 2$ mm.

inelastic image. This order of increase in image contrast may be sufficient to show low contrast, high resolution detail previously masked by the inelastic background. The choice of the defocus difference between two images is not critical, but 50 nm defocus difference seems to be an optimum value for a substantial change in the elastic images and a small variation in the inelastic images. Small defocus differences of 25 nm would work in ideal conditions, but in practice the smaller difference curves would be affected by noise, for example, from the substrate, which has not been included in these calculations.

An example of image subtraction in bright-field microscopy is shown in Fig. 4.13 for two images taken at 50 and 100 nm (the optimum elastic defocus)

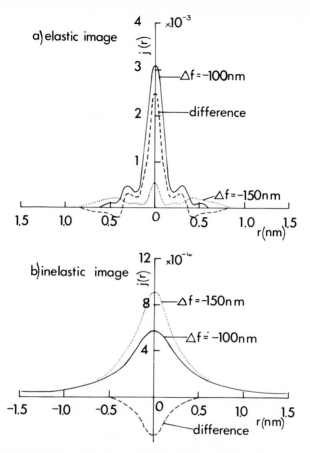

Fig. 4.12 Illustration of image subtraction for coherent dark-field microscopy. The image profiles for both (a) elastic and (b) inelastic scattering are shown for $\Delta f = -100$ nm and $\Delta f = -150$ nm. The respective difference curves are shown by dashed lines. Microscope parameters are the same as for Fig. 4.11.

underfocus. The elastic difference curve displays several side peaks which will lead to a deterioration in image resolution; the reduction in the elastic contribution to the total image is $\sim 28\%$, compared with the 72% decrease in the inelastic image.

In bright-field conditions the background contribution arises mainly from the unscattered electrons, and a reduction of even 72% in the inelastic background will not significantly alter the image contrast. However, as a result of image subtraction, the contribution from the unscattered electrons will be almost completely cancelled. The gain in contrast will then be quite substantial, but this gain will have little to do with the removal of a significant proportion of the inelastic scattering.

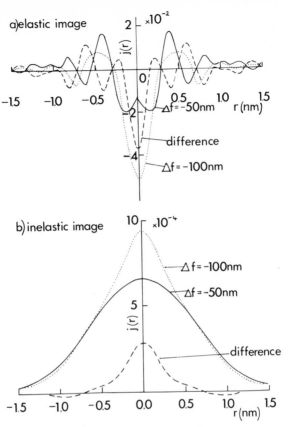

Fig. 4.13 Illustration of image subtraction for bright-field microscopy. The image profiles are shown for (a) elastic, (b) inelastic scattering for $\Delta f = -100$ nm and $\Delta f = -150$ nm. The respective difference curves are shown by dashed lines. Microscope parameters are the same as for Fig. 4.11 with the omission of a dark-field center stop.

In case subtraction seems an attractive way of removing the background due to unscattered electrons, it should be stated that the subtraction of two large numbers to give a small difference (the elastic signal) will give a large error, with a consequent decrease in signal to noise, S/N, in the difference image; for example, if there are two images in bright-field with optical densities 1.30 and 1.10 (with a background of unity) and each image has a 10% noise component, the difference image has a large contrast with a signal of 0.20 on a nearly zero background but a noise contribution of ±0.17; that is, S/N is only just greater than unity as compared with the original S/N value of ~2.5 (signal = 0.30; noise is 10% of 1.30).

In dark-field microscopy this reduction in signal to noise on subtraction of two images is not so serious because the background arises mainly from the inelastic scattering; for example, using the results in Fig. 4.11 the two images (elastic + inelastic) would have relative optical densities of 1.16 and 0.80, respectively, for $\Delta f = -50$ nm and $\Delta f = 0$, and in this case the difference image would have an optical density of 0.36 ± 0.14, that is, $S/N \simeq 2.5$. These simple calculations neglect noise arising from other factors such as radiation damage; any two pictures of the same area of the specimen will have different noise components due to increasing radiation damage. To obviate this, it is recommended that the high resolution image at defocus Δf_E be taken first, followed by the second image at $\Delta f_E \pm 50$ nm, an image which essentially contains the low resolution information to be eliminated by image subtraction.

There are specimens where radiation damage leaves the high resolution information virtually unchanged; such an example would be a heavy atom-labeled DNA strand, where the positions of the heavy atoms are not significantly affected by the radiation dose, and the DNA structure has probably been destroyed before the first micrograph has been taken. For unstained small proteins there is evidence that even after the initial radiation damage, the ashes of the molecule retain a three dimensional configuration that is fairly stable (Ottensmeyer et al., 1975). As an alternative to taking two micrographs of the same specimen area, it may be possible to take pairs of images from different parts of the specimen grid. This suggestion suffers from two disadvantages: (1) isolated macromolecules do not often have a single orientation on the substrate, although groups or aggregates of macromolecules may have a preferential orientation; (2) focus gradients across a grid square may be large enough to affect the focus changes used in image subtraction. In the case of crystalline specimens, where adjacent areas of specimen are likely to be identical, there is no problem in taking several micrographs under minimum radiation dose conditions for each field of view.

PRACTICAL PROBLEMS IN IMAGE SUBTRACTION

It will be evident that the image subtraction method requires an accurate determination of defocus, in both bright-field and dark-field microscopy, although in the latter case the elastic image resolution is less sensitive to defocus errors. Ideally, a method is needed for estimating defocus while the specimen is in the electron microscope, with a quantitative measurement of Δf available from a subsequent analysis of the electron micrograph. In bright-field microscopy choosing focus from the granular appearance of a thin carbon substrate is fairly routine, but this phase contrast structure is not at all clear in dark-field microscopy. This section evaluates the problems of choosing optimum elastic defocus in both bright-field and dark-field microscopy; from theoretical calculations for the resolution function for the elastic image, figures are given for the optimum defocus Δf_E and the precision required in its measurement. The second problem in image subtraction, namely the alignment of two similar images to within the resolution limit inherent in the elastic image, is also examined. Subtraction of two aligned images may be affected by computer after the normalization of the two images (Welles *et al.*, 1974) or by photographic superposition of the positive of one image with the negative of the second image (Krakow *et al.*, 1976).

Determination of Defocus in Bright-field Microscopy

Focus in bright-field microscopy is normally determined by the disappearance of the carbon substrate granular structure as the objective lens current is changed from overfocus to underfocus. Astigmatism is corrected at the same time by a correction for the streaking of this grain structure. The explanation of this behavior of the image of a virtually structureless carbon film can be derived from the theoretical expression for the bright-field image intensity, Eq. (4.19); the granularity arises almost totally from the spherical aberration defect and defocus of the objective lens. In fact it will be shown that the empirically determined "in-focus" image corresponds to ~50 nm underfocus; thus Δf_E, if set with reference to this "in-focus" image, should be a further 50-nm underfocus (for $\Delta f_E = -100$ nm) and not a further 100-nm underfocus.

In order to determine defocus from a micrograph, we measure its optical transform or calculate its Fourier transform using a computer (Thon, 1971). The Fourier transform of Eq. (4.19) is given by (Lenz, 1971):

$$J(\nu) = \delta(\nu) + 2\Phi(\nu) \sin [\gamma(\nu)] B(\nu) \qquad (4.21)$$

where Φ represents the elastic scattering from the carbon substrate; Φ has very little structure so that the transform J is mainly affected by the behavior of $\sin \gamma$, with a cut-off frequency $\nu_{\max}$ determined by the objective aperture function $B(\nu)$. Thus, the transform of a bright-field micrograph comprises an intense center spot, $\delta(\nu)$, at $\nu \simeq 0$, and a term $\Phi \sin \gamma$, where Φ represents the object structure and $\sin \gamma$ is dependent only on the microscope parameters. There will also be a large steep background at small spatial frequencies due to the low resolution inelastic scattering from the carbon film.

In the case of the optical transform, we essentially measure the intensity $|J(\nu)|^2$ proportional to $|\Phi(\nu)|^2 \sin^2 [\gamma(\nu)]$ (cutting off at $\nu = \nu_{\max}$). Since

$$\sin [\gamma(\nu)] = \sin [2\pi/\lambda (C_s \nu^4 \lambda^4 /4 + \Delta f \nu^2 \lambda^2 /2)] \qquad (4.22)$$

the transform depends on C_s and Δf, and at certain spatial frequencies, ν_0, $\sin^2 \gamma = 0$; these zeros, ν_0, in the transform will be determined uniquely by C_s and Δf. Thus, in a focus series (C_s is constant) we will see ring patterns in the transforms with maxima and minima according to the variation in Δf. The optimum defocus will correspond to that Δf for which $\sin^2 \gamma \simeq 1$ for a large range of spatial frequencies. Clearly this is not possible for overfocus ($\Delta f > 0$), since the spherical aberration and defocus terms add to give an oscillatory $\sin^2 \gamma$, but for underfocus ($\Delta f < 0$), the defocus term can partially cancel the oscillations in $\sin^2 \gamma$ caused by the C_s term. It is not possible to make $\sin^2 \gamma \simeq 1$ for all spatial frequencies, but by a suitable choice of Δf, namely Δf_E, the first minimum of $\sin^2 \gamma$ can be extended out to 2-3 nm^{-1} (resolution 0.5-0.3 nm). Equation (4.22) excludes an astigmatism term which depends on x and y (Lenz, 1971; Thon, 1971); in the presence of astigmatism the circular rings become elliptical, particularly near focus, where $\Delta f \simeq 0$.

Figure 4.14 shows a focus series for carbon with the corresponding optical transforms (diffractograms) from in-focus to $\sim$300 nm underfocus (Thon, 1971). Corresponding to $\Delta f = 0$, there is a large region of $\sin^2 \gamma$ that has very small values, because the spherical aberration term is small for low spatial frequencies (up to 1 nm^{-1}). This corresponds to the micrograph shown in Fig. 4.14a, where the optical transform shows only a small annulus in $\sin^2 \gamma$ at about 1 nm resolution, and any information on Φ transferred to the image is limited to this range of spatial frequencies. However, although $\Delta f = 0$ does give a low contrast image, it is not the minimum contrast image; for small underfocus values, 30-50 nm, this region of small $\sin^2 \gamma$ can be extended to spatial frequencies of $\sim$1.5 nm^{-1} (0.7 nm resolution), and this gives the minimum image contrast condition observed in bright-field microscopy.

Figure 4.14b corresponds to optimum defocus with a large region over which $\sin^2 \gamma \simeq 1$; the low-intensity region for small spatial frequencies (up to 0.5 nm^{-1}) is unavoidable if a significant range of spatial frequencies is to be transferred to the image with maximum contrast. Further underfocus produces images shown in Figs. 4.14c and 4.14d, where the central region for which $\sin^2 \gamma \simeq 1$ contracts, and $\sin^2 \gamma$ gives several rings in the optical transform; thus large underfocus values are undesirable because in the range of spatial frequencies of interest in high resolution electron microscopy, 0.5-3 nm^{-1} (2-0.3 nm resolution), certain information on the object structure is lost from the image, corresponding to $\sin^2 \gamma$ having small or zero values.

It will be seen from the optical transforms shown in Fig. 4.14 that the larger spatial frequencies have lower intensities. This decay of $\sin^2 \gamma$ is due to two factors: (1) the structure factor for carbon, $|\Phi|$, decreases with increasing angle of scattering ($\theta = \lambda \nu$); and (2) owing to the limited coherence of the incident beam, the $\sin^2 \gamma$ curve is multiplied by a gaussian function which decays to zero for large spatial frequencies (Cowley, 1976). The limited coherence is due to the energy spread (chromatic coherence) and the angular divergence (spatial coherence) of the incident electron beam (Hibi and Takahashi, 1971).

Chromatic coherence can be improved by using a lower emission temperature for the normal heated filament or by using a field emission electron source. The latter type of electron source will also improve the spatial coherence of the electron beam because the source size is very small; in conventional electron guns, spatial coherence can be improved by using a smaller condenser 2 aperture ($\sim$50-100 μm) and working well away from the condenser 2 cross-over point.

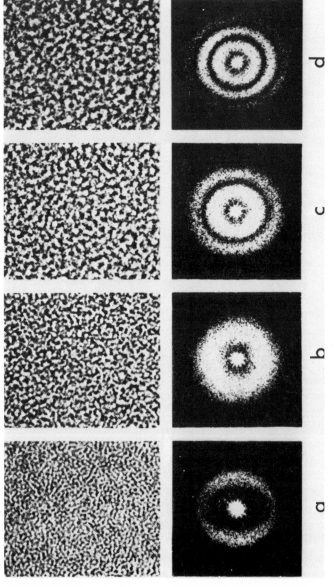

Fig. 4.14 Focus series of a carbon film in bright-field microscopy together with the corresponding optical diffraction patterns: (a) in focus, (b) optimum defocus (about 100 nm underfocus), (c) underfocus by about 200 nm, (d) underfocus by about 300 nm. $E_0 = 100$ keV, $C_s = 4$ mm, $M = 1,500,000$. (*From Thon, 1971.*)

The optical diffraction pattern will in any case cut off at a spatial frequency ν_{max} determined by the objective aperture size $d_{obj} = 2f\lambda\nu_{max}$; the results shown in Fig. 4.14 cut off at $\sim\nu_{max} = 2$ nm^{-1} (0.5 nm resolution).

Clearly, a smaller objective aperture would eliminate the higher spatial frequencies of $\sin^2\gamma$ and make focus discrimination much more difficult. Conversely, although a large objective aperture, say, 60 μm giving a $\nu_{max} = 4$ nm^{-1}, would give excellent focus discrimination, the specimen may not display such a high resolution. In this case all the high resolution detail will arise only from the substrate and contribute to the noise in the image.

These experimental results are confirmed by the theoretical calculations shown in Fig. 4.15 of $\sin^2\gamma$, corresponding to scans across the diameters of the focus series diffractograms of Fig. 4.14. The corresponding resolution functions $q(r)$ are shown on the right-hand side of Fig. 4.15. Clearly optimum defocus (b) not only gives the largest region of $\sin^2\gamma \simeq 1$ but also the sharpest resolution function $q(r)$. In fact, a measure of the image contrast can be obtained from the

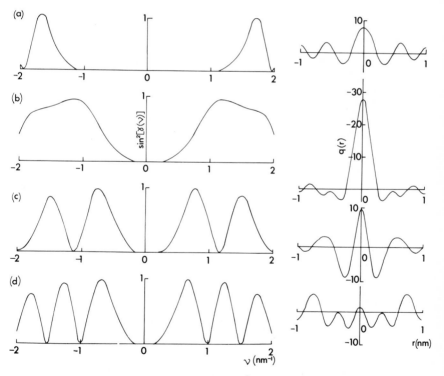

Fig. 4.15 Theoretical results for the intensities $\sin^2[\gamma(\nu)]$ of the optical transforms of the carbon film focus series in Fig. 4.14. The corresponding resolution functions are shown on the right-hand side of the figure. Coherence envelope on $\sin^2\gamma$ corresponds to an illumination angle of $\alpha_c = 0.001$ rad with a thermal electron energy spread of $E_f = 1$ eV.

value of $q(r)$ at $r = 0$; the contrast is a maximum for Δf_E, decreasing either side of this optimum defocus to a minimum for $\Delta f = -50$ nm and $\Delta f = -140$ nm. We now have a clear relation between an experimental quantity, the optical transform of a bright-field micrograph, and the elastic image resolution functions of the previous section.

It is clear that the optical diffractometry described above serves only as a postdiagnostic method after the image has been recorded, but it also gives a clear indication of how to take bright-field images at optimum elastic defocus. The procedure for taking bright-field images for image subtraction would then be as follows: astigmatism should be corrected a high magnification (500,000×) with a thin-foil objective aperture of the correct diameter in place. At lower magnification, "in-focus" is determined from the granular appearance of a thin carbon substrate. The accuracy with which "in-focus" can be determined depends on several factors, including the grain size of the viewing screen, the thickness of the carbon film, and the coherence of the incident electron beam. The conditions for maximum beam coherence have been mentioned already, and it is clear that phase contrast effects are enhanced by using a coherent electron beam (Hibi and Takahashi, 1971).

The carbon film must be thin for two reasons; first, Eq. (4.21) for the optical transform is valid only for thin films ($t \simeq 10$ nm), and thicker films will give additional contributions to the image transform depending on $\cos^2 \gamma$, with a consequent shift of the "in-focus" position by an unpredictable amount; second, a thick substrate will scatter a larger fraction of the electrons inelastically and thus contribute a significant background to the image. The carbon film thickness should be ideally less than 5 nm, thinner films being more difficult to prepare unless supported on a perforated plastic film.

With the determination of the "in-focus" position, the objective lens is under-focused by a further 50 nm approximately to Δf_E, and the specimen area of interest is moved into the electron beam, either by moving the specimen stage or by deflecting the illumination (see Unwin and Henderson, 1975, for details of minimum irradiation techniques), and the image recorded immediately; a second image at $\Delta f_E \pm 50$ nm is taken either from the same specimen area or from a new unirradiated area, if the biological structure has a clearly defined orientation on the substrate. The pairs of micrographs are then examined by optical diffractometer to select the images that have the correct defocus values, and to reject those images which are not at optimum defocus and may addition-ally show astigmatism or specimen drift by the ellipticity of the optical diffrac-tion rings. The optical diffraction patterns will also give an indication of the coherence by the decrease in intensity of the larger spatial frequency rings.

Figure 4.16 shows a focus series of groups of nine hexons from adenovirus, with a range of defocus from -50 nm ("in-focus") to 200 nm underfocus; optimum defocus corresponds to Fig. 4.16b, but this image is far from ideal because the substrate thickness is at least 10 nm and can be seen to give a signifi-

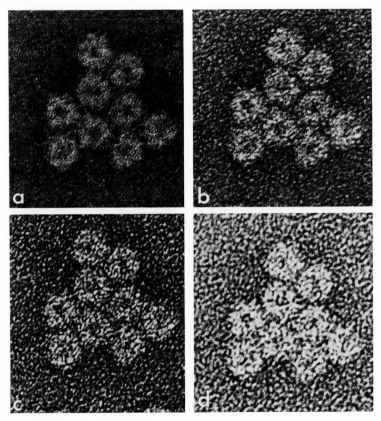

Fig. 4.16 Bright-field focus series of adenovirus groups of nine hexons, stained in sodium silicotungstate: (a) "in focus," (b) optimum defocus (–100 nm), (c) 150 nm underfocus, and (d) 200 nm underfocus. E_0 = 80 keV, M = 1,000,000 (*Courtesy of J. V. Heather*).

cant background structure in the image. However, the example shown in Fig. 4.16 does indicate that it is possible to take micrographs well within 25 nm of optimum defocus. In Fig. 4.17 calculations for the microscope conditions used in Fig. 4.16 show the sensitivity of $\sin^2 \gamma$ and the corresponding resolution function $q(r)$ to an error in the optimum defocus of ± 20 nm. At the optimum defocus, Fig. 4.17b, the transfer function extends to 2.7 nm^{-1} (0.4 nm resolution) and gives the sharpest resolution function. The $\sin^2 \gamma$ curves 20 nm on either side of Δf_E still extend to 2 nm^{-1} (0.5 nm resolution), with a slight broadening of the resolution function $q(r)$.

On the basis of these theoretical calculations an accuracy of 25 nm is required for the determination of Δf_E. This does not seem an unrealistic requirement, although such accuracy probably means taking a large number of focus pairs before the ideal pair of micrographs is found. In order to eliminate some of this

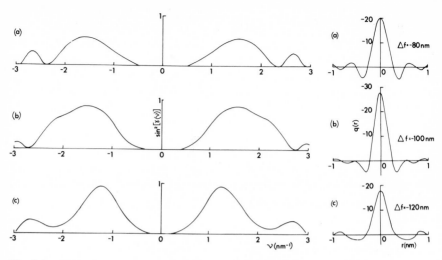

Fig. 4.17 Theoretical results in bright-field microscopy for the optical transform and the corresponding resolution functions for elastic scattering, showing the dependence on defocus ±20 nm from optimum defocus (b). E_0 = 100 keV, C_s = 2 mm, α = 0.01 rad. Coherence parameters: α_c = 0.001 rad, E_f = 1 eV.

trial and error we are developing an on-line focusing device which will directly display the transform $J(\nu)$ using an image obtained from a transmission phosphor in the electron microscope; such a direct display of $\sin\gamma$ will facilitate the correction of astigmatism and an accurate determination of Δf_E.

To complete this section, Table 4.1 shows the theoretical values for Δf_E (±10 nm) in bright-field microscopy for a C_s range of 1.2–2.0 mm, corresponding to values attained in modern transmission electron microscopes; the defocus values corresponding to minimum contrast for a thin carbon film are also given in Table 4.1.

Table 4.1   Optimum Elastic Defocus Δf_E (nm) for E_0 = 100 keV in Bright-field and Dark-field Microscopy

	Bright-field		Dark-field	
C_s(mm)	Δf_E	Δf (minimum contrast)	Δf_E (coherent)	Δf_E (incoherent)
2.0	−100	−50	−70	−100
1.8	−90	−50	−70	−90
1.6	−80	−40	−60	−80
1.4	−80	−40	−60	−70
1.2	−70	−30	−50	−60

Underfocus values are increased by a factor of about 1.06 for E_0 = 80 keV, and 1.15 for E_0 = 60 keV.

Determination of Defocus in Dark-field Microscopy

It is not possible to focus in dark-field in the same way as in bright-field micros-copy; the granular structure of a carbon film is not at all clear in dark-field, using either axial or tilted illumination. Furthermore, there is no objective test of the defocus level, since the optical transform of the image has no simple rela-tion to the state of focus. Reference to Eq. (4.17) for the image intensity j in dark-field shows clearly that the transform will not be simply related to either $\sin \gamma$ or $\cos \gamma$. In fact, the optical diffractograms of dark-field images show only a uniform disk almost independent of defocus, although astigmatism and speci-men drift will show up as distortions of the disk. In most biological applications the accuracy of achieving optimum defocus may not be critical, and it is suf-ficient to focus in bright-field, followed by the appropriate beam tilt to give the dark-field image. However, this change in electron optics can give focus changes of 100–200 nm, well beyond the limits for image subtraction. Even in axial dark-field microscopy using an objective aperture with a center stop (Dubochet, 1973), the electrical charging of the center stop by the intense unscattered beam will cause a significant focus change and gross astigmatism in going from bright-field to dark-field, as the objective aperture is correctly centered.

Cowley *et al.* (1974) have suggested a procedure for focusing in either tilted or axial dark-field using the contrast variation of a hole in a carbon film. Their cal-culations show that in tilted dark-field the intensity around the hole is asymmet-ric, because of the asymmetry of the diffraction pattern with respect to the objective aperture. This asymmetry of the image intensity reverses on going through focus, and the intensity minimum around the hole is calculated to be ~ 30 nm underfocus. Cowley *et al.* (1974) suggest that at a magnification of 100,000X this minimum corresponding to maximum contrast could be deter-mined to within an accuracy of ± 20–30 nm. Optimum defocus in axial or tilted dark-field microscopy is ~ -70 nm for an incident electron energy of 100 keV.

Table 4.1 shows the variation of Δf_E (coherent) for a range of C_s values. These values are slightly larger than those given by Cowley (1976), because he based his values for Δf_E only on the behavior of the resolution function $q'(r)$, that is, choosing Δf_E to make $\cos \gamma \simeq 1$ over as large a range of spatial frequencies as possible. However, the results in Table 4.1 were based on numerical calcula-tions using Eq. (4.17), including the behavior of both the resolution functions $q(r)$ and $q'(r)$. Because the optimum defocus for $q(r)$ is ~ -100 nm and that for $q'(r) \sim -40$ nm ($C_s = 2$ mn), the mean result –70 nm for Δf_E is obtained. This difference of 30 nm, however, between the figures in Table 4.1 and Cowley's (1976) result is not too important in dark-field microscopy, where the contrast appears to be less sensitive to defocus error in Δf_E; an error in Δf_E of ± 40 nm is acceptable.

Probably the best electron-optical configuration for dark-field microscopy is hollow-cone illumination (Dubochet, 1973), since this avoids the problems of (1) electrical charging and critical alignment of an objective aperture with a small (5 μm) central stop, and (2) the interpretation of dark-field images taken with tilted illumination, where the asymmetry of the objective aperture with respect to the diffraction pattern of the specimen can lead to imaging artifacts. However, with the annular condenser aperture used in hollow-cone illumination, the illumination angle on the specimen may be as large as 0.005 rad (cf. normal beam divergence of less than 0.001 rad). This large angle for the illumination gives virtually incoherent imaging conditions; the area of specimen coherently illuminated is $d_{coh} \simeq \lambda/\alpha_c$ (Thon, 1971), which is ~0.7 nm for α_c = 0.005 rad, and detail in the specimen separated by larger distances from 0.7 nm is incoherently imaged, that is, the intensities rather than the amplitudes of the scattering are added.

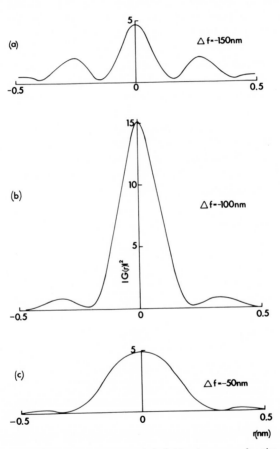

Fig. 4.18 Theoretical curves in incoherent dark-field microscopy for the resolution functions for elastic scattering, showing the dependence on defocus ±50 nm from optimum defocus (b). E_0 = 100 keV, C_s = 2 mm, α = 0.01 rad; hollow cone illumination.

The image intensity j for incoherent imaging of a weak phase object in dark-field microscopy is:

$$j(\mathbf{r}_i) = [\phi(\mathbf{r}_i) - \phi_m]^2 * |G(\mathbf{r}_i)|^2 \qquad (4.23)$$

where $|G(\mathbf{r})|^2 = [q(\mathbf{r})^2 + q'(\mathbf{r})^2]/4$ is the resolution function for incoherent imaging.   As in coherent dark-field images j is related to $[\phi(\mathbf{r}_i) - \phi_m]^2$, but phase contrast effects are completely absent because Eq. (4.23) is an intensity convolution unlike Eq. (4.17), which is an amplitude convolution, although an unattractive one.

Figure 4.18 shows the behavior of the resolution function $|G(\mathbf{r})|^2$ with a variation of ± 50 nm about the optimum defocus Δf_E (b). Clearly the resolution and contrast of this type of dark-field image is less sensitive to defocus variations than the bright-field image near Δf_E.   On the basis of these calculations, the error in the determination of Δf_E can be as much as 40 nm before the image subtraction procedure fails. Because of the low intensity on the viewing screen care must be taken in dark-field microscopy not to find the optimum contrast condition for the inelastic image, rather than the elastic image.   The defocus values of the dark-field images cannot be subsequently verified by optical or computer methods, because the Fourier transform of $|G(\mathbf{r})|^2$, for example, is almost independent of defocus; a comparison of such an experimental transform with the theoretical calculations for the transform, using $\sin \gamma$ and $\cos \gamma$, leads to a defocus discrimination of ± 100 nm at best.   Table 4.1 lists the theoretical results for Δf_E (incoherent), which are very close to the optimum defocus values for bright-field microscopy.

Alignment of Two Images

Before two images can be aligned for image subtraction, the images have to be scaled so that the optical densities of each image represent the same total number of electrons detected. It is impossible to take two successive images which have exactly the same total energy density, that is, the total number of electrons detected by each photographic plate is the same for each image; exposure times for electron images are not that reproducible. It is assumed for image subtraction that, although the spatial distributions of electrons in the images are different (owing to the different resolution of the images), the total number of electrons transmitted by the specimen is the same in each, independent of defocus. Thus, we normalize the two image intensities $j_1(x, y)$ and $j_2(x, y)$ so that over a large region of each image

$$\sum_x \sum_y j_1(x, y) = \sum_x \sum_y j_2(x, y)$$

This normalization is most easily achieved in the computer by the separate addition of the image intensities for each image, and then multiplying one set of image intensities, say j_1, by the ratio of $\Sigma j_2 / \Sigma j_1$. Photographically this type of scaling is not easy to implement without making several photographic copies of one image and evaluating the mean density levels using a microdensitometer.

The basic problem of the alignment of two images for subtraction is the translation and rotation of one image with respect to the second image so that the prominent features of each image coincide. This procedure may be done photographically by superposition of the negative of one image on the positive of the other; the choice of the best alignment, however, is subjective, with the eye choosing the result that shows the best "structure" with the least background. Additional problems arise in subtraction by photographic superposition from the matching of the optical densities of the positive and negative images; it may be necessary to prepare several negatives of one image at different optical densities and use several different film emulsions to match the optical density–exposure characteristics of the negative with the positive image.

Computer alignment of two similar images is more objective (Frank, 1973). In order to illustrate the computational alignment of two images, we examine the translational alignment of two image profiles $f(x)$ and $g(x)$ in one dimension using the cross-correlation function $c(x)$ defined by

$$c(x) = \int_{-\infty}^{+\infty} f(x_0) g(x + x_0) \, dx_0 \qquad (4.24)$$

The shape of $c(x)$ with respect to its peak position and the width of the curve give an indication of how good the correlation (alignment) of two images is, and indicates how far one image has to be moved to give the best alignment.

To see how this works in practice, first consider two profiles $f(x)$ and $g(x)$ that are identical and aligned (Fig. 4.19a). When $x = 0$, in Eq. (4.24), $c(0)$ is the sum of all the products $f(x_0) g(x_0)$ for all x_0 values; then the maxima and minima of both $g(x)$ and $f(x)$ coincide, and both positive and negative variations about the mean density (the x-axis) add to give a large contribution to $c(0)$. If now g is displaced by a small distance x, the maxima and minima will be slightly out of alignment, and some of the prominent features will be subtracted to give zero in the product $f(x_0) g(x + x_0)$; the sum $c(x)$ will therefore be smaller than $c(0)$. Increasing x correspondingly puts $f(x)$ and $g(x)$ further out of register, and it can be seen that generally as x [the displacement of $g(x)$ with respect to $f(x)$] increases, $c(x)$ will decrease. Thus $c(x)$ gives a measure of the similarity of the two image profiles.

If $g(x)$ is now made a low resolution version of $f(x)$, Fig. 4.19b, with the two images still aligned, it can be seen that even for $x = 0$ in Eq. (4.24) the maxima and minima do not exactly coincide, and $c(0)$ is smaller than for the example shown in Fig. 4.19a. Also, displacing $g(x)$ by a small distance x makes little difference to the sum of the products $f(x_0) g(x + x_0)$ because of the original differ-

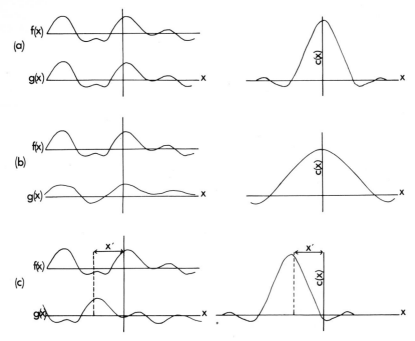

Fig. 4.19 The cross-correlation $c(x)$ and two image profiles $f(x)$ and $g(x)$ when f and g are: (a) the same and aligned, (b) similar and aligned, (c) the same and misaligned by x'.

ences that exist in $f(x)$ and $g(x)$. However, as the displacement x increases, $c(x)$ does decrease because the two profiles become increasingly out of register. So the width and height of $c(x)$ show the similarity between two profiles, a very broad $c(x)$ corresponding to two dissimilar images. If $g(x)$ is out of alignment with $f(x)$ by a distance x', Fig. 4.19c with $f(x)$ and $g(x)$ now identical, then for $x = 0$ the sum of the products $f(x_0)g(x_0)$ will not be very large because maxima and minima will not coincide. However as $g(x)$ is displaced to the right, the two curves approach alignment, and $c(x)$ will increase until $x = -x'$, when $f(x)$ and $g(x)$ will be exactly aligned. The maximum in $c(x)$ then occurs at $x = -x'$, the extent of the original misalignment of the two images. The position of the maximum in the cross-correlation function, $c(x)$, gives the translational mis-alignment of two images, which can then be compensated for using the computer.

Of course, the accuracy with which the peak position of $c(x)$ can be determined depends on the width of $c(x)$, and if $c(x)$ is broad, the alignment cannot be made very accurately. In practice, noise due to the substrate background and damaged specimen will make the cross-correlation less sensitive to image mis-alignment. A large signal-to-noise ratio in both images will blur out the peak in $c(x)$.

This analysis can be extended to two dimensions by calculating the two-dimensional cross-correlation function $c(x,y)$ of two images $f(x,y)$ and $g(x,y)$:

$$c(x,y) = \int_{-\infty}^{+\infty} \int_{-\infty}^{+\infty} f(x_0, y_0) g(x_0 + x, y_0 + y) \, dx_0 \, dy_0 \qquad (4.25)$$

when the maximum in $c(x,y)$ at (x', y') will give the image translational misalignment, which can then be corrected using the computer. In practice the cross-correlation function for translational misalignment is calculated using the fast Fourier transform because the integrals in Eqs. (4.24) and (4.25) become simple products of the Fourier transforms of f and g, with a significant reduction in computing times.

The rotational alignment of two images, which must be considered at the same time as translational alignment, is more difficult because in order to calculate the rotational cross-correlation function, both f and g must be interpolated onto polar coordinate axes (Saxton, 1974). Of course, the eye can assess both these alignment processes in a single step, but the use of cross-correlation functions for alignment is objective, and does not prejudice the final result for the image subtraction. We have made attempts to implement image subtraction by photographic superposition using a whole series of negatives of one image, but we found a strong bias towards choosing the most acceptable results visually. Computer image alignment and subtraction is recommended, provided there is a good image display facility with at least 16 gray levels.

Image Subtraction in Practice

Image subtraction at high resolution is shown in Fig. 4.20 (Welles *et al.*, 1974; Krakow *et al.*, 1976) for SV40 DNA positively stained in uranyl acetate for images taken in dark-field using an axial beam stop in the objective aperture. Figure 4.20a shows the image taken at optimum defocus (~60 nm underfocus),

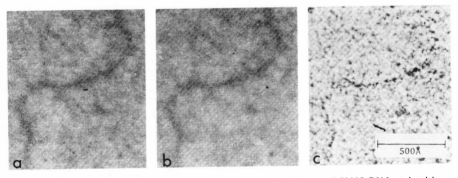

Fig. 4.20 Image subtraction in dark-field microscopy for images of SV40 DNA stained in uranyl acetate: (a) optimum defocus, (b) defocused by 25 nm, (c) computer subtraction of two images. $E_0 = 100$ keV, $M = 370,000$. (*From Welles* et al., *1974*.)

and Fig. 4.20b shows a second image taken 25 nm underfocus. These two images were scaled and aligned by the computer, and the difference image (Fig. 4.20c) was generated by computer. Neither of the two original images (Figs. 4.20a,b) shows either high resolution detail or very good contrast; the difference image shows high resolution detail, such as a loop in the DNA strand (left of center in Fig. 4.20c), which was masked by inelastic scattering in both the original images. The overall result is a definite gain in image contrast and in the signal-to-noise ratio. However, since neither of the original images of the stained DNA was taken under minimum irradiation conditions, it is unlikely that the DNA structure is resolved in the final subtraction. It is most likely that the subtraction has affected the removal of the predominantly inelastic image components of the DNA damage products and the carbon substrate to leave the uranium atoms, which form an essentially elastic image, in clear contrast. Image subtraction, it seems, will be successful in resolving heavy atoms or groups of heavy atoms substituted in biological macromolecules.

A second high resolution application has yet to be tested: if it is assumed that in the dark-field images of unstained protein molecules the protein ashes retain a three-dimensional structure with significant high resolution detail (Ottensmeyer *et al.*, 1975), then image subtraction could be used to remove the background arising from inelastic scattering in the protein and substrate. The experimental problems in this type of dark-field microscopy arise from the poor signal to noise ratio in the images of the isolated protein molecules, and any significant improvement in signal to noise ratio may improve structural analysis of the protein ashes. However, because of the asymmetry of such isolated protein molecules, the pairs of images for image subtraction have to be taken from the same specimen area, with a consequent increase in the total irradiation of the protein molecules.

Negatively stained specimens appear to be intrinsically low resolution specimens (1.5 nm) because the stain does not reflect the morphology of the biological macromolecule at high resolution. The stain mainly acts as a support for the biological material and produces the contrast observed in bright-field microscopy of negatively stained specimens. However, Unwin (1972) has shown by using a phase plate (see following section) that although the negative stain in tobacco mosaic virus (TMV) protein may only accurately reflect the biological structure to 2 nm resolution, the protein itself may have structure preserved to high resolution (0.5–1 nm). In this case we suggest that the image pairs for subtraction could be used to eliminate the low resolution elastic image due to the negative stain in addition to the inelastic scattering from the biological material and substrate. However, in this case the resolution difference between the biological material (representing a small contribution to the total image intensity) and the negative stain (large contribution) is marginal, so that this application of image subtraction is speculative. If we are less ambitious, it should be possible at least to eliminate by using image subtraction a substantial proportion of the inelastic scattering from the substrate.

DARK-FIELD VERSUS BRIGHT-FIELD MICROSCOPY

It is evident from the calculations presented in previous sections that image subtraction can produce a significant contrast enhancement in dark-field microscopy and is only slightly beneficial in bright-field microscopy; the image contrast in dark-field is much more affected by inelastic scattering than the corresponding bright-field contrast, which is limited by the background from unscattered electrons. In this section we wish to quantify the advantages and disadvantages of dark-field microscopy at atomic or molecular resolution and, where there are disadvantages, whether they can be offset by a reduction in the inelastic contribution. Dubochet (1973) has reviewed the experimental configurations for dark-field microscopy, but he did not consider in detail the effects of inelastic scattering on image resolution and contrast. Of the dark-field modes used, both axial and tilted illumination give experimental and image interpretation problems; in the case of axial illumination the electrical charging of the objective aperture stop can cause severe image distortion, whereas the asymmetric position of the objective aperture with respect to the scattered electron beam in tilted illumination can lead to imaging artifacts, depending on the angle and direction of tilt.

The ideal dark-field image is obtained using conical illumination, with a potential image resolution ~ 1.5 times better than the bright-field image at optimum elastic defocus; this can clearly be seen in Figs. 4.17 and 4.18, where a comparison of the corresponding resolution functions shows that the dark-field incoherent function not only has a smaller half-width but also less prominent side lobes. However, there still remains a fundamental problem of image interpretation in dark-field microscopy; namely, whereas bright-field image contrast is linearly dependent on the object structure $[\phi(r) - \phi_m]$, dark-field contrast depends on $[\phi(r) - \phi_m]^2$ (Cowley, 1976).

Because in dark-field microscopy we can only measure $[\phi(r) - \phi_m]^2$, it is not possible to tell the difference between a group of heavy atoms (large $\phi > \phi_m$) on a substrate/biological macromolecule and a hole (or low density region) in the substrate/biological macromolecule (small $\phi < \phi_m$). This ambiguity in distinguishing between high density and low density regions in the image has not been too much trouble when the specimen has comprised only a few heavy atoms in a model organic compound (Henkelman and Ottensmeyer, 1971; Whiting and Ottensmeyer, 1972; Hashimoto et al., 1973; Ottensmeyer et al., 1973), but may be a serious disadvantage in the images of biological specimens, stained or unstained (Massover and Cowley, 1973; Ottensmeyer et al., 1975).

In incoherent imaging conditions, where effectively each specimen point scatters independently, the definition of a mean potential ϕ_m taken over a large area of specimen is not really meaningful. In this case image contrast is probably proportional to ϕ^2 at each image point, and incoherent dark-field images, obtained by using hollow cone illumination, may be easier to interpret than images taken by either axial or tilted illumination. Certainly the images shown by Thon and Willasch (1972) using conical illumination can be interpretated as a direct projection of the atom density of the organo-metallic compound they used. At lower resolution (>2 nm) there does not seem to be any interpretation problem in dark-field images, because the image formation can be described by the differential scattering of electrons from high and low density regions of the specimen (Dupouy, 1967; Dupouy *et al.*, 1969; Johnson and Parsons, 1969). Besides the difficulty in image interpretation, there is the problem in dark-field microscopy of increased specimen irradiation necessary to give an image with the same signal-to-noise ratio as in bright-field microscopy.

In bright-field microscopy the number of electrons imaged is dependent on $1 + 2\phi$ (neglecting lens aberrations), whereas in dark-field microscopy the number of electrons is only proportional to ϕ^2; thus to keep electron noise in the photographic image to the same level as in bright-field, the total electron dose to the specimen may have to be increased by a factor of 10 or more in dark-field microscopy. The frequently quoted advantage of dark-field microscopy is the higher contrast than that obtainable in bright-field microscopy; but taking into account inelastic scattering in the specimen and substrate contributing to the background, the contrast in dark-field may in fact be less than in bright-field microscopy. The main point here is that because of the small dark-field signal, ϕ^2, even small contributions to the background from the substrate (inelastic and elastic scattering) and biological material (inelastic scattering) may lower the image contrast significantly.

In order to clarify some of these points on image contrast and signal to noise ratio in the image, this section gives results for two types of specimens: (a) unstained biological material of thickness 5 nm ($\sim$40 atoms thick) and (b) a group of heavy atoms of thickness 0.5 nm including the effects of a substrate on image contrast. These calculations do not take into account lens aberrations or defocus, and so the contrast figures may be a factor of 2 in error; but the relative bright-field and dark-field figures given should be correct.

Table 4.2 shows a comparison of the contrast, C, and S/N in both bright-field and dark-field images for the two types of specimens described above; the columns on the right-hand side of the Table for dark-field show C and S/N for three situations: all the inelastic scattering is included in the image, the inelastic scattering is removed by energy selection, and 70% of the inelastic scattering is removed by image subtraction. These calculations for C and S/N are also shown for increasing substrate thickness, where it is assumed that all the scattering from the substrate (elastic and inelastic) contributes to the background in the image; only the elastic scattering from the specimen contributes to the signal part of the image.

The calculations in Table 4.2 are made for two electron radiation doses at the specimen: $n_0 = 100$ and 1,000 electrons nm^{-2}. 6.2 electrons nm^{-2} corresponds to 1 coulomb m^{-2} at the specimen (or 620 electrons Å^{-2} = 1 coulomb cm^{-2}

Table 4.2   A Comparison of the Contrast, C (%), and Signal to Noise, S/N, in Bright-field and Dark-field Microscopy for Two Electron Irradiation Conditions, $n_0 = 100$ and $n_0 = 1,000$ electrons nm^{-2} (6.2 electrons nm^{-2} = 1 coulomb m^{-2}). $E_0 = 100$ keV, $\alpha = 0.01$ rad

(a) Specimen: unstained biological material ($Z = 6$) of thickness 5 nm

Substrate thickness (nm)	n_0	Bright-field		Dark-field						
		C	S/N	+ inelastic			no inelastic		30% inelastic	
				C	S/N	F^+	C	S/N	C	S/N
0	100	20%	1.9	15%	0.36	30				
	1,000	20%	6.1	15%	1.1	30	100%	3.0	36%	1.8
1	100	20%	1.9	12%	0.33	35				
	1,000	20%	6.1	12%	1.0	35	82%	2.7	30%	1.6
2	100	19%	1.9	10%	0.31	40				
	1,000	19%	6.0	10%	0.97	40	69%	2.5	26%	1.5
5	100	19%	1.9	7%	0.26	55				
	1,000	19%	5.9	7%	0.81	55	53%	2.2	18%	1.3
10	100	18%	1.8	5%	0.21	80				
	1,000	18%	5.8	5%	0.68	80	36%	1.8	12%	1.0

F^+—factor required in electron dose to give the same signal-to-noise ratio as in the bright-field image.

(b) Specimen: heavy atoms or a small group of atoms ($Z = 79$) of thickness 0.5 nm

Substrate thickness (nm)	n_0	Bright-field		Dark-field						
		C	S/N	+ inelastic			no inelastic		30% inelastic	
				C	S/N	F^+	C	S/N	C	S/N
0	100	30%	2.9	65%	1.1	7				
	1,000	30%	9.2	65%	3.6	7	100%	4.4	86%	4.2
1	100	30%	2.9	47%	0.95	9				
	1,000	30%	9.1	47%	3.0	9	91%	4.3	71%	3.8
2	100	29%	2.9	36%	0.83	12				
	1,000	29%	9.0	36%	2.7	12	83%	4.1	60%	3.4
5	100	28%	2.8	22%	0.67	19				
	1,000	28%	8.9	22%	2.1	19	71%	3.8	42%	2.9
10	100	27%	2.7	14%	0.53	25				
	1,000	27%	8.7	14%	1.8	25	56%	3.3	29%	2.4

in cgs units). The normal image photographed after careful focusing on the specimen area of interest may correspond to a radiation dose of as high as 62,000 electrons nm^{-2} (10,000 C m^{-2}), and such high doses are to be avoided. The least that one can achieve in biological microscopy is minimum radiation conditions corresponding to $n_0 \simeq 1,000$ electrons nm^{-2} (160 C m^{-2}), when the specimen is only exposed to the electron beam for the time required to take a photograph. Low (or subminimal) dose images correspond to ~100 electrons nm^{-2} (16 C m^{-2}) when, because of the low signal to noise ratio in the image, some type of image processing is required (Unwin and Henderson, 1975). The two electron doses used in the calculations in Table 4.2 correspond to low and minimal radiation conditions.

We shall show a sample calculation in detail so that it will be evident how the results in Table 4.2 are derived. For specimen (a), 5 nm of unstained biological material ($Z = 6$), the respective contributions from the unscattered and elastically and inelastically scattered electrons to the image are ($\alpha = 0.01$ rad, $E_0 = 100$ keV): $U = 0.902$, $E(\alpha) = 0.009$, $I(\alpha) = 0.053$ for an incident beam of unit intensity. Now in bright-field microscopy, the elastic signal is proportional to 2ϕ, that is, $2\sqrt{E(\alpha)}$ [since ϕ^2 is effectively the intensity of the elastic scattering $E(\alpha)$]. The background term in bright-field is the total fraction of the incident beam that is transmitted $T(\alpha) = U + E(\alpha) + I(\alpha) = 0.964$; $C \simeq 2\sqrt{E(\alpha)}/T(\alpha) = 0.19/0.964 = 20\%$.

The signal to noise ratio is calculated as follows: the electron noise in the image depends on the square root of the total number of electrons detected; if the incident electron dose is n_0 (measured in electrons per nm^2 at the specimen) the total number of electrons detected is $n_0 T(\alpha)$, and the corresponding electron noise is $N = \sqrt{n_0 T(\alpha)}$ (assuming 100% detection efficiency); with $n_0 = 100$ (low dose image) $N = \sqrt{96.4}$. The signal in bright-field is proportional to $2\sqrt{E(\alpha)}$ for unit intensity and $S = 2n_0\sqrt{E(\alpha)}$ for an intensity of n_0. Thus for $n_0 = 100$ (no substrate) $C = 20\%$ and $S/N = 1.9$. Increasing the electron dose to $n_0 = 1,000$ (minimum irradiation), the contrast is, of course, unchanged, but S/N increases to 6.1.

Now given that the image contrast is high enough (>5% for visual detection), the minimum signal to noise ratio that is acceptable for image interpretation is ~2–5, depending on the nature of the specimen; for example, with single isolated atoms a larger S/N would be required than for a group of heavy atoms in a particular geometrical arrangement, or for an isolated protein molecule, S/N would have to be significantly greater than for a macromolecule with a regular arrangement of protein subunits. Thus, S/N expresses the confidence in a particular structure that has been selected from the image field; a low S/N indicates a low probability that the structure can be distinguished from the noise, while a high $S/N \simeq 5$ gives a greater than 98% probability that a genuine structure can be selected from a large field in the image.

In dark-field microscopy the signal is proportional to ϕ^2, so that the elastic signal now depends on $E(\alpha)$ and the background $T(\alpha)$ excludes the unscattered contribution. For the example given above, $C = E(\alpha)/T(\alpha) = 15\%$, and for $n_0 = 100$ the total number of electrons detected (per nm^2 area of specimen) is 6.2; hence $S = 0.9$, $N = \sqrt{6.2}$, giving $S/N = 0.36$, which is significantly less than in the corresponding bright-field image. In order to increase S/N to the bright-field S/N, the total number of electrons incident on the specimen has to be increased by a factor of about 30, so that S increases to 27, while N only increases to $\sqrt{186}$. A similar result is obtained for $n_0 = 1,000$, when the S/N in dark-field microscopy is still well below the level required for reliable structure determination.

Dark-field microscopy of unstained biological macromolecules $\sim$5 nm in thickness is not clearly better than bright-field microscopy, since the image contrast is comparable and the signal to noise ratio in the dark-field image can only be improved by a substantial increase in the radiation dose to the specimen. The situation becomes even less favorable in dark-field when the contribution to the image from a carbon substrate is added to the background. For a substrate thickness greater than $\sim$2 nm, not only does the dark-field contrast fall below the visual threshold, but the signal to noise ratio becomes significantly less than unity, giving a very small probability that an observer can select a genuine structure from the noisy image. For this reason Table 4.2a also gives the contrast and S/N improvement that could be expected from first removing all the inelastic scattering from the dark-field image by energy selection, and second eliminating 70% of the inelastic contribution by image subtraction. These results are shown only for $n_0 = 1,000$ because lower radiation doses give an unacceptably low S/N value. With the reduction in the inelastic background, contrast is significantly enhanced, and the signal to noise ratio is just acceptable, provided that the substrate is very thin (1-2 nm). The examination of unstained specimens thicker than 5 nm in dark-field is unlikely to gain significantly over bright-field image contrast, for the inelastic scattering will increase relative to elastic scattering as the specimen thickness is increased (Ottensmeyer and Pear, 1975).

With the second type of specimen (Table 4.2b), namely a small group of heavy atoms on substrate, the dark-field image contrast is significantly better than in bright-field, provided that the substrate thickness is less than 2 nm. This improvement in contrast in dark-field is due to the increase in elastic scattering $[E(\alpha) = 0.020]$ relative to the inelastic scattering $[I(\alpha) = 0.011]$ for large Z. The signal to noise ratio in dark-field is acceptable for an electron dose of $n_0 = 1,000$ electrons nm^{-2}, and only a factor of 7 increase in electron dose is required to obtain a S/N as high as obtained in bright-field microscopy. A further gain in contrast and S/N can be obtained by a reduction of the inelastic contribution to the image.

The results in Table 4.2 show that for heavy atom labeling of molecules dark-field microscopy has a considerable advantage in image contrast, with or without the reduction in inelastic background. However, the advantages of dark-field microscopy for the examination of unstained biological macromolecules are not evident, unless the substrate thickness is very small and additionally some effort is made to reduce the inelastic contribution to the image. As stated above, the results in Table 4.2 do not include the effects of lens aberrations and defocus; they were omitted in order to simplify the evaluation of image contrast for the different types of specimen. Lens aberrations would reduce the contrast figures in Table 4.2 by a factor of ~1.5–3.0, depending on the precision with which optimum defocus can be determined; contrast and S/N figures depending on C_s and Δf would make Table 4.2 even more complicated, and a simple comparison between bright-field and dark-field images would not be possible.

In the resolution of single heavy metal atoms in simple organo-metallic compounds both bright-field (Formanek *et al.*, 1971; Baumeister and Hahn, 1973; Hahn and Baumeister, 1973; Parsons *et al.*, 1973) and dark-field images (Henkelman and Ottensmeyer, 1971; Thon and Willasch, 1972; Whiting and Ottensmeyer, 1972; Hashimoto *et al.*, 1973; Ottensmeyer *et al.*, 1973) have been equally convincing; however, for this type of specimen, radiation damage may not be a severe limitation because of the stability of the specimen as compared with a normal biological material.

It is evident that the choice of optimum elastic defocus is more objective in bright-field microscopy, and that scattering effects from the substrate are not so severe as in dark-field microscopy. However, to date there are no published bright-field images of unstained biological macromolecules that are comparable with the dark-field images of the proteins myokinase and protamine (Ottensmeyer *et al.*, 1975) or even showing the detailed core structure of the ferritin molecule (Massover and Cowley, 1973; Massover *et al.*, 1973). The reservations about the interpretation of dark-field images in terms of the specimen structure (Massover and Cowley, 1973; Beer *et al.*, 1975; Cowley, 1976) are to some extent offset by the definitive results obtained for test specimens, particularly for hollow cone illumination (Thon and Willasch, 1972).

In many cases the dark-field images do seem to represent a projection of the atom density, although, in practice, for heavy atom–labeled molecules there may be some ambiguity in distinguishing between the high density (heavy atom) and low density (holes or clefts in the molecules) areas of the specimen. Contrast in dark-field microscopy can be improved by using crystalline substrates which scatter electrons only at certain discrete angles (Bragg diffraction), and the effect of the substrate can be virtually eliminated by tilting the incident beam so that the scattering from the crystal is stopped by the objective aperture (Hashimoto *et al.*, 1973); the preparation of such thin crystals with the correct orientation on the specimen grid is not without its frustrations.

The least that can be done to facilitate dark-field microscopy is to prepare thin carbon substrates, 1–2 nm in thickness, with a smooth structure (Whiting and Ottensmeyer, 1972), although perforated plastic films may be needed to support such thin carbon films. In addition, image subtraction applied to dark-field images can significantly improve image contrast in unstained and heavy atom-labeled macromolecules and therefore favor the use of high resolution dark-field microscopy in preference to bright-field microscopy for these types of specimens.

The final item in this section concerns the use of phase plates in the objective aperture plane to enhance image contrast. This is an additional imaging mode to the dark-field and bright-field techniques discussed above, and we should briefly discuss the merits of phase plates with respect to image subtraction. The simplest form of phase plate comprises a thin carbon film placed on the normal objective aperture; the film has a small hole (5 μm diameter) in the center to allow the unscattered beam to be transmitted unaffected by the phase plate while the scattered electrons are phase-shifted by transmission through the phase plate (Thon, 1971). This technique is analogous to the use of a Zernike phase plate in light microscopy for enhancing phase contrast. The thickness of the carbon film phase plate is chosen to give a $\pi/2$ phase shift in the scattered electron beam, and using Eq. (4.10) with $\phi \simeq \pi/2$, $t \simeq$ 20-25 nm with a mean value for the potential V of 8-10 volts (Jonnson and Parsons, 1973).

In the diffraction plane a phase of $\pi/2$ is added to the scattered wave $\Phi(\nu)$ in Eq. (4.13) but not added to the central beam represented by $\delta(\nu)$. In the subsequent analysis this changes $\sin \gamma$ to $\sin(\gamma + \pi/2) = \cos \gamma$, and the bright-field image intensity is now:

$$j(\mathbf{r}_i) = 1 + 2[\phi(\mathbf{r}_i) - \phi_m] * q'(\mathbf{r}_i) \qquad (4.26)$$

that is, the resolution and contrast depend on the behavior of $q'(\mathbf{r}_i)$ with changing defocus, or the image transform $J(\nu) = \delta(\nu) + 2\Phi(\nu) \cos \gamma$ depends not on $\sin \gamma$ as in conventional bright-field microscopy but on $\cos \gamma$. Now for small spatial frequencies $\cos \gamma \simeq 1$, whereas $\sin \gamma \simeq 0$. Thus, the phase plate certainly enhances the contrast of the lower spatial frequencies; however, since $\cos \gamma$ and $\sin \gamma$ can both be made nearly constant at larger spatial frequencies (>0.5 nm^{-1}), there is very little gain in contrast for high resolution detail. In fact for the optimum defocus for a phase plate image (~ -40 nm) the maximum spatial frequency that $\cos \gamma$ can be extended to is 2.5 nm^{-1} (0.4 nm resolution) as compared with the first zero in $\sin \gamma$ at ~ 3.0 nm^{-1} (0.3 nm resolution for $\Delta f = -100$ nm).

However, there are two clear disadvantages of the phase plate: first, the central hole in the phase plate cannot easily be made smaller than 5 μm, which means that spatial frequencies from 0 up to ~ 0.3 nm^{-1} (3 nm resolution) are not phase-shifted, and the contrast is still determined by the behavior of $\sin \gamma$. Hence this type of phase plate improves image contrast in only a rather restricted resolution range, ~ 1-3 nm. The second disadvantage is the more serious, namely, the additional scattering of electrons in the phase plate, which will give a further background in the image; for example, with a carbon phase plate of 20 nm thickness 23% of the electrons are scattered inelastically and 11% scattered elastically. The effect of this scattering in the phase plate can be reduced by an additional aperture below the objective (Johnson and Parsons, 1973), but with this modification the number of elastically scattered electrons contributing to the image is reduced by about 34%. With these limitations it is possible to recommend the use of a carbon phase plate for only a restricted resolution range near 2 nm; the main application would then seem to be for the examination of negatively stained specimens ($\sim$1.5 nm resolution) and perhaps thin sections, either stained or unstained. There seems to be no gain in using this type of phase plate if the resolution in the specimen is better than 1 nm.

The second type of phase plate is electrostatic, namely, a thin electrically charged thread across the center of the objective aperture, which has none of the disadvantages of the carbon film phase plate (Unwin, 1972). The electron illumination is carefully adjusted so that the central thread reaches a stable electrical charge which generates an electrostatic field in the region of the objective aperture that compensates for the spherical aberration defect (Unwin, 1972); this gives for a frequency range 0.5-2 nm^{-1} (2-0.5 nm resolution) a wave aberration function that is almost$\simeq$ ideal for phase contrast, that is, $\gamma(\nu) \simeq \pi/2$ for $\Delta f = 0$.

The type of images obtained from an electrostatic phase plate is not easy to explain, because a certain fraction (which is variable) of the unscattered and inelastically scattered electrons is stopped by the central thread so that image formation is neither dark-field nor bright-field; in addition the central thread causes diffraction effects in the image (Unwin, 1972). However, for the biological microscopist the interesting result is that this type of phase plate can be used to enhance the contrast of the biological material and decrease the contrast from the negative stain.

Using negatively stained TMV stacked disk protein, Unwin (1972) shows phase plate images clearly representing the actual biological structure to a resolution of $\sim$1 nm, while the normal bright-field image shows contrast only from the negative stain, which does not accurately represent the structure of the biological specimen beyond a resolution of 2 nm. So while the negative stain may give a high resolution representation of the actual morphology of the biological specimen, it does seem to act as a support for the biological structure to at least a resolution of 1 nm. No form of conventional microscopy, even if used with image subtraction, can achieve such a relevant result for the biologist using negatively stained specimens.

In order to explain the loss in contrast of the stain with an enhancement of contrast for the biological material, we consider a model for the specimen, where the stain scatters electrons strongly while the biological material still satisfies the weak phase approximation. The electron wave for electrons scattered by the stain is:

$$\psi_s(\mathbf{r}) = 1 + i\phi_s(\mathbf{r}) - \epsilon_s(\mathbf{r}) \qquad (4.27)$$

where the "absorption" term, ϵ_s, includes the effects of large phase shifts (terms in ϕ_s^2 when ϕ_s is not significantly less than unity—see Lenz, 1971) and the actual attenuation of the incident beam of unit amplitude by scattering (unscattered amplitude <1). The electron wave for the biological material still satisfies the weak phase approximation, $\phi_b \ll 1$:

$$\psi_b(\mathbf{r}) = 1 + i\phi_b(\mathbf{r}) \qquad (4.28)$$

In both Eqs. (4.27) and (4.28) it is assumed that the amplitude of the un-scattered beam is unity so as to simplify the following analysis, although in prac-tice this amplitude can be significantly less than unity (Unwin, 1972).

The contributions to the image intensity from stain and biological material can be calculated using the analysis leading to Eq. (4.19) for the normal bright-field image:

$$j_s(\mathbf{r}_i) = 1 + 2\phi_s(\mathbf{r}_i) * q(\mathbf{r}_i) - 2\epsilon_s(\mathbf{r}_i) * q'(\mathbf{r}_i) \tag{4.29}$$

and

$$j_b(\mathbf{r}_i) = 1 + 2\phi_b(\mathbf{r}_i) * q(\mathbf{r}_i) \tag{4.30}$$

where now q and q' are the resolution functions for the phase plate. Since the Fourier transform $\sin\gamma$ and $\cos\gamma$ of q and q', respectively are approximately $\sin\gamma = 1$ and $\cos\gamma = 0$ (with $\Delta f = 0$) for spatial frequencies $\nu = 0.5\text{-}2 \text{ nm}^{-1}$, high resolution information in ϕ will be imaged with maximum contrast. Thus in the case of the biological material, Eq. (4.30), $j_b(\mathbf{r}_i) \simeq 1 + 2\phi_b(\mathbf{r}_i)$ attains its maximum value for the resolution range 0.5-2 nm.

If the negative stain shows no detail in this resolution range (that is, $\Phi_s(\nu)$ and $E_s(\nu)$ are zero or small for $\nu > 0.5 \text{ nm}^{-1}$), then the only contribution to image contrast from the negative stain arises from $\nu < 0.5 \text{ nm}^{-1}$ (>2 nm resolu-tion); in this case the two terms in Eq. (4.29) may cancel out because $\Phi_s \sin\gamma \simeq E_s \cos\gamma$ with γ increasing from 0 at $\nu = 0$ (when $\sin\gamma \simeq 0$, $\cos\gamma \simeq 1$) to $\pi/2$ (when $\sin\gamma = 1$, $\cos\gamma = 0$). This type of contrast enhancement of the biological material with respect to the negative stain is facilitated by reduction in the am-plitude of the unscattered beam by the central thread of the phase plate.

Under certain conditions the electrostatic phase plate can produce a contrast reduction for low resolution detail (>2 nm) and a contrast enhancement of high resolution detail down to 0.5 nm. This type of contrast enhancement seems to work only if the low resolution material scatters electrons strongly, with the example in this case being the negative stain. If both high and low resolution components of the specimen scatter weakly or scatter strongly, this type of phase plate may not achieve such remarkable results. Of course, in conven-tional bright-field imaging we can choose a defocus Δf in the expression for γ, Eq. (4.22), such that the ϕ_s and ϵ_s terms in Eq. (4.29) do effectively cancel out, but unfortunately this defocus, approximately 130 nm overfocus, is far from the optimum defocus for spatial frequencies greater than 0.5 nm^{-1} (2 nm resolution).

The electrostatic phase plate will probably offer no advantage over conven-tional microscopy for the examination of unstained specimens, for although the contrast of the biological material will be significantly enhanced, so will the con-trast from the carbon substrate. Similar considerations apply to heavy atom-labeled macromolecules, where the phase contrast enhancement for the heavy atoms and macromolecules may be offset by an increase in background from the substrate; note that in the case of single or small groups of heavy atoms the competing effects in Eq. (4.29) will be absent because ϵ_s will be much less than unity, and the phase contrast term ϕ_s will dominate. However, this disadvantage may be offset by the fact that the phase plate produces the ideal phase contrast function, $\sin \gamma \simeq 1$, in a resolution range 0.5-2.0 nm.

IMAGE SUBTRACTION FOR LOW RESOLUTION IMAGES

As the thickness of the specimen increases some of the elastically scattered electrons may be additionally inelastically scattered, with a resultant increase in the ratio of inelastic to elastic contribution to the image (Crick and Misell, 1971; Burge, 1973). This increase in inelastic scattering has two effects: (1) phase contrast effects from the elastic scattering are reduced; (2) elastically scattered electrons which are in addition inelastically scattered are now affected by chromatic aberration. The overall result of increasing specimen thickness is an increase in the angular spread of the transmitted electron beam, because now electrons are scattered an average of two or three times in the specimen. Because of this plural scattering of electrons, image resolution and contrast may no longer be limited only by lens aberrations but also by the spreading of electrons in the specimen.

There is also a basic resolution limit in thick biological specimens due either to specimen preparation (embedding and staining for sections) or superposition of detail from the top and bottom of the specimen, such as for a whole virus particle (Henkelman and Ottensmeyer, 1974; see also Fig. 4.3); in the latter case the detail in the specimen may actually be preserved to 1-2 nm resolution, but in general it is not possible to separate superimposed detail unless the particle has some high symmetry properties. In this low resolution situation (>2 nm) there is no adequate theoretical treatment of image formation.

Scattering contrast, considering only intensities of elastic and inelastic scattering, will be a valid description of image formation when all electrons are inelastically scattered; in this situation there can be no phase contrast effects from the elastically scattered electrons because they will have been scattered inelastically (incoherently) at least once. Incoherent imaging conditions occur for specimen thicknesses greater than 100 nm. However, even for thinner specimens, particularly if structural detail is only preserved at 2 nm resolution, phase contrast effects are very weak. In fact, if we repeat the calculations of the elastic image profiles with $\nu_{max} = 0.5$ nm^{-1} (2 nm resolution) using the weak phase approximation, we find that the optimum defocus increases to $\sim$-400 nm with the phase contrast in bright-field reduced by a factor of 5-10. Consequently, at low resolution inelastic scattering limits image contrast to a greater extent than at 0.5 nm, where bright-field phase contrast dominates. Additionally, as the specimen thickness increases, a substantial fraction of the inelastically scattered electrons will also be elastically scattered; the angular spread of the inelastic scattering will increase with a consequent increase in chromatic aberration, which depends on θ in Eq. (4.1), $r = C_c E \theta / E_0$.

We will present in this section calculations showing how inelastic scattering increases with specimen thickness, and show how the angular spread of the transmitted electron beam affects image resolution, with a comparison of the resolution limit due to the chromatic aberration in the inelastic image. These calculations are made using the scattering contrast model with a geometrical evaluation of the image resolution; provided phase contrast effects are not important, these results should give an indication of the factors limiting image resolution and contrast for biological sections and large particles, such as whole viruses. Contrast enhancement by image subtraction cannot be based on the previous calculations because it was assumed that the specimen was thin enough for the incident electrons to be scattered either elastically or inelastically; in addition it was assumed that the elastic image contained genuine information on the structure of the specimen to a resolution of 0.5 nm ($\nu_{max} = 2$ nm^{-1}). We will attempt to assess the prospects for the enhancement of contrast by image subtraction for low resolution specimens.

Figure 4.21 shows the relative proportions of inelastic (I) and elastic (E) scattering in different materials from carbon (corresponding to an unstained section) to gold (corresponding to a heavily stained specimen). These results give an indication of how the total inelastic scattering increases relative to the elastic scattering as the specimen thickness increases. Since these results are for no objective aperture, it is evident that the ratio I/E will increase even further when an objective aperture is used. This is a slight oversimplification because elastic scattering will broaden the angular distribution of the inelastically scat-

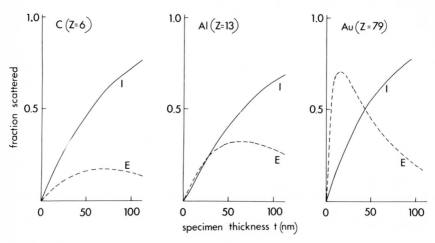

Fig. 4.21 The relative proportions of elastic (E) and inelastic (I) scattering in different materials as the specimen thickness increases (no objective aperture): (a) carbon, (b) aluminum, (c) gold. $E_0 = 100$ keV.

tered electrons, but generally an objective aperture will differentially enhance the inelastic contribution to the image.

A second effect of increasing specimen thickness is the significant reduction in the unscattered component, which means that the bright-field phase contrast cannot be calculated from the weak phase approximation. In fact, because of this attenuation of the incident beam, the elastic image, arising from the interference of the unscattered and elastically scattered electrons, has a reduced contrast. The results in Fig. 4.21 give some indication of the specimen thickness limit above which phase contrast will be canceled by "absorption" effects. Referring to Eq. (4.27), the attenuation of the amplitude of the unscattered electron beam is approximately $(1 - \epsilon_s)$, with a corresponding intensity reduction of $(1 - \epsilon_s)^2$. For 20 nm of carbon $(1 - \epsilon_s)^2 = 0.74$, giving $\epsilon_s \simeq 0.14$, which is significantly less than unity. However, for $t = 50$ nm $(1 - \epsilon_s)^2 = 0.36$, $\epsilon_s \simeq 0.40$; such a large absorption term can effectively cancel the phase contrast term in Eq. (4.29), particularly if the structure is preserved only to relatively low resolution. Even if the cancellation of the phase contrast can be offset by choosing an optimum defocus, the interpretation of an image, arising from a combination of phase and absorption effects, is not going to be simple.

In thick specimens image resolution may be limited by the spread of the electron beam at the exit face of the specimen. For an angle of scattering θ, the spread of the electron beam at the exit face will be $d = 2\theta t$ (geometrical approximation). As a measure of this quantity we have chosen θ corresponding to the semi-angle $\alpha_{1/2}$ (rad), which contains 50% of the scattered electrons for both inelastic and elastic scattering (Misell, 1973c). This is an arbitrary definition for the image resolution as limited by beam broadening, and the actual profile of the beam should be calculated using the actual angular distributions; however, we would still have the problem of defining the resolution in terms of the half-width or some other measure of the width of this profile.

Figure 4.22 shows the variation of $\alpha_{1/2}$ for carbon, aluminum, and gold with increasing specimen thickness. In general the value of $\alpha_{1/2}$ for inelastic scattering is smaller than $\alpha_{1/2}$ for elastic scattering, but the differences are small for thick specimens (when the inelastic beam is broadened by further elastic scattering). For gold the value of $\alpha_{1/2}$ increases very rapidly with thickness, indicating that beam spreading will limit resolution for heavily stained specimens at a smaller specimen thickness than for unstained specimens. Figure 4.23 shows how $d_{1/2} = 2\alpha_{1/2}t$ increases with specimen thickness of carbon. Evidently, for thicknesses less than 50 nm beam spreading for both elastic and inelastic scattering is less than 0.5 nm, and is comparable to the resolution limit due to spherical aberration. However, $d_{1/2}$ increases as t^2, so that for an unstained specimen of thickness 100 nm the elastic image has a resolution significantly greater than 2 nm, well above the spherical aberration limit. Stained specimens will reach this resolution limit for thinner specimens.

The inelastic image resolution is not, however, limited by beam spreading but

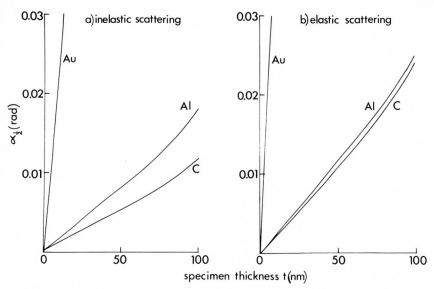

Fig. 4.22 The variation of the semi-angle $\alpha_{1/2}$ (rad), which includes 50% of the (a) inelastically and (b) elastically scattered electrons for carbon, aluminum, and gold with increasing specimen thickness. $E_0 = 100$ keV.

by chromatic aberration. As a measure of the chromatic aberration defect we use Eq. (4.1), $r = C_c E \theta / E_0$ (Marton et al., 1955; Cosslett, 1956), with $\theta = \alpha_{1/2}$ and $E = E_{mp}$, the most probable energy loss (~25 eV for biological materials). The variation of the chromatic aberration with specimen thickness is shown in Fig. 4.23 (curve C.A.). Clearly the chromatic defect determines the inelastic image resolution. These results are slightly lower than Cosslett's (1956) estimates of chromatic aberration based on the mean energy loss but about the same as the results based on the energy half-width of the energy loss distribution of the inelastically scattered electrons (Cosslett, 1969). For thick specimens with a resolution limit of perhaps 2 nm there is a similar situation to that at high resolution. The final image comprises an elastic image, limited in resolution by specimen preservation and beam spreading, superimposed on a large inelastic background of low resolution. Image contrast will again be limited by inelastic scattering, although at a lower resolution.

The relative contributions of the elastic and inelastic scattering to the final image will depend on specimen thickness. For thin sections chromatic aberration will not limit image resolution because the elastic image, enhanced by phase contrast at optimum defocus, will dominate the total image. However, for thick sections (50–100 nm), the elastic image with reduced phase contrast will be submerged in a low resolution inelastic image. The results for the chromatic defect shown in Fig. 4.23 should serve only as a guide because the chromatic defect should properly be calculated from a detailed knowledge of the angular

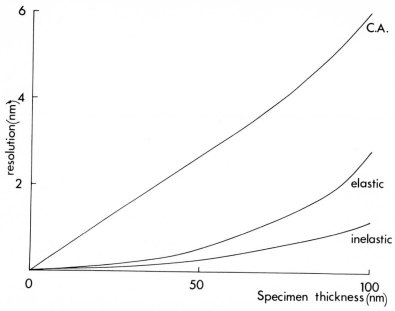

Fig. 4.23 The image resolution as limited by beam spreading for inelastic and elastic scattering, $d_{1/2} = 2\alpha_{1/2}t$, with increasing specimen thickness, t (nm). The resolution limit due to chromatic aberration (C.A.) in the inelastic image is also shown. $E_0 = 100$ keV, $C_c = 2$mm.

and energy distributions of the inelastically scattered electrons (Crick and Misell, 1971; Nagata and Hama, 1971; Dupouy *et al.*, 1973); these distributions would, however, seldom be available for biological specimens.

For the type of specimen considered in this section, image subtraction can be used to enhance image contrast, but under slightly different conditions to those for high resolution. First, we disregard the exceptional cases of high resolution detail superimposed on a large inelastic background (Nagata and Hama, 1971); in the electron microscopy of sections and large particles, it is unlikely that specimen substructure will be preserved to a better resolution than 1–2 nm.

For the elastic image there are two definite conditions for maximum image contrast: (1) phase contrast at low resolution, when ϕ is much greater than ϵ, for fairly thin (20–50 nm) unstained biological specimens; (2) "absorption" contrast at low resolution, when ϵ is much greater than ϕ, for thick unstained specimens (>50 nm) or thin stained specimens (20–50 nm). In the first case, we consider in Eq. (4.29) the maximization of the $\phi * q$ term, assuming that the resolution is 1–2 nm ($\nu_{max} = 1$–0.5 nm^{-1}); this gives an optimum elastic defocus of ~ -200 to -400 nm and a phase contrast term that is several times lower than phase contrast at high resolution. In the second case, $\epsilon \gg \phi$, we maximize the $\epsilon * q'$ term in Eq. (4.29), and this gives an optimum defocus of 0 to -200 nm for the resolution range 1–2 nm; the optimum defocus in this case is not very critical because cos γ, the transform of q', $\simeq 1$ at low spatial frequencies for a large defocus range. Of course, in intermediate situations when the phase and "absorption" terms in Eq. (4.29) are comparable, we cannot give a precise figure for optimum defocus giving maximum image contrast; observation of maximum contrast will be in the defocus range of 0 to -400 nm, but the interpretation of the elastic image will be a problem.

While we are achieving optimum imaging conditions for the elastic image, it will be impossible, because of the large energy spread of the inelastic scattering in thick specimens, to find a single optimum defocus for the inelastic image; optimum inelastic defocus will be about -500 nm for 25 eV loss electrons and $-1,000$ nm for 50 eV loss electrons (Fig. 4.5). In order to implement image subtraction at low resolution we suggest that the condition for maximum contrast be determined by observation of the biological specimen, while the second image is taken some 200 nm further underfocus. The elastic image taken some 200 nm from its optimum defocus should show only minimal contrast; the inelastic image profile will of course vary over this defocus range of 200 nm or more, but the differential effect that was used at high resolution should still be valid.

We cannot expect as large a contrast enhancement as obtained at high resolution for two reasons: (1) the inelastic contribution to the image is relatively larger; (2) the elastic image contrast is lower than that obtained for a weak phase object. In fact, image subtraction cannot work for thick unstained sections where the image under any defocus conditions is virtually an inelastic image, because all the elastic scattering has been depleted by inelastic scattering in the specimen. In this situation neither the energy-selecting electron microscope (Henkelman and Ottensmeyer, 1974) nor the scanning transmission electron microscope (Crewe, 1971) could produce a significant contrast enhancement; the latter technique would fail because the angular discrimination, used to separate inelastic and elastic scattering in thin specimens, would be unable to discriminate between any elastic scattering and a significant proportion of large-angle inelastic scattering.

Figure 4.24 shows an example of image subtraction at low resolution for dark-field images of mouse muscle section (Krakow and Siegel, unpublished result). Figure 4.24a shows the image at optimum elastic defocus, where the muscle fibers show poor contrast in comparison with the sarcoplasmic reticulum. An image taken several hundred nm further underfocus (Fig. 4.24b) shows an image with virtually no structural detail in the muscle fibers, although the sarcoplasmic reticulum can still be seen. The subtraction is effected photographically by superimposing the photographic negative of (b) (Fig. 4.24c) with the optimum defocus image (Fig. 4.24a); the final result (Fig. 4.24d) shows the muscle fibers with improved contrast, while the sarcoplasmic reticulum shows reversed con-

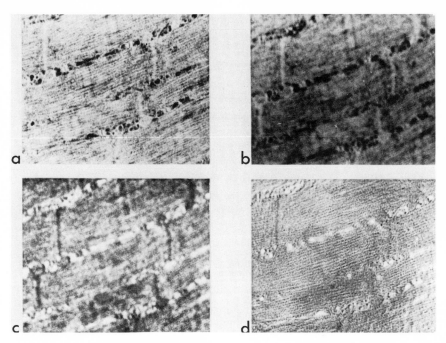

Fig. 4.24 Image subtraction at low resolution for dark-field images of sectioned mouse muscle: (a) optimum defocus, (b) several hundred nm underfocus, (c) photographic reversal of (b), and (d) photographic superposition of (a) and (c) after alignment, $E_0 = 80$ keV, $M = 30,000$. (*Courtesy of Dr. W. Krakow.*)

trast from the original image (Fig. 4.24a). While the original mouse muscle preparation may not be the best that can be achieved by the biologist, there is a marked improvement in the image clarity after image subtraction.

CONCLUDING REMARKS

A considerable amount of space in this review has been given to basic image formation in the transmission electron microscope, particularly the description of image formation by inelastically scattered electrons. For most purposes inelastic scattering and its role in image formation are treated very briefly; here we have examined how inelastic scattering limits image contrast, and we have suggested a particular method for enhancing image contrast by the reduction of this inelastic contribution. Image subtraction, using a high resolution elastic image superimposed on a low resolution inelastic image, is considered to be a method of contrast enhancement, rather than a method for improving image resolution. As an additional bonus it may also be possible to see higher resolution detail than is observable in the original image. However, considerable care should be taken in the interpretation of apparent high resolution detail.

The ultimate resolution observable in the images of biological specimens may be near to the resolution of the microscope (0.3 nm or better), but we should be interested only in image detail that truly reflects the structure of the biological specimen. Thus although we may resolve in negatively stained specimens particles as small as 1 nm, these particles of stain may be in no way related to the original biological structure. Similarly, if we examine an unstained protein molecule, we may see in the image substructure such as clefts or holes corresponding to a resolution of 0.5 nm; however, if the protein molecule has been damaged to the extent that the ashes no longer represent the true morphology of the original molecule, then the statement that the image resolution is 0.5 nm ΠΛs no biological significance (Beer *et al.*, 1975).

Contrast enhancement by image subtraction is based upon the assumption that an electron microscope image comprises two components, namely, an elastic image which can represent the object structure to high resolution and an inelastic image reflecting the object structure only at a lower resolution. In the ideal situation where the specimen structure is preserved to high resolution, there is no doubt that image subtraction can lead to more information on the structure. However, care must be taken in the interpretation of the different image obtained by image subtraction if, for example, the specimen is damaged to such an extent that the high resolution detail represents radiation damage fragments only. To emphasize this point we have tried image subtraction on the groups of nine hexons from adenovirus; the result of subtraction was certainly the contrast enhancement of some fine detail, but this detail was quite unrelated to either of the original images. We can have some confidence in the results from image subtraction if the high resolution detail consistently relates to the overall shape of the molecule.

Currently the application of image subtraction at high resolution is represented by only one example, namely, uranium-labeled DNA (Welles *et al.*, 1974; Krakow *et al.*, 1976) where there is a significant improvement in contrast of the uranium atoms; the DNA structure, however, is not at all evident. It is hoped that this result will encourage the biologist to collaborate with the physicist to try this method using the best images that can be obtained from good specimen preparations. Less than the best in specimen preparation and in image recording may only give, after image processing, a result that could have been obtained from a good specimen under optimum imaging conditions (minimal radiation and accurate focusing).

Image subtraction will be particularly effective in the contrast enhancement of (1) heavy atoms used to label macromolecules and (2) unstained biological molecules such as enzymes and virus protein subunits. There are some reservations about its application to negatively stained specimens, where the image contrast is dominated by the negative stain; the negative stain in many cases may not be an accurate representation of the morphology of the biological molecule, although in the image negative stain particles as small as 1 nm may be resolved. Image subtraction will achieve maximum gain in dark-field microscopy, where the background in the image is due mainly to inelastic scattering in the specimen and substrate.

In bright-field microscopy phase contrast effects, arising from the interference of the elastically scattered and unscattered electrons, give an image with a modulated elastic signal on a large background due to unscattered electrons. Calculations for both unstained and stained specimens show that the removal of 70% of the inelastic image contribution by image subtraction improves dark-field image contrast to an extent which recommends this imaging mode in preference to bright-field microscopy in terms of contrast alone. There are, however, some additional problems in dark-field microscopy, namely, image interpretation (Cowley, 1976), an increase in radiation damage over bright-field microscopy for a given signal to noise ratio in the image (Beer et al., 1975), and focusing to within ±40 nm of optimum elastic defocus (Cowley et al., 1974). The use of hollow cone illumination should improve image interpretation, and provided that we can accept lower signal to noise values (2 for unstained specimens, 4 for heavy atom-labeled macromolecules), there will be no increase in radiation damage.

The obvious disadvantage of image subtraction is the requirement for two images, when a second image recorded will represent a further degradation in the structure of the specimen. This problem can be avoided if the biological structure is periodic (Unwin and Henderson, 1975) or lies on the substrate in a well-defined orientation, as is the case for multiple subunit structures such as the groups of nine hexons; in these cases the two images can be recorded from different areas of the specimen grid, although we must expect focus gradients when the two areas are separated by a grid square or more. In the case of heavy atom-labeled macromolecules radiation damage does not seem to be quite as severe a problem, provided that we expect to resolve only the heavy atoms and not the macromolecule itself; this is a satisfactory situation if the specific labeling by heavy atoms is used to resolve the base sequence in, say, DNA.

In comparison with other methods of contrast enhancement, image subtraction is preferred to the carbon phase plate (Johnson and Parsons, 1973), principally because of the loss in image contrast by scattering in the phase plate; the competition with the electrostatic phase plate (Unwin, 1972) is somewhat more severe, particularly for negatively stained specimens. Photographic contrast enhancement (Farnell and Flint, 1970) is easier to use than image subtraction; however, the enhancement of contrast by photographic methods is a nonlinear process, emphasizing the high density areas while suppressing the low density regions of the image, so that image interpretation may be a problem at high resolution. However, at lower resolution for stained sections this type of photographic enhancement produces results similar to those obtained from energy-selecting microscopes (Henkelman and Ottensmeyer, 1974), enhancing the high density (stained regions) of the specimen which scatter elastically while suppressing the low density regions of the section which scatter inelastically. In image subtraction at low resolution the incorrect choice of defocus could produce image pairs that eliminate all the low resolution image detail on subtraction, leaving high resolution detail which cannot be true to the object structure. We have given approximate defocus values for the application of image subtraction to sectioned specimens or large particles such as viruses, but the precise optimum defocus value will depend on the specimen preservation and thickness.

Specialized instruments such as the energy-selecting electron microscope (Henkelman and Ottensmeyer, 1974) and the scanning transmission electron microscope (Crewe, 1971) have some clear advantages over contrast enhancement by image subtraction. The energy-selecting microscope separates elastic and inelastic images on a simple energy difference basis, and does not rely on the correctness of a theoretical model for removing inelastic scattering from the image. The results to date show a consistent increase in image contrast, but in many cases energy-selected images show no additional structural detail. The energy selecting microscope has not yet been applied to the observation of specimens which may retain structural integrity to high resolution; it is in this latter application that image subtraction may compete with the energy-selecting microscope.

The scanning transmission electron microscope achieves a natural separation of the inelastic and elastic scattering in terms of their angular behavior, provided that the specimen is thin; additionally the inelastic image is not affected by chromatic aberration because there are no lenses after the specimen. In dark-field the scanning transmission microscope is more efficient in electron collection than the conventional transmission microscope because all electrons scattered outside an angle of about 0.01 rad are collected by the annular detector; in the conventional microscope used in dark-field only those electrons scattered within the objective aperture are collected to form the image. The scanning transmission electron microscope would, however, not produce an effective separation of the inelastic and elastic scattering in a fairly thick section (50 nm), for the inelastic electrons will also be scattered into fairly large angles. However, the total (elastic + inelastic) image should show a better resolution and higher contrast than is obtainable in the conventional microscope, as the inelastic image resolution is no longer limited by chromatic aberration (Stroud *et al.*, 1969).

The scanning microscope may not produce a significant contrast enhancement for negatively stained specimens, for the low resolution (small angle, $\theta = \lambda \nu$) elastic scattering (1.5–2 nm) will pass through the center of the annular detector with the inelastically scattered electrons, while the large angle elastic scattering from the stain will contain no structurally significant information. On the whole, image subtraction cannot compete with the results that could be obtained by scanning transmission electron microscopy principally because tnese images are unaffected by chromatic aberration.

As stated at the beginning of this review, if you have access to either a high resolution scanning transmission microscope or an energy-selecting microscope capable of high resolution imaging, then contrast enhancement by image subtraction will be of little interest. This chapter was written for the large number of biological electron microscopists who may not have access to such instruments, and it is hoped that this detailed account of the basis of image subtraction will encourage them to try the method.

I am grateful to Professor M. A. Hayat for the opportunity to contribute to this volume. I should like to thank Janet Misell for preparing the typescript and correcting my errors, and Elaine Brown for her care in drawing all the diagrams.

References

Baumeister, W., and Hahn, M. H. (1973). Electron microscopy of thorium atoms in monomolecular layers. *Nature* **241**, 445.

Beer, M., Frank, J., Hanszen, K.-J., Kellenberger, E., and Williams, R. C. (1975). The possibilities and prospects of obtaining high-resolution information (below 30 Å) on biological material using the electron microscope. *Quart. Rev. Biophys.* **7**, 211.

Burge, R. E. (1973). Mechanisms of contrast and image formation of biological specimens in the transmission electron microscope. *J. Microscopy* **98**, 251.

Castaing, R. (1971). Secondary ion microanalysis and energy-selecting electron microscopy. In: *Electron Microscopy in Material Science* (Valdrè, U., ed.), p. 103. Academic Press, New York.

Colliex, C., and Jouffrey, B. (1970). Images filtrées obtenues avec des électrons ayant subi des pertes d'énergie dues à l'excitation de niveaux profonds. *C. R. Acad. Sci. Paris* **270B**, 673.

Cosslett, V. E. (1956). Specimen thickness and image resolution in electron microscopy. *Brit. J. Appl. Phys.* **7**, 10.

Cosslett, V. E. (1969). Energy loss and chromatic aberration in electron microscopy. *Z. angew. Phys.* **27**, 138.

Cowley, J. M. (1976). The principles of high resolution electron microscopy. In: *Principles and Techniques of Electron Microscopy*, Vol. 6 (Hayat, M. A., ed.). Van Nostrand Reinhold Company, New York and London.

Cowley, J. M., Massover, W. H., and Jap, B. K. (1974). The focussing of high resolution dark field electron microscope images. *Optik* **40**, 42.

Crewe, A. V. (1971). High resolution scanning microscopy of biological specimens. *Phil. Trans. Roy. Soc. Lond.* **B261**, 61.

Crick, R. A., and Misell, D. L. (1971). A theoretical consideration of some defects in electron optical images. *J. Phys. D: Appl. Phys.* **4**, 1.

Dubochet, J. (1973). High resolution dark-field electron microscopy. In: *Principles and Techniques of Electron Microscopy*, Vol. 3 (Hayat, M. A., ed.). Van Nostrand Reinhold Company, New York and London.

Dupouy, G. (1967). Contrast improvement in electron microscopic images of amorphous objects. *J. Electron Microscopy* **16**, 5.

Dupouy, G., Marais, B., and Verdier, P. (1973). La résolution; cas des objets très minces. Rôle de l'aberration chromatique. *C. R. Acad. Sci. Paris* **276B**, 887.

Dupouy, G., Perrier, F., Enjalbert, L., Lapchine, L., and Verdier, P. (1969). Accroissement du contraste des images d'objets amorphes en microscopie électronique. *C. R. Acad. Sci. Paris* **268B**, 1341.

Egerton, R. F., Philip, J. G., Turner, P. S. and Whelan, M. J. (1975). Modification of a transmission electron microscope to give energy filtered images. *J. Phys. E: Sci. Instrum.* 8, 1033.

El Hili, A. (1967). Influence de la difussion inélastique sur la formation des contrastes de diffraction en microscopie électronique. *J. Microscopie* 6, 693.

Farnell, G. C., and Flint, R. B. (1970). Method for increasing the photographic contrast of electron micrographs. *J. Microscopy* 92, 145.

Formanek, H., Müller, M., Hahn, M. H.. and Koller, Th. (1971). Visualisation of single heavy atoms with the electron microscope. *Naturwissenschaften* 58, 339.

Frank, J. (1973). Computer processing of electron micrographs. In: *Advanced Techniques in Biological Electron Microscopy* (Koehler, J. K., ed.), p. 215. Springer-Verlag, New York.

Hahn, M., and Baumeister, W. (1973). Möglichkeiten und Grenzen elektronenmikroskopischer Abbildung einzelner Atome in sub- und supramolekularen Systemen. *Cytobiologie* 7, 224.

Hashimoto, H., Kumao, A., Hino, K., Endoh, H., Yotsumoto, H., and Ono, A. (1973). Visualisation of single atoms in molecules and crystals by dark field electron microscopy. *J. Electron Microscopy* 22, 123.

Henkelman, R. M., and Ottensmeyer, F. P. (1971). Visualisation of single heavy atoms by dark field electron microscopy. *Proc. Nat. Acad. Sci. USA* 68, 3000.

Henkelman, R. M., and Ottensmeyer, F. P. (1974). An energy filter for biological electron microscopy. *J. Microscopy* 102, 79.

Hibi, T., and Takahashi, S. (1971). Relation between coherence of electron beam and contrast of electron image of biological substance. *J. Electron Microscopy* 20, 17.

Isaacson, M., Langmore, J. P., and Rose, H. (1974). Determination of the nonlocalisation of the inelastic scattering of electrons by electron microscopy. *Optik* 41, 92.

Johnson, H. M., and Parsons, D. F. (1969). Enhanced contrast in electron microscopy of unstained biological material. I. Strioscopy (dark-field microscopy). *J. Microscopy* 90, 199.

Johnson, H. M., and Parsons, D. F. (1973). Enhanced contrast in electron microscopy of unstained biological material. III. In-focus phase contrast of large objects. *J. Microscopy* 98, 1.

Krakow, W., Welles, K. B. and Siegel, B. M. (1976). Image processing of dark-field electron micrographs. *J. Phys. D: Appl. Phys.* 9, 175.

Lenz, F. (1971). Transfer of image information in the electron microscope. In: *Electron Microscopy in Material Science* (Valdrè, U., ed.), p. 540. Academic Press, New York.

Marton, L., Leder, L. B., Mendlowitz, H., Simpson, J. A., and Marton, C. (1955). Les pertes d'énergie des électrons et leur rôle en microscopie électronique. In: *Les Techniques Récentes en Microscopie Electronique et Corpusculaire*, p. 175. C.N.R.S., Toulouse.

Massover, W. H., and Cowley, J. M. (1973). The ultrastructure of ferritin macromolecules. The lattice structure of the core crystallities. *Proc. Nat. Acad. Sci. USA* 70, 3847.

Massover, W. H., Lacaze, J.-C., and Durrieu, L. (1973). The ultrastructure of ferritin macromolecules. I. Ultrahigh voltage electron microscopy (1–3 MeV). *J. Ultrastructure Res.* 43, 460.

Misell, D. L. (1971). Image formation in the electron microscope. II. The application of transfer theory to a consideration of inelastic electron scattering. *J. Phys. A: Gen. Phys.* 4, 798.

Misell, D. L. (1973a). On the electron microscopy of biological specimens. *J. Phys. D: Appl. Phys.* 6, L1.

Misell, D. L. (1973b). Image resolution in high-voltage electron microscopy. *J. Phys. D: Appl. Phys.* 6, 1409.

Misell, D. L. (1973c). Some aspects of inelastic and elastic electron scattering in the scanning electron microscope. In: *Scanning Electron Microscopy* (Johari. O., ed.), p. 225. ITT Research Institute, Chicago.

Misell, D. L. (1975). The resolution and contrast in biological sections determined by inelastic and elastic scattering. In: *Physical Aspects of Electron Microscopy and Microbeam Analysis* (Siegel, B. M., and Beaman, D. R., eds.), Chapter 5. John Wiley and Sons, New York.

Misell, D. L., and Atkins, A. J. (1973). Image resolution and image contrast in the electron microscope. III. Inelastic scattering and coherent illumination. *J. Phys. A: Gen. Phys.* 6, 218.

Misell, D. L., and Burge, R. E. (1973). Limitations of image information due to inelastic electron scattering. In: *Image Processing in Electron Microscopy* (Hawkes, P. W., ed.), p. 168. Academic Press, London.

Misell, D. L., and Burge, R. E. (1975). Contrast enhancement in biological electron microscopy using two micrographs. *J. Microscopy* 103, 195.

Nagata, F., and Hama, K. (1971). Chromatic aberration on electron microscope image of biological sectioned specimen. *J. Electron Microscopy* 20, 172.

Ottensmeyer, F. P., and Pear, M. (1975). Contrast in unstained sections: a comparison of bright and dark field electron microscopy. *J. Ultrastructure Res.* 51, 253.

Ottensmeyer, F. P., Schmidt, E. E., and Olbrecht, A. J. (1973). Image of a sulfur atom. *Science* 179, 175.

Ottensmeyer, F. P., Whiting, R. F., Schmidt, E. E., and Clemens, R. S. (1975). Electron microscopy of proteins: a close look at the ashes of myokinase and protamine. *J. Ultrastruct. Res.* 52, 193.

Parsons, J. R., Johnson, H. M., Hoelke, C. W., and Hosbons, R. R. (1973). Imaging of uranium atoms with the electron microscope by phase contrast. *Phil. Mag.* 27, 1359.

Saxton, W. O. (1974). A new computer language for electron image processing. *Computer Graphics Image Proc.* 3, 266.

Stroud, A. N., Welter, L. M., Resh, D. A., Habeck, D. A., Crewe, A. V., and Wall, J. (1969). Scanning electron microscopy of cells. *Science* 164, 830.

Thon, F. (1971). Phase contrast electron microscopy. In: *Electron Microscopy in Material Science* (Valdrè, U., ed.), p. 571. Academic Press, New York.

Thon, F. and Willasch, D. (1972). Imaging of heavy atoms in dark field electron microscopy using hollow cone illumination. *Optik* 36, 55.

Unwin, P. N. T. (1972). Electron microscopy of biological specimens by means of an electrostatic phase plate. *Proc. R. Soc. Lond.* A329, 327.

Unwin, P. N. T., and Henderson, R. (1975). Molecular structure determination by electron microscopy of unstained crystalline specimens. *J. Mol. Biol.* 94, 425.

van Dorsten, A. C. (1971). Contrast phenomena in electron images of amorphous and macromolecular objects. In: *Electron Microscopy in Material Science* (Valdrè, U., ed.), p. 627. Academic Press. New York.

Welles, K., Krakow, W., and Siegel, B. M. (1974). Image processing by computer, heavy-light atom discrimination, non-linear effects, image difference technique. In: *VIIIth International Congress of Electron Microscopy*, *Canberra*, Vol. I, p. 320. Australian Academy of Science, Canberra, A.C.T.

Whiting, R. F., and Ottensmeyer, F. P. (1972). Heavy atoms in model compounds and nucleic acids imaged by dark field transmission electron microscopy. *J. Mol. Biol.* 67, 173.

5. INTERFERENCE PHENOMENON ON OSMIUM TETROXIDE–FIXED SPECIMENS FOR SYSTEMATIC ELECTRON MICROSCOPY

Karl Hermann Andres and Monika von Düring

Institut für Anatomie II, Ruhr-Universität, Bochum, Federal Republic of Germany

INTRODUCTION

After histological specimens are fixed with an aldehyde for optimal preservation of structure, they have such a high molecular density that the penetration by osmium tetroxide and Araldite into them is severely limited. This problem is overcome by using very small (1–2 mm) specimens. However, topographical relationship is lost among very small tissue specimens. As knowledge of cytological and histological structures has grown, so too has grown the question of where these structures are situated with regard to the organ or organism as a whole. This question applies especially to the central nervous system, where topographical relationship of nerve cells and nerve fiber systems and their connections are most important (Andres and v. Düring, 1974).

246

A prerequisite for systematic electron microscopy is fixation with glutaraldehyde by vascular perfusion (Andres, 1967), as well as embedding of complete organ slices for semithin sectioning. Once the osmium tetroxide–fixed tissue slices are embedded, structures on their surface, apart from their overall contour, cannot be distinguished. However, the method presented in this chapter explains how it is possible to see both the cut surface of a section and the natural surface of the organ, as well as to identify its structures. This allows the prepared tissue block to be adjusted to a definite position for sectioning. It is possible therefore to examine a specific area from the embedded organ or organ slice by light and electron microscopy and at the same time to keep its topographical relationship to the whole. This method bridges the gap between the macrostructure and fine structure of an organ. In addition, the method allows a precise removal of the desired specimen out of the organ for electron microscopic analysis. Finally, with the interference reflecting light (IRL) method it is possible to recognize poorly fixed specimens for elimination.

APPARATUS

The apparatus for the IRL method (Fig. 5.1) consists of the following components:

1. Two Zeiss heavy duty microscope lights with metal halide short arc lamps (CSI 250 Watt) (Phillips)
2. Two metal concave mirrors (f. 11 cm)
3. One adjustable turntable
4. One stand with fine adjustment
5. One 35-mm single-lens reflex camera with focusing finder, and various objectives and extension tubes

Heavy-duty microscope lights are necessary to ensure that the specimen dish (4 cm diameter) is evenly illuminated. A high light intensity is essential, as a large proportion of the light is absorbed by the black, osmium tetroxide–fixed specimen. The table and specimen dish should also be made of a black, matte material so that the section is not irradiated from the periphery.

To reduce convection in the specimen dish, each lamp is fitted with six heat filters, while the beam of light itself is encased in two adjustable concave metal mirrors with ribbed casing, which further absorbs the heat. Since the specimen dish must be evenly lit, it is placed between the lights on an adjustable turntable. The beam of light itself can be focused above on the two movable axes of the mirrors, which, unlike the hot lamp casing, can be continuously focused and repositioned, as they remain cool.

A 35-mm single-lens reflex camera with a 6X focusing finder is used for this work. The camera is fixed on a stand with fine adjustment for focusing. A direct magnification of up to 12X is made possible owing to the wide-angle lenses (17 mm, 24 mm, and 35 mm) and a normal lens (50 mm) mounted in

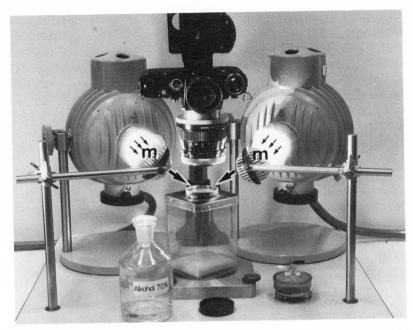

Fig. 5.1 The camera and lights set up in their optimal position for photography of the osmium tetroxide–fixed specimens. Arrows indicate the direction of the beam of light. Movable metal mirrors (*m*) direct the light onto the specimen.

retrofocus on an extension tube. This method of photography with lenses mounted in retrofocus gives the necessary distance of 4-5 cm between the specimen and the camera lens. The high optical quality of new camera lenses allows the use of apertures of 4.5 or 5.6 so that a high resolution of specimen structures and a relatively short exposure time can be obtained. The film used is Agfa Ortho 25 high resolution film which is developed in Agfa Rodinal 1:100. The quality of the film allows an enlargement of the prints up to 10X or more.

FIXATION

Fixation by glutaraldehyde perfusion directly into the blood vascular system is essential for the detailed analysis of organs or organ systems. The fixation solution is colored with methylene blue, which acts as an indicator of the speed of the fixation in addition to enabling the stained glands, epithelia, and sensory organs to be easily located and dissected.

PRODUCTION OF INTERFERENCE REFLECTING LIGHT (IRL) PHENOMENA

During the dehydration procedure, interference phenomena become visible in a beam of reflecting light at the 70% alcohol stage. As any contamination of the alcohol will produce its own bright interference, the alcohol should be absolutely free of dust. A bacterial filter has proved to be useful. In order to reduce the possibility of surface reflection, the surface of the section must be completely covered with alcohol. One of the most important requirements in taking the photographs is that vibration on the surface of the alcohol in the specimen dish be completely eliminated. A piece of rubber is placed under the stand to absorb the vibrations.

The reflecting light is partly absorbed by the black specimen surface and partly reflected by interference on specific surface structures in the form of different iridescent colors, which give, on the dark tissue background, a very high contrast. In this way, a specific structural pattern of the tissue is illuminated and gives the equivalent of a photograph of low magnification of the histological specimen. As the intensity of the structures is dependent upon the angle of incidence, it is useful to turn the specimen dish to find the optimal position.

NATURAL SURFACE

The surface of formalin-or glutaraldehyde-fixed specimens of vertebrates exhibits a diffuse reflection in higher stereo-microscopical magnification which does not allow fine details to be recognized. After fixation of the specimen with osmium tetroxide and illumination in the IRL apparatus, details of the surface structures can be recognized up to a magnification of 100X or above. In this manner a dark-field effect is achieved.

In such photographs one can clearly see the sharply defined borders between different epithelia (Figs. 5.2B, 5.8B), small papillae, sensory hairs, and glandular pores of the skin (v. Düring, 1973). Ciliated epithelia exhibit an interference spectra whose color is dependent upon the thickness of the cilia. This phenomenon is particularly noticeable in the olfactory mucosa of different vertebrates (Andres, 1968, 1969, 1975); for example, in fish it is blue, in *Amphibia* and reptiles it is yellow, and in birds the olfactory mucosa is of copper color. The olfactory mucosa of mammals, however, exhibits a silver grey iridescence due to the exceptional thinness (80 nm) of its cilia. Vertically positioned structures such as taste buds appear black as they absorb the light (Figs. 5.2, 5.3). The IRL method is ideally suited for scanning electron microscopy, as the surface structure can be seen clearly, ensuring as well that the surface is intact.

SECTION SURFACE

Structural details can be seen not only on the natural surface but also on the cut surface of sections by using the IRL method. Sections appear similar to a photograph of low magnification of a histological section. If the surface of the section needs to be examined with the SEM (Fig. 5.4), the IRL method is ideally suited for selecting section segments.

To achieve the maximum effect with this method for systematic analysis of

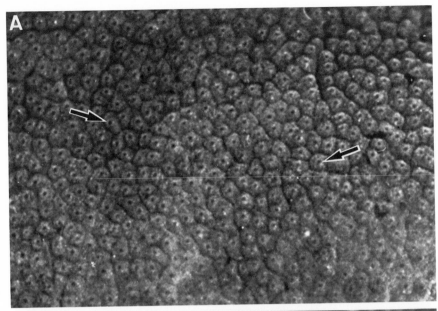

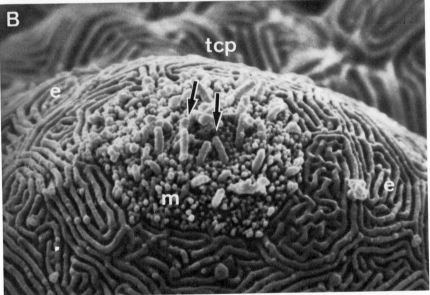

Fig. 5.2 (A) Interference photography shows in detail the surface structure of an organ. For example, in the palatal organ of the goldfish (because of high magnification, ×100) it is possible to see the dense arrangement of taste buds. The pores of the taste buds (*arrows*) are the black spots on the tips of the buds. (B) A scanning electron micrograph of the same region with the surface of the taste bud; sensory cell processes (*tcp*), microvilli of supporting cells (*m*), and the surface epithelium (*e*), which shows microridges. ×4,000.

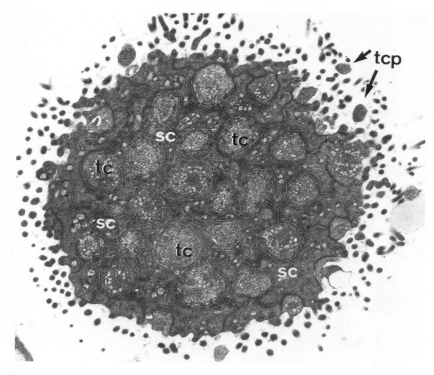

Fig. 5.3 Tangential section through the pore of the taste bud taken from the same specimen as in Fig. 5.2A; taste cell (*tc*), sensory cell-processes (*tcp*), and supporting cell (*sc*). ×8,100.

organs, it is advisable to use slices of not more than 1–2 mm in thickness. A small apparatus, based on the same principle as a bread slicer with a razor blade as knife, is used. With this apparatus the problem of the specimen's being displaced during sectioning has been overcome, so that the surfaces of all sections are parallel to one another. Thus, when the slices are photographed, all structures on their surfaces are in focus, which also eliminates the necessity of refocusing between each slice.

However, for irregularly shaped specimens a special method for sectioning has been developed. The specimen is first placed in the perspex oblong container (Fig. 5.5), and then gelatin (which is on the point of setting) is poured over it and allowed to set. Thus, an even surface is produced that allows the specimen to be cut without further difficulty. The gelatin-embedded specimen is taken out of the case and placed in an oblong frame, which has spaces at 1- or 2-mm intervals along both its sides and top. These spaces act as aids in the cutting procedure, which is then carried out with the vibration cutter. Finally, the slices are pushed out of the frame, and the gelatin is carefully removed.

Owing to the consistency of the gelatin, we have developed the use of a

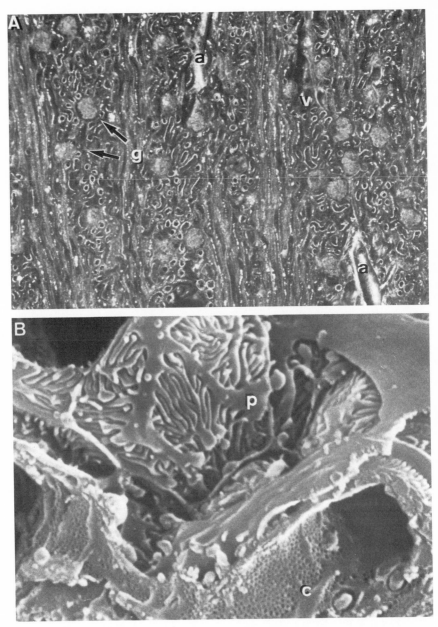

Fig. 5.4 (A) A slice taken from the cortex of rabbit kidney. Structures such as glomeruli (g) and convoluted and straight tubules are easily distinguished in the IRL method. Also shown: arteries (a) and vein (v). ×70. (B) A scanning electron micrograph of part of a glomerulus showing the podocytes (p) and the fenestrated endothelium of the capillaries (c). ×6,300.

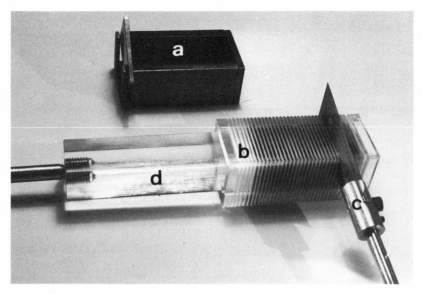

Fig. 5.5 Case (*a*), in which the gelatin will be poured over the specimen. Oblong frame (*b*), which is the same size as the gelatin block from the form above; the frame shows the 1-mm spaces that act as guides for the razor blade in the vibrocutter. Knife holder of the vibro-cutter (*c*) and pusher (*d*).

vibrocutter which vibrates at 50 Hz. A normal electric shaver has been adapted for this purpose. When the slices are placed in a 70% alcohol stage during the dehydration, they are photographed as described above on both sides before the dehydration and embedding procedures are continued. In this way the slices are documented.

Since cytoplasmic granules of certain sizes and density exhibit varying colors, the distribution of granules and their size allows different parts of the organ to be recognized, e.g., the hypophysis, adrenal medulla, and cortex (Fig. 5.6). The distribution of the islets of Langerhans is particularly clear in the pancreas, as the endocrine part of the gland has a denser granule structure. Connective tissue networks in different skin and organ sections are seen clearly, as they exhibit a silver-white iridescence. The denser the collagen fibers are, the more silver-white is their iridescence. Cartilage and bone have characteristic white patterns (Karduck and Richter, 1975; Decker *et al.*, 1976), whereas striated muscle is easily distinguished by its many iridescent colors. The striations seen on the liver vein (Fig. 5.7A) are the arrangements of smooth muscle fibers as seen through the thin endothelial wall.

When the specimen shows autolytic changes prior to fixation, the interference spectra of the surface cannot be seen. Pathologically altered specimens (swelling, infection, scar-tissue) (Fig. 5.7B) can also be used with this method, as the inter-ference spectra for the normal tissues are either altered or absent.

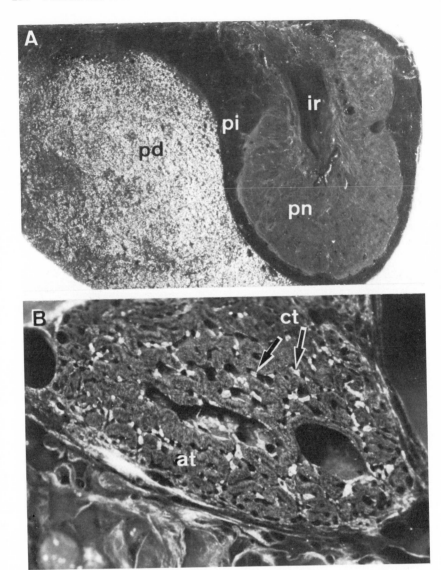

Fig. 5.6 Examples showing the differentiation of tissue in endocrine glands as seen in the IRL method. (A) A median section through the hypophysis (cat). The three regions: pars distalis (*pd*), pars intermedia (*pi*), and pars nervosa (*pn*) are easily differentiated owing to the different interference spectra of their tissues. Also shown: infundibular recess (*ir*). ×11. (B) From these sections through the adrenal body (caiman) one can see the different interference spectra of the chromaffin tissue (*ct*) and adrenocortical tissue (*at*), once again enabling a precise excision of the specimen for further examination. Note the richly anastomotic sinusoids. ×56.

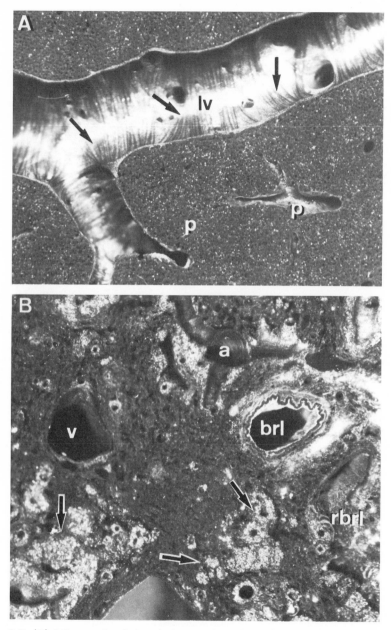

Fig. 5.7 (A) A liver section (caiman) photographed in the IRL method. The striations on the liver vein (*lv*) show the arrangement of smooth muscle fibers (*arrows*) seen through the thin endothelial wall. Liver parenchyma (*p*) and portal vein (*p*). The white spots in the liver parenchyma are due to pigmentation in the Kupfer cell. ×40. (B) A slice of lung (Rhesus monkey) showing signs of silicosis (*arrows*) after inhaling stone dust containing silicon dioxide. Bronchiole (*brl*), respiratory bronchiole (*rbrl*), lung artery (*a*), and lung vein (*v*). ×28.

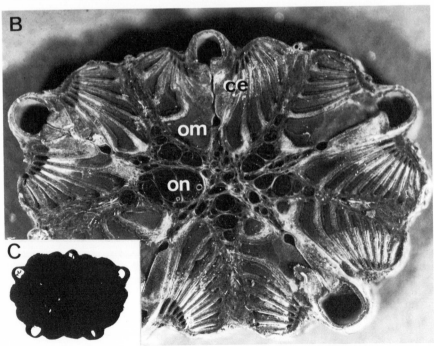

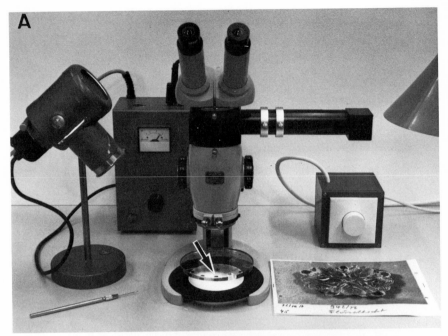

PRECISE EXCISION OF SPECIMEN FOR TEM

With the help of a camera lucida placed in a stereo-microscope, the photograph of the slice taken in the IRL method can be superimposed onto the embedded specimen. With the zoom system of the stereo-microscope combined with the zoom system of the camera lucida, the photograph as well as the slice can be enlarged so that they can be superimposed onto one another. In addition, a dimmer adaptor for the light source ensures optimal illumination (Fig. 5.8). The slices are embedded in Araldite or Epon using a shallow plastic dish as mould. Care should be taken to use a dish that is perfectly smooth; otherwise the undersurface of the Araldite or Epon will be rough. Should this, however, happen, its surface can be made clear once more by applying liquid paraffin.

A specific area can be selected for electron microscopy and marked on the slice, which is then placed on a hotplate (at 80°C) until the Araldite or Epon becomes pliable. The selected area is then cut out without disturbing the remainder of the specimen, which can be labeled and stored for further investigation. The block is mounted on an Araldite base in the required position, using a quick-setting resin (Techovit, Kulzer). The above-described method using Araldite or Epon has produced the best results.

TEM ANALYSIS OF CNS WITH IRL METHOD

The IRL method is a remarkable aid in the analysis of CNS and sensory organs, as it bridges the gap between macroscopy and electron microscopy. The pattern of nerve fiber networks and the direction of the nerve fiber pathway are characteristic for each section of the nervous system. Owing to the iridescent colors that these tissue structures show in the IRL method, it can be used as a means of diagnosis. For example, fibers which run parallel to the surface of the section are blue/yellow in color, whereas bundles of nerve fibers in cross section appear black as they absorb the light (Figs. 5.8B, 5.10, 5.11C,D). The direction of obliquely cut nerve fibers can be determined by adjusting the turntable until they are reflected in the beam of light. Neuropils of molecular layers appear diffuse without a specific interference spectrum (Fig. 5.10), and small nerve cells are also not recognizable in well-fixed specimens.

Fig. 5.8 (A) A stereo-microscope with camera lucida, which is used to superimpose interference photographs (Fig. 5.8B) onto the embedded and osmium tetroxide–fixed specimen (*arrow*) (Fig. 5.8C). (B) A slice of the nasal cavity of *Polypterus bichir* photographed by the interference reflecting light method (IRL). Olfactory nerve bundles (*on*), olfactory mucosa (*om*), and ciliated epithelium (*ce*) of the olfactory folds. ×16.2. (C) The osmium tetroxide-fixed slice as seen after embedding in Araldite. The specimen is completely opaque with only the outline visible. ×4.7.

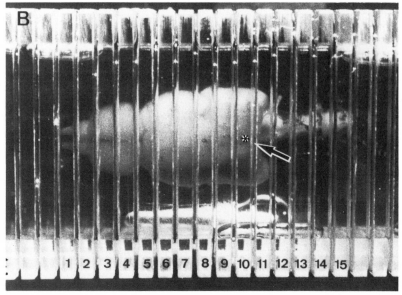

Fig. 5.9 (A) Perspex oblong container. It is possible to achieve an ideal position for the specimen by adjusting the screws on whose tips the specimen rests. (B) The brain of *Gymnotus carapo* in gelatin in the oblong frame prior to sectioning. The numbers show the sequence of slices (see Fig. 5.10). Projection to the dorsal surface of the pacemaker nucleus from Fig. 5.11 (*). ×5.

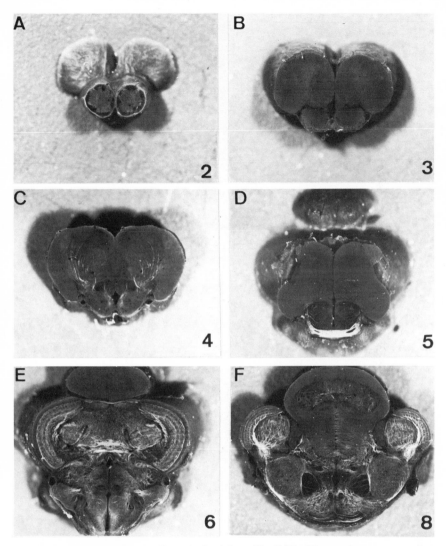

Fig. 5.10 Examples of brain slices obtained from Fig. 5.9B. The numbers correspond to those seen in Fig. 5.9B. With the IRL method it is possible to obtain a histological low magnification picture of the surface of the slices. ×8.

For the study of the CNS it is extremely important to adjust the brain so that exact frontal, sagittal, or horizontal plane sections can be made. For this purpose the brain is placed in the container (Fig. 5.9A) so that it rests on the tips of the screws. The screws have been fitted with fine needles in their tips so that the brain rests in fact only on these fine needle tips. By adjusting these screws it is possible to orientate the brain in the desired plane for sectioning, as well as for the coordinate system of a stereotaxic instrument.

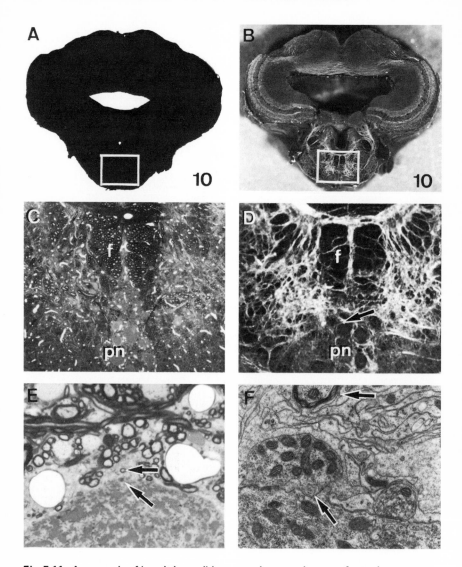

Fig. 5.11 An example of how it is possible to examine a precise area of a section, e.g., pace-maker nucleus of medulla oblongata (*Gymnotus*). Slice No. 10 from the specimen in Fig. 5.9B (*). (A) Osmium tetroxide–fixed slice embedded in Araldite. (B) The same slice as in (A) above in the IRL method. ×7. (C) Semithin section of the area enclosed in the white square in (A) and (B). Fasciculus longitudinalis medialis (*f*) and pacemaker nucleus (*pn*). ×40. (D) Higher magnification of the area enclosed in a white square in the IRL method. Fascic-ulus longitudinalis medialis (*f*), pacemaker nucleus (*pn*). ×40. (E and F) Semithin and sub-sequent ultrathin section of the pacemaker nucleus. Arrows indicate the same axon and synapse; capillaries perfused. E, × 900; F, × 9,000. From the electron micrograph in Fig. 5.11 it is possible to retrace each individual step as far as Fig. 5.9B, where one can see the topographical relationship of the fine structure to the whole. See Fig. 5.9B (*).

In order to show how in practice it is possible to connect the macroscopical structure of the brain with its microscopical and electron microscopic structure, we shall take *Gymnotus* brain. The fine structure of the pacemaker nucleus of this specimen and its topographical relationship to the complete brain are described (Figs. 5.9B, 5.11). The brain is sectioned as previously described and photographed in the IRL method. In this way an atlas of the brain of the animal is produced, thus providing an excellent basis for a systematic morphological analysis. Examples from such a brain atlas are shown in Fig. 5.10, where the numbers correspond to the numbers in Fig. 5.9B. In Fig. 5.11 one can see how, as described, it is possible to isolate the area of the pacemaker nucleus and examine its fine structure. Subsequent semithin and ultrathin sections make it possible to analyze and localize synaptic contacts on the perikaryon and at the same time to locate its position in the complete brain (compare Fig. 5.11 with Fig. 5.9B).

Although we have described the systematic application of the IRL method to the study of brain, it is equally suited to the systematic examination of any type of tissue.

We thank Mrs. Christine Parker for her assistance in preparing the chapter. The method discussed above was developed by us in the course of our research, which is supported by the German Research Council, SP "Receptor Physiology" and SFB 114 grants.

References

Andres, K. H. (1967). Zur Methodik der Perfusionsfixierung des Zentrainervensystems von Säugern. Tag. d. Niederl. u. Dt. Elektr. Ges. Aachen 1965. *Mikroskopie* **21**, 169.

Andres, K. H. (1968). Neue Befunde zur Feinstruktur des olfaktorischen Saumes. *J. Ultrastruct. Res.* **25**, 163.

Andres, K. H. (1969). Der olfaktorische Saum der Katze. *Z. Zellforsch.* **96**, 250.

Andres, K. H. (1975). Neue morphologische Grundlagen zur Physiologie des Riechens und Schmeckens. *Arch. Oto-Rhino-Laryng.* **210**, 1.

Andres, K. H., and v. Düring, M. (1974). Interferenzphänomene am osmierten Präparat für die systematische elektronenmikroskopische Untersuchung. *Mikroskopie* **30**, 139.

Decker, S., Rehn, J., v. Düring, M., and Decker, B. (1976). Morphologischexperimenteller Beitrag zur Kenntnis der Vorgänge bei der Verpflanzung von autologer Beckenkammspongiosa bei Hunden. *Arch. orthop. Unfall-Chir.* **85**, 303.

v. Düring, M. (1973). The ultrastructure of lamellated mechanoreceptors in the skin of reptiles. *Z. Anat. Entwickl. -Gesch.* **143**, 81.

Karduck, A., and Richter, H. G. (1975). Effects of highly dosed ultrasound on the growing rabbit's larynx. *Arch. Oto-Rhino-Laryng.* **211**, 71.

6. COMPUTER PROCESSING OF ELECTRON MICROGRAPHS

P. W. Hawkes

C.N.R.S. Laboratory of Electron Optics, Toulouse, France

INTRODUCTION

The computer is a comparatively recent addition to the tools at the disposal of the electron microscopist, and the techniques being developed for processing the images obtained with conventional transmission electron microscopes are still in their infancy. The most successful and best-known application of the computer has been in the field of three-dimensional reconstruction from projections, but this by no means exhausts the possibilities of computer processing. The role of the computer is rather different for the two other main types of electron microscope, the scanning transmission instrument (STEM), general accounts of which are to be found in Zeitler (1976) and Reimer and Pfefferkorn (1977), and the conventional scanning electron microscope (SEM), described in Oatley (1972), Holt *et al.* (1974), Hayat (1974-1978) and Goldstein and Yakowitz (1975), for example, and in an excellent earlier review by Hayes and Pease (1968).

The difference arises because a scanning image consists of a sequence of electrical signals, each generated by a single "point," or small area, on the specimen, which is converted into visible form by a cathode ray tube. The signal may represent any one of several beam–specimen interactions—x-ray production, back-scattering, fluorescence, for example—and it is often possible to display simultaneously several types of image, representing different aspects of the specimen. Since these signals are electrical, they can be sent directly to a computer, via a standard interconnection or "interface," and useful operations can be performed on them as they emerge from the microscope. Commercial scanning electron microscopes are therefore available already linked to the type of circuitry used in the computers in such a way that simple controls enable the user to perform some kinds of image processing almost effortlessly. Most of this chapter is concerned with conventional transmission electron microscopy, but we also give some account of the processing of scanning micrographs.

In the early days of electron microscopy, the operation of the instrument was so complicated that a technician was permanently on hand to assist all but the most technically inclined users. Computer processing is passing through a similar phase, for it is extremely unlikely that the biological user of the microscope will be prepared to master the programming techniques necessary for any kind of digital processing. Just as microscope manufacturers have striven to simplify the routine operation of electron microscopes, however, so user-orientated computer programs and, indeed, whole computer systems are becoming available; and we can confidently anticipate that, in a few years, the microscope user who believes that his images can benefit from computer processing will be able to try out the techniques with only a vestigial knowledge of computing. Nevertheless, in order to use these techniques sensibly and to decide whether particular images can profitably be treated, some familiarity with the underlying ideas and the restrictions is essential. Our aim in this chapter is precisely this: to explain in terms that will be comprehensible to the microscope user with little or no knowledge of computing the types of treatment that the computer can or should soon be able to perform, the types of image with which it must be provided, and the type of information it can yield. In the later part of the chapter, we discuss some specific techniques in fair detail, but the subject is in such a rapidly expanding and developing state that this material will inevitably become dated rather quickly. For this reason, we concentrate throughout on the underlying principles and general ideas, which should survive longer.

An important practical aspect of image processing is the intimate relation between the performances of each of the links of the chain leading from the electron–specimen interaction in the electron microscope to the array of measured intensities, representing the micrograph in a form that the computer can manipulate, stored on magnetic tape (or some other suitable medium). We shall return to this in due course, but it is worth stressing from the outset that unless the microscope magnification is high enough, the film or plate response will limit the real resolution obtainable; again, if the adjustment of the microdensitometer used to measure the intensities is not matched to the desired resolution, the results obtained with the computer may be at best true but not what had been hoped for, and at worst misleading. The problem is, of course, very different in the scanning instruments where the image is built up point-by-point and is available directly for processing; here it is the relation between beam brightness, scanning rate, and resolution that must be respected.

Electron Image Formation and Digital Processing

The electron micrograph that is finally obtained after the specimen has undergone all the manifestly drastic processes of preparation and the more subtly harmful effects of microscopy is imperfect for any of a wide variety of reasons. The preparation processes will have disturbed the structure to some extent; if the specimen is stained, positively or negatively, the image will contain information mostly about the stain, and the latter may not have embraced all the details of the original structure; the electron beam may well have destroyed some of the fine detail of the specimen. Even if all of these effects are negligible, the electron image will not necessarily represent the structure faithfully, since two intimately related but effectively distinct mechanisms are involved in contrast formation.

Roughly speaking, coarse detail is amplitude contrast and fine detail is phase contrast. By this, we mean that electrons deflected in the specimen by heavy atom stain through reasonably large angles will be intercepted by the objective aperture; their absence at the image produces a dark region, just as the absorption of light in a semitransparent specimen in light microscopy produces amplitude contrast. These electrons are capable of producing an image that is a faithful representation of the distribution of heavy atom stain in the specimen down to a resolution of a nanometer or a few nanometers. Below this, the contrast is again a consequence of the deflection of the electrons by the atoms of the specimen, but now the deflections are not large enough for the electrons to strike the objective aperture. Nevertheless, their point of arrival at the image plane varies with the amount by which they are deflected—a consequence of the spherical aberration of the objective lens—and by exploiting this fact, and choosing the objective focus carefully, a very satisfactory phase contrast can be achieved. Unfortunately, this phase contrast differs from that familiar in light microscopy in that there is no equivalent of a high-precision phase plate; genuine phase-shifting devices do exist, but they are not routinely available and are unlikely ever to be as satisfactory as their light microscope counterparts (Müller, 1976; Willasch, 1975).

Instead, we have the combination of an aberration, over which the user has no control, and the defocus, which can be varied at will; this is analogous to using a phase plate of varying thickness, the thickness variations of which can be altered by turning the defocus knob. In consequence, the phase contrast image will not always provide a faithful representation of the phase variations at the specimen: some image detail may be true, some may be artifact, and some object structure may be lost altogether. Image imperfections of this kind are especially important at high resolution and can be very misleading: it is perfectly conceivable that the larger structures seen in a clear, crisp, and hence convincing-looking micrograph do represent the object structure faithfully, while some of the fine detail, however sharp, does not.

In order to understand how this aberrant behavior comes about, and what hope we have of distinguishing between true and false image details, we need more knowledge of the mechanism of phase contrast imagery. More mathematical versions of the qualitative account that follows are to be found in the chapters by Humphreys (1976), Burge (1976), and Cowley (1976) in Volume 6 of this series and in Lenz (1971), Hanszen (1971), and Erickson (1973). The relationship between the electron distribution at the specimen and the image is straightforward only for a class of specimens of limited interest, namely, specimens that only deflect the incident beam electrons through extremely small angles and absorb very little energy in the process. Such specimens are known as weakly scattering objects or weak-phase, weak-amplitude objects. Very often, the amplitude effect is ignored, and we then speak of weak-phase objects. For this class, weakly scattering specimens, the image *intensity* is related straightforwardly (by the linear mathematical operation known as convolution) to the specimen phase and amplitude distributions.

Convolutions possess the important property that if a distribution is equal to the convolution of two others, as is the case here, the spatial frequency spectrum* or Fourier transform of the distribution is equal to the ordinary product of the Fourier transforms of the other two. This means that if we consider not the image itself but its Fourier transform, the intensity at each point is equal to the intensity at the corresponding point in the Fourier transform of the phase distribution multiplied by a number determined by the microscope parameters (defocus, spherical aberration coefficient, and wavelength), for a weak-phase object. For a weak-phase, weak-amplitude object, the result is essentially the same except that a second term, consisting of the Fourier transform of the amplitude distribution again multiplied by an instrument function, must be added. Under these conditions, the microscope acts as a *filter*, imaging object detail in certain (well-defined) size ranges faithfully, imaging detail in other ranges with reversed contrast, and suppressing the remainder.

If the specimen is not a weakly scattering object, and very few specimens are, the mechanism just described still operates, but the image intensity contains an extra term, which destroys the simple proportionality between its spectrum and that of the object phase and/or amplitude. The problem of working back from the image intensity to the object phase and amplitude is now extremely complicated. As we have seen, all electron–specimen interactions result in deflection of the incident electrons. Virtually no electrons are absorbed in the specimen, though of course many lose more than a negligible amount of energy as they pass through it: these are the inelastically scattered electrons. This means that the best possible information that the microscope could give us would be a map or picture of the deflections of the electrons as they left the specimen: this would correspond very closely to a map of the phase variations of the specimen, that is, the variations of refractive index for a specimen of uniform thickness.

Unfortunately, we cannot record *directions* of arrival of the electrons at the

---

*An attempt to explain this important concept in nontechnical terms is to be found in Hawkes (1972).

image, but only their *positions*, the points at which they strike the recording medium: plate or film or perhaps image intensifier. Since direction is intimately related to phase, this is conventionally expressed by saying that we can only record intensities and not phases, although at high resolution (and for some specimens, such as ferromagnetics, at quite modest resolution) it is the phase distribution that is of interest. We stress high resolution because at lower resolution the objective aperture acts as a crudely direction-sensitive device and hence converts direction (phase) information into variations of image intensity. This problem (our inability to record phase directly) arises in several branches of physics, and is traditionally known simply as the phase problem. It can be solved, in principle at least, by computer processing of pairs (or larger numbers) of electron micrographs taken under slightly different conditions.

The final type of inadequacy from which electron micrographs may suffer is that of being essentially two-dimensional. In most types of specimens, the ultimate object of electron microscopy is to establish the three-dimensional structure of the material under scrutiny. The fact that the electron micrograph gives only a projection through the specimen can be circumvented in various ways. The simplest is the difficult and tiresome procedure of serial sections: the same area is observed in a series of thin sections, and the three-dimensional structure is built up by following the changes from section to section. Even ignoring the obvious difficulties, this method is restricted to stained sectioned material, and the resolution is limited by the section thickness. The alternative that has received the most attention is computer-aided reconstruction using a series of views through the same section taken at different angles. In practice, the method has been most successful when applied to specimens with high symmetry or where several views of nearly identical particles appear in the same micrograph. In these cases, several views are in effect available simultaneously. This method already has a large literature and is sufficiently self-contained to merit a separate chapter. Therefore, only a brief account is given here.

Preparation Artifacts and Radiation Damage

In the previous section, we saw that the computer has a useful role to play in remedying the intrinsic deficiencies of the conventional transmission electron microscope: its filtering effect in spatial frequency terminology, its inability to record phase information in general, and its reduction of a three-dimensional structure to a two-dimensional image. Can the computer help us to eliminate or correct image imperfections of other kinds? All the information about the specimen acquired by the electron beam is coded in the latter as the distribution of electrons and their directions of motion. There is therefore nothing that a computer, or indeed any other device, can do to identify or remedy preparation artifacts or to recognize the effects of damage, given a single image. It is, on the other hand, at least conceivable that a sequence of images taken with increasing exposures could yield some information about the increase of damage with dose. It has even been suggested that one could project some way back from such a sequence to, or towards, the undamaged structure; in the types of specimen that would warrant such an effort, however, the more fragile structures are often lost so very rapidly that there would be little left from which to extrapolate back (Frank *et al.*, 1974; Hoppe, 1975; for extensive discussion of radiation damage, see Glaeser, 1975, Isaacson, 1975, 1976, Parsons, 1975, Reimer, 1975, Baumeister and Hahn, 1976b, Baumeister *et al.*, 1976, Scherzer, 1976, and, more generally, Hoppe *et al.*, 1970, Huxley and Klug, 1971, Beer *et al.*, 1975, and Cosslett and Hoppe, 1977).

If interesting structures are irretrievably damaged by normal electron doses, can we invoke the aid of the computer to "improve" an image obtained with a very low electron dose, so low that the direct image is of extremely poor quality? If the specimen is plane and periodic, the work of Unwin and Henderson on the unstained purple membrane of *Halobacterium halobium* and beef liver catalase shows that we can successfully exploit the periodicity of the specimen to provide, in effect, a very large number of low-dose images from a single minimum-exposure picture. These can be combined to give a good high-resolution image of the fragile structure (Unwin and Henderson, 1975; Unwin, 1975). Kuo and Glaeser (1975) have come to the same conclusion. Furthermore, it is possible to take a sequence of very low-dose pictures at different tilts, so that the total dose can be tolerated by the specimen, after which the computer is used to synthesize a "true" image (Henderson and Unwin, 1975; Hoppe *et al.*, 1976a,b).

Electron Microscope Parameters

An electron micrograph does not consist only of information, reliable or misleading, about the specimen. If the substrate is a thin layer of amorphous carbon, or any amorphous substance, the image will be seen against an irregular speckle pattern, which is in fact an interference pattern formed by electrons deflected by the atoms of the substrate. This pattern may well obscure the image of the specimen if the latter is not significantly more intense than the background; it is for this reason that considerable efforts are being made to find alternative supports, either crystalline—since these would contribute a very well-defined background, which could be subtracted (e.g., Baumeister and Hahn, 1974, 1976a, Hahn and Baumeister, 1974, using vermiculite, and Mihama and Tanaka, 1976, using beryllium oxide)—or amorphous but less troublesome (e.g., Dorignac, 1974, using boron and Johansen, 1976, and Kölbel, 1976, using carbon).

Nevertheless, these speckle patterns do contain very useful information about the microscope imaging parameters. We have mentioned that the microscope "filters" the Fourier transform of the specimen phase and/or amplitude; the filter factor is known as the electron microscope transfer function, and it is essential that it be known accurately in any attempt to reconstitute filtered images. The transfer function is determined by the coefficient of spherical aberration, C_s, the defocus, Δf, and the wavelength, λ (typically 3.7×10^{-12} m = 0.037 Å at 100 kV). The coefficient C_s does not of course vary (though it may vary somewhat from the value given by the manufacturer), but Δf is best obtained from the micrograph itself; this is a simple task provided that a part of the image contains a speckle pattern, reasonably free of superimposed specimen.

Scanning Electron Microscope Images

The computer is likely to be useful to the scanning electron microscopist not so much for correcting recondite imperfections by sophisticated programs but for helping him to make the best use of the immense amount of information produced by such microscopes. This point is already fully appreciated by the manufacturers of commercial instruments (Paden *et al.*, 1973). The Cambridge Scientific Instruments "Stereoscan 180," for example, can provide many different types of images (the secondary electron image, the back-scattered electron image, the absorbed electron image, selected-area and large-area channeling patterns, the transmission image, various X-ray images, the cathodoluminescence image, the magnetic contrast image, and the selected-area electron diffraction pattern). Any of these can be displayed directly or after a variety of types of processing: contours may replace changes in grey level, thus rendering small gradations easily visible; the contrast may be expanded when the range in the original image is such that some detail would otherwise be invisible; the image may be differentiated, a technique that emphasizes edges and makes changes in image brightness just as easy to see against a bright background as against a dark field. Nevertheless, it is intuitively obvious that more information about the specimen is being generated by the microscope than the user can readily digest.

The role of the computer here is to reorganize the information contained in the various types of images in such a way that it can be assimilated and appreciated more easily. Already the Stereoscan 180 has two displays as standard, so that the same image can be examined before and after processing, for example, or two different types of images can be compared by eye; with a computer, a much wider range of calculations could be performed. At the simplest extreme, the different images could be subtracted from one another, to bring out the differences, and many more refined comparisons (cross-correlation, for example) can equally well be envisaged. Much of the work on computer processing of scanning electron microscope images has appeared in the proceedings of the annual symposia organized by the IIT Research Institute in Chicago. From these and other sources, we draw attention particularly to the pioneering work of White *et al.* (1968), MacDonald (1968), Simon (1969), McMillan *et al.* (1969), and White *et al.* (1970); to the series of surveys by Newbury and Joy (1973), Joy and Verney (1973), Fiori *et al.* (1974), and Newbury (1975); and to the more technical material in Matson *et al*. (1970), MacDonald and Waldrop (1971), White *et al*. (1971, 1972a,b), Finlay and Brown (1971), Herzog and Everhart (1973), Baggett and Glassman (1974), Yew and Pease (1974), Herzog *et al*. (1974), Schmiesser (1974), Lewis and Sakrison (1975), and Oron and Gilbert (1976).

A similar situation arises with the scanning transmission instrument, where once again the wealth of information provided in the various images is difficult to appreciate. Furthermore, the simple annular detector normally used in these instruments will be supplemented in the future with detectors of more complex geometry—sets of concentric rings, for example (Rose, 1974). This proliferation of STEM signals, each providing an image or a "subimage" from which the complete picture is to be synthesized, cannot be fully exploited without the participation of a computer. A full discussion of the role of the computer in scanning electron microscopy is to be found in the companion series on scanning (Hayat, 1974–1978).

EQUIPMENT

In this section, we discuss the equipment needed to digitize the image produced by a conventional transmission electron microscope and to reconstitute a half-tone picture from a digitized image. We also explain the type of computer and peripherals that are required for the processing itself. Self-contained image analysis devices, such as the Quantimet, are not considered here, but are briefly described at the end of the section on types of image processing.

Microdensitometry[*]

Electron micrographs are recorded as an essentially continuous intensity (current) distribution on photographic film or plates. The computer must be provided with a discrete digitized version of this photographic record, that is, the intensity must be sampled at each point of a square or rectangular array; in reality, the transparency is averaged over a very small square or circular area, and the image is regarded as a mosaic of these elementary areas. It is clearly of considerable importance that this area be chosen correctly for, if it is too large, fine detail will not survive, and if it is too small, one piece of "information" on the micrograph will be represented by several measured intensity values, some of which will be redundant. Since a serious difficulty with computer processing is that the number of measurements which can be handled in a reasonable time is very limited, it is clearly desirable to choose the sampling interval—the size of the elementary areas—as large as possible without loss of useful resolution. This point is discussed very carefully by Dainty and Shaw (1974), and a good account is also to be found in Frank (1973).

---

[*]We draw attention to the issue of *Optical Engineering* devoted to microdensitometers: **12**, No. 6 (1973).

The exact choice is determined by the grain size of the emulsion used, or better, by its modulation transfer function (Zeitler and Hayes, 1965; Zeitler, 1968; Dainty and Shaw, 1974); the latter gives a very precise description of the resolving capacity of the emulsion, but is unfortunately not available for all the emulsions that are used in electron microscopy (see, for example, the work of Burge and Garrard, 1968; Burge *et al.*, 1968; Jones and Cosslett, reported in Cosslett *et al*, 1974, which contains many references to earlier work; and Glaeser, 1976). Generally speaking, measurements of micrograph transparency are made at intervals of 25 or 50 μm apart, although for some applications very much finer sampling may be necessary. There is a limit to the accuracy with which the transparency can be measured: that is, to the number of grey levels that can be distinguished. Provided that the number is not too small, this is not a serious limitation; for beyond a certain number, the worth of the extra values becomes debatable. In practice, the maximum number of grey levels is tailored to the properties of the computer. A very common figure is 256 (i.e., 0–255), since each value can be coded as a group of eight binary digits (bits), and small computers work with groups (words) of twelve or sixteen bits; large computers have longer words, but eight bits can usually be packed compactly into them. (Note: Decimal 255 = binary 11 111 111.)

The measurements are made with the aid of a microdensitometer, of which a number of models are available commercially. These models differ in the size of sampling area and hence the distance between adjacent samples, the step-length; in the rate at which measurements are made; in the number of grey levels that can be distinguished; and in the type of pictures that can be measured (flat plates or flexible film). The fastest microdensitometers are drum scanners, for which the micrograph must be recorded on (or transferred to) film; the latter is wrapped around a drum, which spins at a uniform speed and, at the end of each revolution, moves one step-length along its axle. Each revolution thus yields a row of measurements, and the instruments are designed so that a predetermined square or rectangular area is measured. Typical drum scanners are the "Scandig," manufactured by Joyce-Loebl/Tech-Ops and the Optronics "Photoscan." Both offer a range of sampling rasters, from 12.5 $\times$ 12.5 μm^2 up to 200 $\times$ 200 μm^2; the maximum rate is 55,000 measurements per second, but it cannot always be attained, since the drum cannot rotate faster than some maximum value (8 rev/s in the "Photoscan"). The number of grey levels is 256, the zero level being recalibrated between each revolution to ensure consistency; the optical density, which can be modified to some extent by the user, ranges between zero and two or three, a factor of a hundred or a thousand, respectively, in transparency, since optical density (see Dainty and Shaw, 1974) is a logarithmic measure.

The measurements are recorded either on magnetic tape or sent directly to a computer. The choice is largely dependent on the local conditions: if the computer is used only for image processing (a "dedicated" computer), and all the processing is performed on this one machine, there is little point in recording the image on magnetic tape only to transfer the latter to the tape-transport of the computer; the digitized image may as well be dispatched directly into the computer. If, on the other hand, many users are digitizing images, for subsequent treatment on different computers or on some large central computer, it is sensible to separate the digitization stage from the computing stage. The choice does not always rest with the user; the latest "Scandig," model 3, is available with an interface to a Data General Nova computer as standard.

The other family of microdensitometers consists of the flat-bed instruments, which accept images on plates or film; they are rather slower than the drum-scanners but may achieve much higher precision. The Optronics "Specscan," for example, has a sampling raster of 1, 5, 10, . . ., 100 μm, an optical density range from zero to four, and 256 grey levels; the maximum data rate is 2, 000–5, 000 measurements per second (depending on the exact model). The Joyce-Loebl 3CS also goes down to 1 μm in resolution, and the optical density range varies from 0.2 to 6.0; when interfaced to a computer, the latter can be used to generate any desired scanning pattern and to perform calculations on the measurements as they are collected. The maximum measuring rate is 20 points per second. Yet another such instrument, comparable to the "Specscan" in performance, is the Perkin-Elmer* 1010 (which, like the others mentioned here, exists in several versions).

We have already seen that the choice of sampling raster is determined by the properties of the emulsion. The latter thus dictates the threshold beneath which small detail will not appear in the microdensitometer record, and the electron microscope magnification must therefore be chosen high enough to ensure that fine detail of interest in the specimen is enlarged beyond this threshold. Conversely, it is important not to magnify beyond the minimum tolerable value, to keep the electron dose administered to the specimen as low as possible, for the higher the dose, the less chance the fine detail has of survival. The compromise that must be made may be very unpalatable, since for many specimens we find ourselves in a dilemma with no easy way out: the fine detail cannot survive the dose that is necessary to record the image, at least conventionally. When this situation arises, radical solutions must be sought, as we shall see.

Microdensitometers of the caliber of the "Photoscan" and the "Scandig" are extremely expensive devices, and, together with the photographic step, they represent a series of potential sources of noise and error between the electron image and the computer. We may therefore inquire whether it is possible to connect the microscope directly to the computer without passing through this intermediate stage. Clearly, we can at best mask the microdensitometry stage, for the conventional transmission electron microscope produces the whole image simultaneously, whereas the computer can only receive data sequentially: some device,

*Marketed in Europe by Joyce-Loebl.

whether or not we call it a microdensitometer, must measure the current that arrives at each "point" of the plane normally occupied by the photographic plate in the microscope and transmit these measurements one by one to the computer. Nevertheless, direct measurement of this current distribution does avoid the need for photographic processing, although care must of course be taken to ensure that it does not introduce new problems just as unwelcome; a number of experimental systems have been built (Glaeser *et al.*, 1971; Goldfarb and Siegel, 1973, 1975; Siegel, 1976; Brüders *et al.*, 1976) and others are planned. The essential components of the chain leading from the electron beam to the computer are an image intensifier (Herrmann *et al.*, 1972; Herrmann and Krahl, 1973; Coleman and Boksenberg, 1976), a vidicon television camera, and a storage device, either a fast intermediate unit such as a video disc (Glaeser *et al.*, 1971) or the computer itself (Siegel, 1976).

Commercial systems for coupling a conventional transmission electron microscope to a computer are not yet available, though there is every likelihood that such interfaces will soon come on the market (Herrmann *et al.*, 1976a,b).

Computing

The ideal computer for image processing has a large memory, fast processors for performing otherwise slow operations or sequences that need to be repeated many times (Fourier transformation, for example), and one or more visual display units (cathode ray tubes) for inspecting the image, or modified forms of it, at various stages of the processing. In practice, these requirements are usually satisfied in one of two extreme ways. If a large modern computer is available, it may be convenient to write the digitized image on magnetic tape and then submit it to a sequence of programs, already written and filed in the computer disc memory. This procedure has the immense advantage that even very large memory requirements can usually be satisfied in one way or another, and the actual computing time will be shorter than that with a less sophisticated machine; nevertheless, the real time that elapses between submission of the computing job and reception of the results may well be considerable. Alternatively, a mini-computer may be dedicated wholly or largely to image processing. This solution has two attractive features: first, although the calculations may take longer, the turnround time may well be shorter, owing to the lack of competition for the computer facilities. Second, and more important, the user has more direct control over the progress of the processing; he can scrutinize the image on a monitor as it proceeds and, in consequence, change the course of events whenever necessary. (This is, of course, possible too with a large computer, but the visual display possibilities are likely to be less convenient.) Ideally, of course, a dedicated mini-computer is used routinely, with a large computer also available for tasks beyond the capacity of the small computer. In practice, many of the functioning electron image processing centers have a dedicated mini-computer (a Digital Equipment Co. PDP 8/E in the Electron Microscope Section of the Cavendish Laboratory in Cambridge; a PDP 11/20 in the School of Applied and Engineering Physics at Cornell University; a PDP 11/70 in the Laboratories and Research Division of the New York State Department of Health at Albany, N.Y.; a SEMS/CII Mitra 125 in the Laboratoire d'Optique Electronique du C.N.R.S. in Toulouse).

A certain number of peripheral devices are essential for an image processing computer. A digitized picture usually contains measurements from regions of the image that will subsequently be jettisoned or, at least, excluded from the processing, since it is difficult to window the microdensitometer exactly over the feature of interest unless a computer-driven instrument such as the Joyce-Loebl 3CS is available. At least one and better two magnetic tape decks must therefore be provided, since magnetic tape offers the most convenient means of storing images awaiting processing or simply of archiving images that may be needed in the future. A convenient configuration might therefore consist of a central processor with its own core memory, supplemented by two magnetic tape decks used for both temporary and permanent storage and, in particular, for digitized images and large programs and for receiving the processed images at the end of the calculation; in addition, at least one large magnetic disc is essential, because of the rapid access to any information written on it and also because the system programs that actually execute the user programs will require it.

In addition to these storage devices (discs and tape-decks), a fast floating-point processor, now a standard accessory, is virtually indispensable, as it dramatically reduces the time taken to perform certain very common calculations. Certain other processors, which are rapidly becoming standard attachments, are also highly desirable, especially a unit for performing the Fourier transform.

Next, and as important as any other part of the computer system, visual displays must be available. A range of display screens are offered as routine accessories with most mini-computers; they differ in the size of the image, the number of grey levels possible, and general convenience of use. We shall see in the section on picture-handling languages that provision is made for examining images and related visual material at every stage of a calculation; it is clearly preferable if these intermediate images can be generated virtually instantaneously on an oscilloscope screen rather than slowly and cumbersomely on paper.

Finally, a number of routine accessories are usually added. A teletype—that is, a typewriter connected to the computer—is indispensable, since it is the normal mode of communication between user and computer for relatively short commands. Some means of entering longer programs other than by magnetic tape is commonly provided; it may take the form of a card reader, a paper-tape reader attached to the teletype, a fast paper-tape reader incorporated in the computer, a magnetic cassette reader again incorporated in the teletype, or a relatively cheap type of disc, which being less rigid than traditional discs is known as a floppy disc. Complementing such an input facility, some means of printing-out programs for long-term storage or frequent use is also very convenient; for cards, a separate punch is required, but all the other readers mentioned above also act as punches or, in the case of cassettes and floppy discs, as write-devices. If the computer is dedicated entirely to image processing, a visual display may be adequate for examining the results, combined with the paper record from the teletype. It is more normal to provide a line-printer, however, in which case any kind of computing task can be performed. Figures 6.1 and 6.2 show the computer configurations adopted in Cambridge and Toulouse.

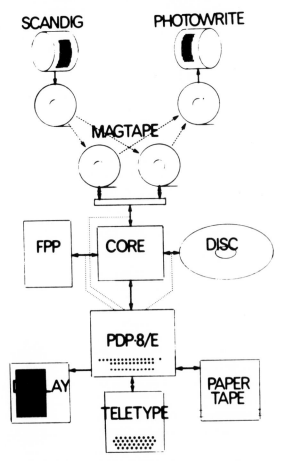

Fig. 6.1 The image processing equipment used in the Electron Microscope Section of the Cavendish Laboratory, Cambridge (Horner, 1976).

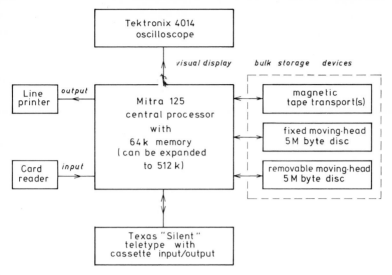

Fig. 6.2 The computer configuration adopted for image processing in the Laboratoire d'Optique Electronique du C.N.R.S. in Toulouse.

Filmwriting

It is clearly highly desirable that the visual quality of the processed image be at least as good as that of the original. Even though the aim of the processing may have been to answer a specific question—is some given structure genuine, or is it an artifact introduced by the mode of image formation, for example—most users will expect a reply in the form of an image that contains the desired information rather than a bald quantitative answer or even a statement, to the effect that the structure is spurious, say, typed out on the teletype. Moreover, it is not likely that image processing will be used mostly for answering specific questions, although such questions may well arise when we are concerned with the ultimate limits of microscopy. The detection of single atoms or atom clusters seen against a noisy substrate raises questions that can be very simply expressed, and a reply in the form of a confidence limit is then as good as any other. Thus, we might wish to know the probability that an observed triangular structure represents a trio of, say mercury atoms. A computer reply "Not better than 30% " would be perfectly adequate here. For more mundane specimens, however, it is still best to present the eye with an image that contains a minimum of misleading structure and is as similar as possible in general appearance to those with which it is familiar, and then interpret the resulting picture visually.

Very few instruments are available that produce a high quality negative from a picture stored in digital form on magnetic tape or in a computer memory. One such device is the Optronics "Photowrite," which in all essential respects is the converse of the "Photoscan" drum-scanner. Like the latter, the "Photowrite" can be connected directly to a computer or fed from a magnetic tape. A combined instrument is also marketed by Optronics, uniting the functions of microdensitometer and filmwriter: it is known as a "Photomation." These film-writers are very expensive ($58,670 in 1976 for a "Photowrite" with 25-, 50-, and 100-μm apertures connected to a tape-deck; $104,570 for a "Photomation" connected to a tape-deck) and are not yet in widespread use. Various alternatives are possible. A few grey levels can be generated on images displayed on cathode ray tubes; such images are tolerable for visual scrutiny though far below the standard of the original micrograph. Alternatively, the computer may produce a contour map (Fig. 6.3) or overprint characters on a line-printer in such a way that the illusion of a continuous picture is created, particularly if the overprinted image is itself photographed and printed in such a way as to smooth out imperfec-

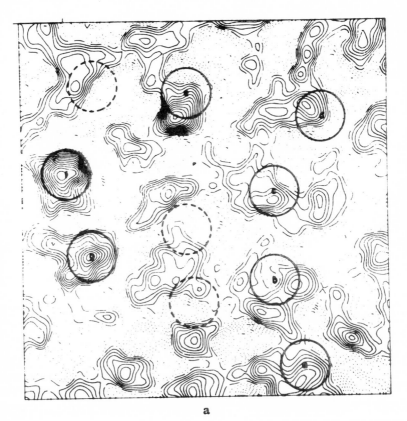

a

Fig. 6.3 Contour plot of a reconstructed image. $Ta_6 Br_{12}$ ions have been molecularly dispersed over a thin carbon foil (3–5 nm thick); phase (a) and amplitude (b) images have been computed from a series of eight micrographs and their positions are circled; dashed circles indicate doubtful ions. The distance between tantalum atoms (c) is 0.29 nm.

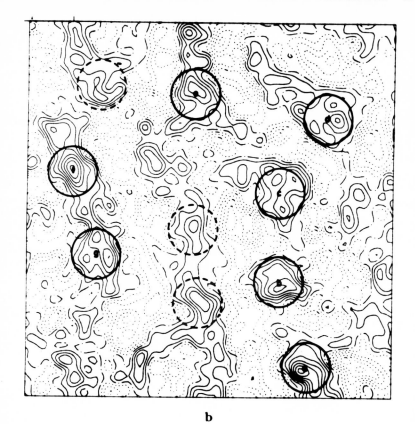

b

c

Fig. 6.3 (*Continued*)

tions. This is an abysmally poor way of displaying an image after going to a great deal of trouble to improve it, however. For certain types of images, with well-defined and preferably isolated features, such as diffraction patterns from periodic structures, the oscilloscope screen or the line-printer can give a very satisfactory representation not as an intensity distribution but as a "mountain range" seen in perspective. The high points of the image—for example, the spots of the diffraction pattern—are then seen as pinnacles rising above a plain, flanked perhaps by foothills. This form of display is well known in scanning electron microscopy, where it is known as y-modulation (e.g., Goldstein and Yakowitz, 1975).

TYPES OF IMAGE PROCESSING

The type of processing that can usefully be attempted depends critically on the nature of the specimen and on the way in which the image has been obtained. It is obviously vain to attempt to recover high resolution information from a fragile specimen that has been exposed to an electron dose large enough to destroy all the fine detail. It is not quite so obvious, though equally true, that high resolution information cannot be retrieved from an image taken with inadequate source coherence: that is, recorded in a microscope with too large an effective source or too wide an energy spread in the beam incident on the specimen. In a few laboratories, in which electron image processing has a high priority, attention is routinely accorded to such points. This chapter is, however, primarily addressed to biologists using modern commercial microscopes, and it is important to realize that certain conditions must be fulfilled if processing is to have a reasonable chance of success. We shall have more to say about this as we discuss the individual procedures, but, generally speaking, any precautionary measures that are offered by the manufacturer with a view to improving image quality should be adopted: the most stable gun supply, the smallest brightest source (pointed or lancet rather than hairpin filaments or, best of all, a field-emission tip), an anticontamination device, and any attachment designed to facilitate minimum exposure microscopy.

In addition, any extraneous interference with the microscope is to be avoided, always on the principle that the better the direct image, the more chance the processing has of yielding interesting results. The user rarely has much choice here, but should always have in mind that most types of image processing are difficult and delicate operations. Experience alone will establish which of them are worth attempting routinely and which should be embarked upon only in desperation, when some vital point can be resolved in no other way. It is scarcely ever worth attempting to process an image obtained in poor microscope conditions with the attitude that all the imperfections can be removed by the computer.

In the following paragraphs, we describe succinctly most of the types of electron image processing that have been proposed, concentrating as before on the principles rather than the details of the procedures. We begin with some very simple tricks that can be played with the computer for image embellishment; like aesthetic surgery, some of these may appeal to the user's sensibility without rendering his results any more useful scientifically—though perhaps they may be more publishable. We then examine the problems of weak (uncommon) objects and strongly scattering (normal) objects in turn. Next, we consider the important case of very fragile but periodic specimens and three-dimensional reconstruction from specimens with and without high symmetry. Finally, we discuss briefly the ability of the computer to reduce the labor involved in routine tasks such as particle counting and sizing.

Improvement of Image Appearance: "Aesthetic" Processing

This brief section illustrates one of the fringe benefits of an image processing facility, although the techniques described here are scarcely important enough to justify its existence. The point we wish to make is that once an image has been converted into digital form, a number of simple operations, some of which are routine stages of the photographic printing process, can be carried out method-ically and repeatably. For example, an important micrograph, though otherwise perfectly acceptable, may have been disfigured by nonuniform illumination in the electron microscope. This defect would be remedied in the dark-room by "dodging," but can be corrected in the computer by regarding the micrograph as the desired image superposed on a nonuniform background, the latter varying smoothly from bright on one side to dark on the other. The computer is then required to establish the direction of this nonuniformity and its extent; once this is known, the nonuniform background is simply subtracted, leaving an image with the same structural content but a more pleasing appearance. A similar im-provement can in principle be made to micrographs of sections marred by chatter of the microtome knife; striations due to chatter are often rather irregular, how-ever, and it may be less trouble to cut fresh sections, unless the material is in short supply for some reason or the feature in question is particularly elusive.

Somewhat less trivially, the plate or film may contain a very wide range of optical densities, ranging from very dark to very light. In the photographic print-ing process, the dark areas have to be protected from overexposure either manually or with the aid of an electronic device such as the LogEtronics (Bonner, 1965; Prosch, 1974; Jakobs and Katterwe, 1976). In the computer, however, the levels can easily be adjusted so that both light and dark zones reproduce satisfac-torily, although considerable care must then be taken in interpreting the micro-graph, since the relation between image blackness and specimen "density" will have been distorted, though not destroyed.

Finally, in this group of types of image modification of mainly aesthetic inter-est, we include some simple ways of removing noise. Most images exhibit some unwanted granularity, which arises in a variety of ways but conveys no information about the image, or at least, none that the eye can interpret. If the dominant size range of this granularity is, on the average, appreciably smaller than that of the smallest specimen detail of interest to us, we can eliminate most of the unwanted background by taking the Fourier transform of the image—a simple task for the computer—and removing the outer region of it, since this corresponds to small spacings (high spatial frequencies). We note in passing that the outer region must not be cut off too abruptly, since sudden changes in a Fourier transform have repercussions over all of the corresponding image, but attenuated gradually over a finite zone.

Noise removal is particularly interesting in the case of periodic objects, since the Fourier transform of a periodic object consists not of a continuous half-tone but of sharp spots; any structure in the transform that does not form a regular array is therefore noise and can be eliminated altogether, just as in the exactly analogous technique of optical filtering, where a mask containing holes for the diffraction spots is placed in the Fourier transform plane of a lens and the noise is intercepted by the mask (Klug and De Rosier, 1966; Gibbs and Rowe, 1976; Donelli and Paoletti, 1977). This technique is a good example of the precept that the more knowledge we have of a specimen, even if it appears very vague, the more chance we have of devising successful procedures. Thus the mere fact of periodicity is sufficient to enable us to improve image visual quality dramatically. So useful is such translational and rotational filtering that it takes us outside the realm of purely aesthetic processing.

The work of Amos and Klug (1972) on the fine polyheads of bacteriophage T4, Aebi *et al.* (1973) on the T-layer of *Bacillus brevis*, and Steven *et al.* (1976a,b) on the P23 surface shell of bacteriophage T4 polyheads provides convincing examples of the practical usefulness of identifying linear periodicities with the computer and of cleansing the image of all noise that does not conform to the pattern. Rotational periodicities have been identified and exploited in the same way by Crowther and Amos (1971b); there are distinct advantages over the optical method associated with Markham (Markham *et al.*, 1963, 1964; Horne and Markham, 1972).

Removal of Electron Optical Artifacts

In this section, we describe a class of procedures that enable the microscopist to improve the reliability of his images by ensuring that any contrast that has been reversed by the microscope is restored to its proper sense and, a more difficult task, by replacing image detail that has been filtered out by the microscope; the latter task involves combining at least two micrographs. Such techniques are far removed from the relatively simple image-cleaning operations described in the preceding paragraph; they have no photographic counterpart, and, in the general case where missing detail is replaced, they cannot be performed other than with the computer. The simpler task, that of eliminating contrast reversals and the contrast distortions that accompany them can be performed with the aid of an optical bench and suitable filters (Baumeister and Hahn, 1973; Stroke et al., 1974; Kübler et al., 1976). Optical processing falls outside the province of the present chapter (see Gibbs and Rowe, 1976), but we mention in passing that, although optical reconstruction requires very considerable experimental effort, the computer has, ironically, removed one of its major inconveniences. Until recently, the fact that every micrograph would need a slightly different filter and that such filters were rather difficult to make reduced the technique to a mere "tour de force"; now, however, it is possible to produce such filters by computer, provided that the latter is equipped with a high resolution microfilm output. An ingenious idea due to Lohmann et al. (1967) and Lohmann and Paris (1968) has been used by Burge and Scott (1975, 1976) to produce the kind of filters needed for electron microscopy in this way.

Contrast reversals and related image defects can only be corrected with any degree of confidence in the result if the cause of the defect is well understood, and if a given type of defect is always due to the same cause. This second condition is vital, for if it is not satisfied, a variety of "true" specimen structures could all give rise to the same image, or to some common feature of the image, and we would not know how to choose the true structure in any given case. This dilemma is known generically as the problem of uniqueness. The present section is restricted to techniques that provide a unique answer for a simple reason; essentially, it is that the range of specimens that can be analyzed is very narrow. In the section on the phase problem, we shall encounter techniques that are capable of succeeding even in more complicated situations. Here, we consider only weakly scattering specimens, for which the recorded image is related linearly, and hence unambiguously, to the phase and amplitude structure of the specimen. Unfortunately this limits the interest of the method to a very small range of biologists studying ultrathin and probably unstained sections (Misell, 1976). Nevertheless, the technique is of somewhat wider application, for it could be applied, as a preliminary step, to images of more everyday material.

The essence of the method is simple. For the restricted class of weak-phase objects, the recorded image is related in a straightforward way to the phase distribution over the object; the Fourier transform of the image is related to that of the object in an even simpler way, however: the intensity at each point of the transform of the image is equal to the product of the value at the corresponding point of the object transform and a variable attenuation factor, fixed by the microscope parameters (spherical aberration, defocus, wavelength, and, to a lesser extent, astigmatism). At first sight, it would seem that we have merely to restore the image intensities to their original values to obtain the true specimen. Although this is correct in principle, there are two related obstacles: first, some parts of the object transform are blocked altogether—they are not merely attenuated but wholly suppressed and cannot therefore be restored to their original value, since we have no information to work on. Furthermore, some of the detail will have been attenuated so strongly that it cannot be distinguished from the unwanted background, the noise, and there is no point in "restoring" detail when we cannot be sure whether it truly represents the specimen or just some speckling due to photographic grain size or the substrate, for example. We must therefore devise a reconstruction scheme that is strongly biased towards detail that we believe for some good reason to be true. This is a well known problem, and although the various solutions differ in detail, they share a common pattern.

Let us suppose that the intensity at some point in the image transform, denoted by I, is related to the corresponding intensity in the specimen transform, S, by the relation:

$$I = FS$$

where F denotes the attenuation: where all information is lost, $F = 0$, and where the true information cannot be distinguished from noise, F is small but not zero. Since we cannot reconstruct S, the specimen transform, by simply solving the equation, $S = I/F$, we write instead:

$$S = \frac{IF^*}{FF^* + N}$$

where N is a quantity determined by the noise and F^* is related in a simple way to F. Thus, where $F = 0$ or $F \approx 0$, the specimen transform will be unhelpful but not positively misleading; elsewhere, $FF^* \gg N$ and we obtain $S \approx I/F$ as required. This type of filtering has been studied in some detail and attempted in practice (e.g., Welton, 1971; Kübler *et al.*, 1976).

We have stressed that the loss or falsification of image information is determined by the microscope parameters: C_s, Δf, and λ. In particular the specific information that is lost completely (at and near $F = 0$) varies if these parameters are altered. We cannot (easily) change C_s, and the variation with λ is somewhat too small to be useful; furthermore, the electron–specimen interactions are altered in a way that is undesirable from our present standpoint if we change the accelerating voltage from say 60 kV ($\lambda = 4.9$ pm) to 120 kV ($\lambda = 3.3$ pm), about the maximum range available on conventional commercial microscopes. The defocus Δf, on the other hand, can be altered very easily indeed, by means of the focus control, so that if the specimen can tolerate the increased dose, two or more images of the same specimen area can be recorded with different values of defocus. Information absent from one image will therefore survive on the other, and the two can in principle be combined and reconstructed to give a faithful representation of the object for all sizes of detail, down to the limiting resolution, of course. This procedure has been much discussed and analyzed, but the experimental difficulties are formidable, for the reason that is by now all too familiar: we need this kind of reconstruction when high-resolution detail is of interest, and it is precisely such detail that is likely to be damaged, particularly in weak-phase specimens.

Very similar techniques are available for the more general and more realistic case of weak-phase, weak-amplitude objects. Here, at least two micrographs taken with different values of defocus must be obtained, as the information contained in a single picture is always inadequate.

We mentioned earlier the importance of working with images as free of unnecessary imperfections as possible. The reason for this can be readily understood if we regard the microscope as a device that filters the information coded in the beam by the specimen. The harmful results of an excessively large source and of fluctuations in beam current are immediately apparent in the attenuating effects that we described above. The larger the fluctuations, the coarser the detail that survives the passage of the beam through the microscope. This problem may be considered in terms of chromatic aberration, where we imagine several images superimposed on one another, each slightly out of focus with respect to the others, so that fine detail is blurred out; or in terms of partial temporal coherence, which has the advantage that the phenomena can be investigated in more detail, at the cost of a considerable increase in mathematical complexity. Too large a source also blurs fine detail, but not in so straightforward a way, for detail in a certain size range survives with little degradation, the range being determined by the defocus as well as the source geometry. We note in passing that this can be exploited by using a thin hollow beam instead of a solid beam of electrons, if the microscope condenser system is equipped with suitable apertures.

The more mundane precautions scarcely need further justification. Clearly every effort to prevent contamination is worthwhile, for any unwanted material adhering to the specimen will make it harder to extract the true specimen from the noise due to the substrate and to this contamination. Any movement of the specimen holder relative to the microscope due to some flaw in the stage mechanism or to external vibrations will blur the image and hence complicate the task of reconstruction; some such movements can be canceled in the computer, with very little loss of image quality, but it is still better to avoid them in the first place.

The Phase Problem

The fact that the techniques described in the preceding section were restricted to weakly scattering specimens rendered them of very limited interest to the majority of biological electron microscopists. A much wider range of specimens could benefit, potentially at least, from the fact that computer processing is capable of indicating how much reliance can be placed on electron microscope images if this restriction could be lifted. It is essentially with this problem in mind that attempts are being made to solve the "phase problem." This problem has already been discussed in this series by Burge (1976), to whose chapter we refer readers interested in the technical details of the various procedures; see also Burge *et al.* (1976). Here we give a more qualitative account.

If the electron beam is deflected in the specimen to such an extent that the weak-phase, weak-amplitude approximation breaks down, which is almost always the case, the fine detail in the image is related to the specimen structure in a way that is again simple to express mathematically; unfortunately, we can no longer predict the original specimen structure from the image unambiguously and cannot predict it at all if only one conventional image is available. Various procedures have been examined, all of which entail using more than one micrograph—generally speaking, the techniques work best when three pictures are available and, subject to certain reservations which usually emerge during practical attempts to use the method, the more the pictures differ from one another, the more chance the method has of success.

The pictures used may consist of an image and its electron diffraction pattern or two images taken with different values of defocus, if only two are used, or two images and a diffraction pattern or three images, if the specimen can tolerate three exposures. Once they are in the computer, the fact that they must be interrelated in an indirect but known way is repeatedly exploited until, eventually, a pair of images representing the phase and amplitude structure of the specimen is obtained. The calculation is iterative—that is, a sequence of operations is performed over and over again until a satisfactory result is reached—and it is regarded as successful if, on performing the sequence again, there is no noticeable change in the result. It is, of course, possible for this stage never to be reached: the procedure has not converged and must be abandoned. This result can occur with only two input pictures but is almost unknown with three. A more serious question is that of uniqueness, which we have already mentioned. Even if the sequence converges, is the answer correct, or could we arrive at any one of several answers, for a given pair (or more) of micrographs? This is an extremely difficult question to answer in general, though an extensive body of theory has now accumulated around it (Ferwerda, 1976).

The difficulty occurs at two levels. Even if the intensity measurements are regarded as "exact," the mathematics needed to establish or disprove uniqueness is complicated and yields answers from which it is not easy to draw general conclusions. In the case of a pair of pictures, consisting of image and diffraction pattern, there is reason to believe that the result will not always be unique. This procedure, the first to succeed in solving the phase problem, is known as the Gerchberg-Saxton algorithm, after the two Cambridge scientists who originated it (Gerchberg and Saxton, 1972, 1973). The later proposal, to use two images with different defocus values (Gerchberg and Saxton, 1973; Misell, 1973a,b), should in general yield a unique result. Unfortunately, the intensity measurements can never be exact, so that the relations that are assumed to exist between the image pairs are never quite true. This element of uncertainty, due to "noise" on the image, invalidates most of the theoretical analysis of uniqueness.

Fortunately, however, it is unlikely in practice that such doubts will arise, for the microscopist who invokes techniques such as these is likely to be studying in depth a specimen about which a reasonable amount is already known. Prior knowledge, not necessarily acquired with the electron microscope, is thus invoked to help in deciding whether the images produced by computer processing of this type are true or false, or, at least, questionable.

What use are these phase and amplitude distributions? What is the relation between them and the specimen structure? These questions can be answered on different levels of sophistication. Most simply, and for most purposes entirely adequately, we can regard the phase image as a faithful map of the projected structure of the object, that is, the structure that would be seen if all the specimens were compressed into a thin layer. The amplitude image should be similar, but may be expected to emphasize features that scatter a large proportion of the incident beam inelastically. It should be possible to use the phase and amplitude distributions more quantitatively, since a great deal is known about the variation of scattering with atomic number, but it is proving very difficult to make any progress on this front. Hoppe (1974) has pointed out that at the highest resolution, misleading effects due to the support film will be much less apparent in the amplitude image than in the phase image, always provided that the specimen generates sufficient amplitude information.

Three-dimensional Reconstruction

The various kinds of computer image processing that we have been describing are essentially two-dimensional. We have regarded the interaction between the electron beam and the specimen as a redistribution of the directions of motion of the electrons, largely ignoring the depth in the specimen at which the interaction occurs; the image-forming process is then regarded as a means of rendering this redistribution visible. The microscopist is, however, naturally interested in the three-dimensional structure of his specimen. Even before computer-aided methods became practical, qualitative and semiquantitative stereological methods were in use, on microscopes equipped with suitable tilting stages, and failing this, serial sections were used to build up a three-dimensional structure when the material was suitable. These methods too have benefited from the computer and are the subject of individual chapters in this series (Lange, 1976; Weibel and Bolender, 1973). In this section, we introduce the reader to techniques that are entirely dependent on the presence of rather sophisticated computing facilities.

If the three-dimensional structure of a specimen is to be explored, several views of it must be available; for example, we could not decide whether a small round disc in an image represented a spherical structure or a cylinder seen end-on without having at least one other view. These different views may be obtained in various ways, and, in particular, they may be real or virtual. The most successful attempts to perform three-dimensional reconstruction have used virtual views, by which we mean that the structures being studied were known to possess a particular brand of relatively high symmetry. A single image of such a specimen then effectively contains as much information as several views of the structure, the exact number being determined by the symmetry. Alternatively, the image may contain many examples of the same structure which happen to have different orientations; this too can give very satisfactory results, particularly when the individual particles exhibit some symmetry. Even so, problems can arise, for it is not at all obvious that the views could all have been obtained from the same original structure, since the latter may have become flattened or otherwise distorted to a different extent in each view. Nevertheless, each of these techniques requires at best only a single exposure. If neither of these procedures is applicable, either the specimen or the incident beam can be tilted to give a range of views. This need not, in principle at least, increase the electron dose to which the specimen is subjected, although the practical problems are formidable.

The techniques that exploit the native symmetry of the specimen were originally developed, both experimentally and theoretically, in the M.R.C. Laboratory of Molecular Biology in Cambridge (De Rosier and Klug, 1968; Crowther *et al.*, 1970; De Rosier, 1971), where the more difficult case of specimens with very little symmetry is now being actively investigated. This work has been reviewed by Amos (1974) and Crowther and Klug (1975), and an elementary introduction to the principles involved is given by Sandler (1975); the different mathematical techniques that have been used for performing the reconstruction are surveyed in Gordon and Herman (1974). Three-dimensional reconstruction programs using these methods are now available in several research centers. The relation between the number of views, real or virtual, and the resolution and the best way of combining these views have generated a large, occasionally repetitious literature, analyzed in Gordon and Herman (1974) and Vainshtein (1977); most users will be content to adopt a processing package, such as that in use in the Biozentrum in Basle, which we discuss briefly in the section on "Languages."

The best type of symmetry, so far as three-dimensional reconstruction is concerned, is helical symmetry; a good example of the practical use of the technique is to be found in the work of Amos and Klug (1975) and Smith *et al.* (1976), who studied the structure of the tail sheath of the T4 bacteriophage. With lower symmetry, the situation becomes more complicated; nevertheless, for fairly small spherical viruses, the protein coats of all of which have icosahedral symmetry, as far as is known, three or four views are sufficient to give a reconstruction containing all the detail that is preserved in the negatively stained specimen. Examples of reconstructions from such material are to be found in the work of Crowther and Amos (1971a), Mellema and Amos (1972), Finch *et al.* (1974), and Crowther *et al.* (1974). Other examples of three-dimensional reconstruction are to be found in Finch and Klug (1971; stacked-disc aggregate of tobacco mosaic virus protein), Mellema and Klug (1972; haemocyanin), Crowther *et al.* (1975; R17 and f2), Jack *et al.* (1975; tomato bushy stunt virus), Mellema (1975; capsids of alfalfa mosaic virus), and Wakabayashi *et al.* (1975; actin-tropomyosin and related complexes). Many further examples are listed in the review articles cited earlier.

The many successful attempts to reconstruct three-dimensional structures using these methods have relied heavily on the high specimen symmetry. It is natural to inquire how the techniques perform when material of lower symmetry is used, and Amos (1976) has examined the case of intact flagellar axonemes, which consist of a highly ordered bundle of microtubules, a central pair of singlets surrounded by nine doublets. If the symmetry of the central pair is disregarded, the structure can be regarded as possessing ninefold symmetry, so that at least eight different views would be necessary to give even 10 nm resolution. In this preliminary attempt, only four views were available to Amos, so that the resolution was not better than 20 nm, but it should be possible to improve on this. (Related work is described in Amos *et al.*, 1976.)

If the specimen has no known or suspected symmetry properties that we can exploit, three-dimensional reconstruction becomes very much more difficult, especially at high resolution. The development of the various new or refined techniques that become necessary has been studied for several years in the Max-Planck-Institut für Biochemie in Munich under the leadership of W. Hoppe (Hoppe *et al.*, 1968, 1970; Hoppe, 1972, 1974; Typke *et al.*, 1976). A recent article by Hoppe *et al.* (1976b) and the relevant contributions to an EMBO Workshop on "Unconventional electron microscopic methods for the study of molecular structures" (published as Cosslett and Hoppe, 1977) give a good idea of the effort involved.

Since several views must be obtained, either the beam or the specimen has to be tilted. Work is in progress on developing beam-tilt circuitry for the electron microscope, which would produce a sequence of images with the desired pattern of views; in a more sophisticated scheme, the image would be sent directly to a storage medium of some kind and built up gradually by deflecting the electron beam over a pattern of different tilts several times (Hoppe and Köstler, 1976). This procedure would ensure that each view showed the specimen damaged by about the same amount. It is necessary because all reconstruction schemes assume that the various views provided come from the same specimen; this would not be the case if the first few images represented a much less damaged structure than the later ones. So far, however, discussion has concentrated on tilting the specimen, and Hoppe *et al.* (1976a,b) have developed a high-precision microgoniometer for this purpose.

For many types of specimens, it is essential to restrict the electron dose administered to the specimen to as low a value as possible; this does not pose any new problems, since the set of tilted images can be obtained with the same total dose as a single minimum-exposure image, although statistical considerations may limit the number of views that can be obtained with a given dose to an unacceptably low figure (Hegerl and Hoppe, 1976; Hoppe and Grill, 1977).

The other problems of high-resolution three-dimensional reconstruction mainly arise in the computing stage. The molecule under examination may lie in slightly different planes for different tilt angles, and allowance must be made for this; some common origin must be recognized in the different views, by the use of correlation functions (briefly described in the section on "Picture-handling Languages"); finally, decisions must be made to deal with gaps in the information available due to the fact that the number of views is never as large as we should like.

These techniques have been used by Hoppe *et al.* (1974) to generate a three-dimensional image of negatively stained yeast fatty-acid synthetase molecules. (This work attracted criticism by Baumeister and Hahn, 1975, which was vigorously refuted by Hoppe *et al.*, 1975.)

Very Fragile Periodic Specimens

Specimens that fall into this category are of sufficient importance to warrant separate discussion, although the problems to be solved fall within the province of earlier sections. The advantage of periodicity is that the number of electrons needed to produce the desired image or diffraction pattern can be spread over a large number of effectively identical unit cells, so that each receives only a very small dose of electrons. The relation between the number of electrons per unit area, the area of each unit cell, and the minimum number of unit cells is discussed in detail by Unwin and Henderson (1975), who were the first to develop a practical method of determining the structures of unstained crystalline specimens. Using doses as small as 50 electrons/nm^2, these researchers succeeded in establishing the two-dimensional (projected) structure of two very radiation-sensitive substances, the specialized part of the cell membrane of *Halobacterium halobium* known as the purple membrane and platelets of beef liver catalase (Unwin and Henderson, 1975; Unwin, 1975).

We have repeatedly stressed the need to take precautions when obtaining images for computer processing; with very sensitive specimens, these precautions become even more draconian. Unwin and Henderson found it necessary to modify their (Philips EM 301) microscope by introducing a shutter mechanism above the specimen and a second viewing screen with binoculars; they also developed a new preparation procedure, in which glucose was used to replace the aqueous medium in the catalase specimen and to strengthen the resistance to vacuum of the purple membrane.

Both images and diffraction patterns were used in the subsequent computer reconstruction, images being measured with a Perkin-Elmer 1010A automatic microdensitometer and diffraction patterns with a Joyce-Loebl IIIC model. A large computer was then required to obtain the Fourier transform of the images, and the information from this, from the electron diffraction pattern, and from a high-dose micrograph was combined to produce two-dimensional images of the two specimens, to a resolution of 0.9 nm in the case of catalase and of 0.7 nm in that of the purple membrane.

Henderson and Unwin (1975) have subsequently succeeded in establishing the three-dimensional structure of the purple membrane, again unstained and exposed to a very low electron dose. For this, 15 diffraction patterns and 18 pairs of micrographs were obtained at various tilting angles between zero and 57°.

We have drawn particular attention to the work of Unwin and Henderson, demonstrating admirably as it does that results of practical interest can be obtained by "averaging" many images of the same structure, each obtained with a very small electron dose. The idea had already been advanced earlier, however, in particular by Glaeser (1971), and a very thorough discussion of the relation between the various parameters involved, including the recording medium, is to be found in Kuo and Glaeser (1975). Attempts are being made to apply the procedure to other substances (Mellema and Schepman, 1976; L-malate hydrolase).

Since there is no shortage of radiation-sensitive specimens that do not form periodic arrays, it is natural to inquire whether there is any hope of extending the techniques of Unwin and Henderson and Kuo and Glaeser to such material. A possible scheme for specimens in the form of irregular arrays of nonoverlapping, nearly identical particles has been proposed by Frank (1975). This scheme uses the properties of auto- and cross-correlation functions to establish the irregular lattice on which the nearly identical particles are situated, after which it is relatively easy to perform the averaging. The method has a much greater chance of success if the individual particles have the same orientation (see also Saxton and Frank, 1977).

Pattern Recognition; Particle Counting; Automation

The techniques described in the foregoing sections have relied heavily on a detailed knowledge of the image-forming process. We now consider the computer not as a device for facilitating image interpretation but as a means of reducing the labor of extracting numerical information from certain types of image. We might, for example, need to count the number of particles per unit area on a large quantity of micrographs, in a medical or genetic application. Alternatively, the lengths of a large number of strands might need to be measured, or the ellipticity of oval structures; it is easy to multiply examples of this kind of measurement, which can certainly be made by hand, but which may entail many hours of tedium with the ever-present possibility of error due to human frailty.

For such problems, the difficulty is to pose the questions in a form suitable for the computer. How is the computer to "recognize" an oval or a strand in a digitized image, particularly as the latter is extremely unlikely to consist of sharply contrasted structures against a noise-free background? Problems of this kind are among those studied in the field of pattern recognition. (Most of the texts on this subject are highly mathematical; Rosenfeld and Weszka, 1976, Rosenfeld and Kak, 1976, and Andrews, 1972, are among the more readable. A survey of typical applications, edited by Rosenfeld, 1976, gives a very vivid impression of the present state of knowledge in this domain, although electron microscopy is not one of the topics covered.) Once the computer has been "taught" how to recognize the structure of interest, it is extremely easy to perform the various simple numerical operations that are commonly required: counting; sizing; measuring length, periphery, maximum and minimum axes of an ellipse; calculation of area; averaging of some property over many particles—this list is by no means exhaustive.

Recognition of some class of particles in an image requires a different kind of approach from those we have met in previous sections, for we may deliberately wish to describe the particles rather vaguely. For example, in searching for oval particles, we might give the range within which the maximum and minimum dimensions fall, to distinguish the particles of interest from others or from specks of dirt or holes in the preparation. Even more vaguely, we might seek all particles bigger than some minimum size and convex, thus excluding kidney- or heart-shaped particles. We might wish to measure the length of strands which are liable to be tangled and knotted; here we would need to incorporate some strategy to enable the computer to follow a given strand, even if knotted, but to prevent it from jumping from one strand to another, even if tangled.

There is no single way of solving all questions of this kind, and techniques for dealing with them are still in a very primitive state. For certain types of measurement, it is best to develop optical equipment that will perform some of the recognition semiautomatically at least, assisted by a computer, after which the latter does any calculations that may be required. For others, it is simplest to measure the image with a microdensitometer and leave all the work of recognition and calculation to the computer; in this case, the latter will usually need to be large, since automatic pattern recognition notoriously requires vast amounts of computer memory. This approach is discussed by Ledley (1970), Nathan (1971), and Billingsley (1971).

Of the special-purpose systems, we restrict this account to a few typical examples, without attempting a comprehensive discussion. An extremely clear and thorough survey has been made by Preston (1976); although particularly concerned with cytology (cf. Prewitt, 1972), this long review contains an excellent account of most aspects of automatic counting and sizing of image structures. Most of the systems in use today were designed specifically for light microscopy; frequently, therefore, the microscope forms an integral part of the chain. For electron microscopy, provision would, of course, need to be made for working with electron microscope plates, film, or micrographs.

One such system is the programmable flying-spot microscope and picture preprocessor described by Eccles *et al.* (1976a), in use at Chelsea College, London. This consists of a small computer (PDP 8/E) linked to a flying-spot microscope; it is a very flexible arrangement, permitting very many extremely useful kinds of search and trace routines. The description by Eccles *et al.* of a typical operation illustrates this vividly:

The field is scanned systematically until an object is encountered, an object being indicated by a fall in transmitted light below a threshold value. There is then a jump from a scan to a search program and the boundary of the object is sought and traced, in the course of which operation the list of points forming the boundary is fed to the on-line computer. When tracing of the boundary is complete (i.e. when the starting-point has been regained), the computer is interrogated for further instructions. The answer to the interrogation might normally depend on the size of the object, but possibly on its shape and position (e.g., whether its boundary includes the edge of the field).

A decision is made within the computer whether to pass on to the next object or to obtain further information about the present one. . . .

The spatial and grey-level resolution of such a system is discussed in a companion paper (Eccles *et al.*, 1976b). The authors point out that a comparatively simple modification to their system would render it suitable for analyzing electron micrographs.

Another interesting system is described by Dunn *et al.* (1975) of the Universities of California and Pittsburgh. These authors were interested in the problem of analyzing the sizes and relative positions of large numbers of nerve fibers; they therefore developed a system for measuring nerve axon diameters and myelin thickness, and hence for calculating statistical information about the individual fibers in nerve bundles. Their system is, however, versatile and can be used to make similar measurements on any micrograph. They use a manual measuring instrument connected to a mini-computer (a PDP-11 with 24 kword of memory and a 1.2 Mword disc); the user traces out the structures of interest in his micrograph with a special stylus, which sends the coordinates of its path to the computer. Here, therefore, all the pattern recognition is performed by eye, which has advantages and disadvantages: the combination of eye and brain is much superior to any computer in recognizing shapes of a particular kind and is not likely to be misled by accidental image defects—apparent breaks in strands that are "clearly" continuous, for example; on the debit side, if a large number of similar structures have to be traced out, the image sent to the computer may be imperfect, through human failure: fatigue, inattention, outside distraction. . . . A number of checking routines, designed to detect such imperfections, are provided with the system of Dunn *et al.* (This last problem is not a trivial one; Preston, 1976, notes that "the National Communicable Disease Center [U.S.A.] has found that 40% of laboratories tested nationwide do unsatisfactory work" in cytological examinations, a task of comparable dullness.)

Several other systems have been built, each with its own special features, some emphasizing the pattern recognition aspect of optical data processing, others the large-scale counting and measuring tasks. We draw attention to a long review article by Mendelsohn *et al* (1968), which, although rather out of date now, gives a good idea of the development of such a system.

Finally, we turn to the systems that are commercially available, from Zeiss, Leitz, Wild Heerbrugg, Bausch and Lomb, and, above all, Image Analyzing Computers (Imanco). The Quantimet series produced by this last firm are general-purpose image analyzers, in which the primary image produced by the microscope (or via an epidiascope, for example) is projected onto a special television camera. The resulting electrical signal is then analyzed, and information can be extracted about the area of selected features, the size range of particles, and many other geometrical and statistical properties of the image. This information is available in a wide variety of forms: visual, on television monitors or a line-printer, or digital, on paper tape (or any other bulk storage device). General accounts of the Quantimet series are to be found in Beadle (1971), Cole and Bond (1972), and Crawley and Gardner (1972). The role of the Quantimet in scanning microscopy is discussed by Braggins *et al.* (1971).

PICTURE-HANDLING LANGUAGES

If the techniques described in the foregoing sections are to be used with any hope of success outside a few highly specialized laboratories, they will have to be made available in a form that requires little familiarity with computing. There are various ways of achieving this. Perhaps the most attractive in the long term is the interactive or conversational program, which asks the user questions, via a teletype or a TV screen, sometimes offering a choice of answers from which the user must select one. For example, the following (wholly fictional) exchange might occur:

Computer: Give the number of the technique required.

  1. Wiener filtering
  2. Phase determination
  3. Three-dimensional reconstruction
  4. Linear averaging
  5. Rotational averaging
  6. Other (see manual)

User : 2
C   : Accelerating voltage in kV?
U   : 100
C   : How many images?
U   : 2
C   : How many diffraction patterns?
U   : 1
C   : State image size
U   : 64 [i.e., 64 X 64 image points]
C   : List names and locations of each Im and Dp
U   : Im 1 "IM.GOLGI.1" MT01 [i.e., on magnetic tape deck 1]
       Im 2 "IM.GOLGI.2" MT01
       Dp 1 "DP.GOLGI.1" MT01
C   : How many iterations?
U   : 25 [i.e., halt calculation after 25 iterations]
C   : Requests?

```
U   : ACF (IM.GOLGI.1) [calculate the auto-correlation function of the first
      Golgi image]
      ACF (IM.GOLGI.2)
C   : Ready
U   : Display ACF (IM.GOLGI.1) [ACF then appears on the monitor]
C   : Ready
U   : Display ACF (IM.GOLGI.2)
C   : Ready
U   : Go [i.e., perform the calculation "2"]
C   : [after a pause] : Calculation completed; error 4% after 25 iterations.
      Requests?
U   : Display image amplitude
C   : Ready
U   : Display image phase
C   : Ready
U   : Display image amplitude + image phase
C   : Ready
U   : Image amplitude to PW [filmwriter]
C   : Ready
U   : Image phase to PW
C   : Ready
U   : Save IM.GOLGI.1, IM.GOLGI.2, DP.GOLGI.1
C   : Disc file (s) created on D01
U   : END
```

Such interactive programs, which cushion the user very effectively against the shock of using the computer, can be designed to accommodate varying degrees of participation on the part of the user. The simpler the questions, the more rigid is the structure; and if all the possible desires or wants of the user are to be catered to, it is better to leave him the option of interrupting the flow of question and answer in order to request other facilities that have been provided by the programmer but are not offered spontaneously.

A more flexible solution is the use of a picture-handling language, sufficiently simple to be easily mastered and almost self-explanatory, but offering a great deal more adaptability in operation. One such language, known as IMPROC (Saxton, 1974), was specifically designed for processing electron micrographs; it has subsequently been modified by W. O. Saxton and M. F. Horner (Horner, 1976) in such a way that it can be used on a minicomputer (in their case, a PDP 8/E) as well as on the IBM 370/165 for which it was first conceived. The new version is known as SEMPER (System for Electron Micrograph Processing and Enhancement of Resolution). We describe a recent version of this language, which is written wholly in Fortran and can hence run on almost any computer, whereas IMPROC relied upon a facility unsuited to mini-computers. A detailed account of the language SEMPER is to be found in the manual. Here we attempt to give some idea of the power of the language, which has been designed with users having little or no training in computing in mind.

What types of operation can we perform with such a language and how easy is it to learn how to use it? Before giving a complete answer to these questions, we describe a typical problem and its solution, using SEMPER, to give the reader a feel for what is involved. Suppose for example that we have two images of the same specimen area taken at different electron doses and wish to subtract them, to see how much has been lost. Clearly, the two pictures must be aligned extremely accurately—by which we mean so accurately that the finest detail that has survived is superimposed when the two images are subtracted. At worst, the original pictures will be rotated with respect to one another, will be misaligned and may even have slightly different magnifications. We ignore the magnification difference here, though it too can be remedied in practice. It is sometimes more convenient to work with a circular image instead of a square or rectangular one, and our first task is then to cut out a circular region from each image. This is done with a MASK instruction, thus:

MASK   FROM   PIC1   TO   PIC1M

PIC1 is the name allocated earlier to the first picture. The effect of this instruction is to mask off a circular region of PIC1 and name the result PIC1M. This basic instruction can be elaborated if the user has some particular preference for the mask radius, for example, or if some special action is required around the edge of the mask; otherwise the language automatically follows the "default" procedure, that is, the procedure laid down in the absence of any special supplementary instructions. Even the latter require a minimum of effort on the part of the user. To impose a radius of 20 units, for example, we have only to add RADIUS 20 to the line above.

The angle between two images is obtained in two stages. The computer will be requested to rotate one image over the other until a strong resemblance is noted. For this, we must first convert the real image into a centro-symmetric structure that is still highly characteristic of the original, so that the center of rotation is common to the two images; we must then devise a measure of resemblance that the computer can easily calculate. These requirements lead us into the realm of correlation functions. A suitable centro-symmetric representation of the image is its auto-correlation function (ACF); we obtain the autocorrelation functions of the pictures by writing

ACF   FROM   PIC1   TO   ACF1

and likewise

ACF   FROM   PIC2   TO   ACF2

The instruction

ORIENT   FROM   ACF1   WITH   ACF2

then yields the angle THETA between the autocorrelation functions by calculating the cross-correlation function around a set of rings.

The cross-correlation of two functions, or rings of sample values of the pictures, has the property of reaching a reasonably sharp maximum when the superimposed images bear the greatest resemblance to each other. One image is then rotated through the angle THETA, after which the relative position of the two images – that is, the translational misalignment–is calculated, again by means of cross-correlation. Finally, one image is shifted by the amount of the misalignment. This alignment step might be achieved as follows:

(1) EXIN   P2   PIC2
(2) XCF   PIC1   TO   PIC2   WITH   PIC2
(3) EXIN   P2   PIC2   UP   Y   RIGHT   X

Line (1) uses EXIN, an instruction that we have not yet met; it simply converts an image in a form suitable for filmwriter or microdensitometer ("external") to a form suited to the computer ("internal"), or conversely. Line (1) thus renames the microdensitometer picture P2 as PIC2. Line (2) forms the cross-correlation function of PIC1 and PIC2 and stores the result as PIC2, thus destroying the record of PIC2 in the computer, though it is of course still available as P2. More important, it yields the coordinates X and Y of the origin of one picture relative to that of the other. Line (3) again brings the microdensitometer picture P2 into the computer but also shifts it upwards by Y and to the right by X so that PIC1 and PIC2 are now exactly aligned. They can now be subtracted and the result examined on the monitor, or if desired, sent to the filmwriter. Subtraction and the other arithmetic operations require such self-explanatory commands as SUBTRACT, ADD, DIVIDE, and MULTIPLY. Visual display is obtained by means of DISPLAY, which draws the image requested on the oscilloscope screen with nine grey levels; CONTOUR, which plots a contour map; or SCAN, which gives an oblique aerial perspective view of the image, represented as a three-dimensional structure–a mountain range in which the height corresponds to the contrast of the image.

This example gives a fair idea of the simplicity of the instructions needed to perform rather complicated operations on the images provided. SEMPER offers many other possibilities, just as easy to master, some of which we now list briefly. FOURIER and IMAGE calculate the diffraction pattern of an image and *vice versa*, respectively; PS calculates the power-spectrum, a function intimately related to the auto-correlation function; WALSH produces an image transform similar in some ways to the Fourier transform. Another series of instructions handles the "filter" involving spherical aberration, defocus, and astigmatism that characterizes the action of the microscope at high resolution. Lattice averaging and linear averaging are available, using LATTAV and AVERAGE. HILBERT provides the transform associated with the use of a half-plane aperture in the microscope. Aesthetic processing is provided by REMOVE with the options DUST, RAMP, and ATTEN. RAMP renders the illumination uniform across a picture that is dark on one side and light on the other, while DUST eliminates points where the local intensity changes so rapidly that they are more likely to be dust than true structure; ATTEN also renders the illumination uniform but also sets the mean illumination at a suitable value, thus boosting the visibility of the low contrast sections.

In addition to these instructions designed for the practicing microscopist, there are a number of simple commands that permit more direct control of and interaction with the computer.

Another highly developed system for analyzing images of crystalline or mainly regular systems is used in the Biozentrum in Basle; this system is dependent on the availability of a large IBM computer and will be found bewildering at first to the microscopist who has no prior experience of computing, although the commands are again relatively simple. We shall not give much description of this system here; we mention it in order to show that more or less accessible systems for performing very complex operations on pictures with relatively little computing effort are slowly beginning to proliferate, and it is safe to predict that this process will gain momentum as more and simpler systems become available.

The various operations provided in the Biozentrum processing program are set in motion by two-letter commands.* Subtraction of an uneven background, to correct for nonuniform illumination, for example, involves the user in the following dialogue:

```
User       : sg [subtract gradient]
Computer : IF YOU WANT THE BACKGROUND SUBTRACTED ENTER '1'
User       : I1
             0 or 1 [if 0 is typed, the background is subtracted; if 1 is typed, the
             background is again subtracted but in addition, the average intensity
             is reduced to zero]
Computer : SLX = a, SLY = b  Q = c [where a, b, and c are numbers; these tell the
             user the direction of the intensity gradient that has been corrected
             and its magnitude]
             LINEAR GRADIENT SUBTRACTED
```

There are a great many possibilities, each initiated by a two-letter command, some of which permit very complex operations on the image. Thus filter (fl) removes nonperiodic detail (noise) from the image of a periodic structure by means of a filter, consisting of rectangular "holes" located on a rectangular lattice; fl is only one of a family of such filters, designed for different symmetries. In a similar spirit, li (linear superposition) superposes rectangular unit cells packed in a rectangular array as in a two-dimensional crystal; li produces the following exchange:

```
User       : li
```

*The following description is taken from the manual, kindly provided by P. R. Smith of the Biozentrum.

Computer : ENTER UNIT CELL SIZE AND NUMBER OF UNIT CELLS IN X
 AND Y DIRECTION
User : ...
Computer : ENTER WANTED NUMBER OF UNIT CELLS IN X AND Y
 DIRECTION IN FINAL ARRAY
User : ...
Computer : MARKHAMED

Other commands enable the user to work with layerlines in various ways and to perform a very wide range of operations on pictures.

Other languages are in use, but most are less oriented toward the convenience of the user unfamiliar with computer jargon and conventions. The powerful language in use at the Jet Propulsion Laboratory in Pasadena for processing pictures from satellites and spacecraft is an example.

CONCLUDING REMARKS

In his opening address at the meeting of the German Electron Microscopy Society held in Berlin in 1975, Professor P. Sitte discussed "Electron microscopy and biology—the fate of a symbiosis." He drew attention to the fact that, although periodic specimens are important in some branches of biology,

> some very important advances in . . . cell biology have been achieved by the application of non-integrating, non-statistical methods like ordinary electron microscopy . . . . What appears to be of primary importance in the field of biological electron microscopy is not so much a further improvement of optical resolution of the EM itself but (a) devices for the observation of aperiodic organic structures at low local beam intensity . . . and finally (d) an improvement in methods of qualitative and quantitative analysis of submicroscopic biostructures. (Sitte, 1976)

In the foregoing sections, we have seen that, although considerable efforts are being devoted to the problems of high resolution, the most interesting results obtained so far by combining the power of the electron microscope with that of the computer have been at modest resolution, hardly a surprising result. More interestingly, we have seen that most of the technical problems of three-dimensional reconstruction have now been solved for specimens with high natural symmetry and that material with low symmetry or none at all is now being investigated.

We can confidently anticipate that such efforts will continue and that other types of computer image analysis, not yet introduced into electron microscopy, will be incorporated if they appear to be useful. A good example is texture analysis, which we have not mentioned, since it is a very new subject and its relevance to electron microscopy is not yet obvious. Nevertheless, it seems reasonable to expect that, since the eye relies so heavily upon textural differences, the computer may well benefit from textural studies also. (A good account of work in this field is to be found in Pressman, 1976, 1977.) Further information about the directions in which digital image processing is progressing in fields other than electron microscopy may be obtained from the numerous texts that have appeared recently, in particular the monographs by Andrews and Hunt (1977) and Rosenfeld and Kak (1976) and the sets of essays edited by Rosenfeld (1976), Fu (1976), and Huang (1975).

In the coming years, progressively more biologists will be able to sample computer image processing, in order to discover whether their particular images can benefit from the procedures available; conversely, those developing the procedures will be better able to appreciate the most useful directions in which to pursue their efforts. If computer processing of biology electron microscopes is to flourish, a genuine "symbiosis" will have to be achieved between computer experts and biologists.

References[*]

Aebi, U., Smith, P. R., Dubochet, J., Henry, C., and Kellenberger, E. (1973). A study of the structure of the T-layer of *Bacillus brevis. J. Supramol. Str.* **1**, 498.

Amos, L. A. (1974). Image analysis of macromolecular structures. *J. Microscopy* **100**, 143.

Amos, L. A. (1976). Three-dimensional image reconstruction of intact flagellar axonemes from a tilt series of electron micrographs. *Proc. Sixth Eur. Cong. Electron Microscopy* (Jerusalem) **1**, 14.

Amos, L. A., and Klug, A. (1972). Image filtering by computer. *Proc. Fifth Eur. Cong. Electron Microscopy* (Manchester), 580.

Amos, L. A., and Klug, A. (1975). Three-dimensional image reconstructions of the contractile tail of T4 bacteriophage. *J. Mol. Biol.* **99**, 51.

Amos, L. A., Linck, R. W., and Klug, A. (1976). Molecular structure of flagellar microtubules. In: *Cell Motility* (Goldman, R. D., Pollard, T. D. and Rosenbaum, J. L., eds.), pp. 847–867. Cold Spring Harbor Laboratory, New York.

Andrews, H. A. (1972). *Introduction to Mathematical Techniques in Pattern Recognition.* Wiley-Interscience, New York and London.

Andrews, H. C., and Hunt, B. R. (1977). *Digital Image Restoration.* Prentice-Hall, Englewood Cliffs, N.J.

Baggett, M. C., and Glassman, L. H. (1974). SEM image processing by analog homomorphic filtering techniques. In: *Scanning Electron Microscopy* (Johari, O., and Corvin, I., eds.), pp. 199–206. IITRI, Chicago.

Baumeister, W., and Hahn, M. (1973). Electron microscopy of thorium atoms in monomolecular layers. *Nature* **241**, 445.

Baumeister, W., and Hahn, M. H. (1974). Suppression of lattice periods in vermiculite single crystal specimen supports for high resolution electron microscopy. *J. Microscopy* **101**, 111.

Baumeister, W., and Hahn, M. (1975). Relevance of three-dimensional reconstructions of stain distributions for structural analysis of biomolecules. *Hoppe-Seylers Z. Physiol. Chem.* **356**, 1313.

---

[*]Wherever possible, we have referred to articles and books that can normally be expected to be accessible to biologists and avoid too heavy reliance on mathematical reasoning. A much shorter account of the material in this section, to which it may provide an introduction, is to be found in Hawkes (1975).

Baumeister, W., and Hahn, M. (1976a). An improved method for preparing single crystal specimen supports: H_2O_2 exfoliation of vermiculite. *Micron* **7**, 247.

Baumeister, W., and Hahn, M. (1976b). Prospects for atomic resolution electron microscopy in membranology. *Prog. Surface Membrane Sci.* **11**, 227.

Baumeister, W., Fringeli, U. P., Hahn, M. H., Kopp, F., and Seredynski, J. (1976). Radiation damage in tripalmitin layers studied by means of infrared spectroscopy and electron microscopy. *Biophys. J.* **16**, 791.

Beadle, C. (1971). The Quantimet image analysing computer and its applications. *Adv. Opt. Electron Micr.* **4**, 361.

Beer, M., Frank, J., Hanszen, K. J., Kellenberger, E., and Williams, R. C. (1975). The possibilities and prospects of obtaining high-resolution information (below 30 Å) on biological material using the electron microscope. *Quart. Rev. Biophys.* **7**, 211.

Billingsley, F. C. (1971). Image processing for electron microscopy. II. A digital system. *Adv. Opt. Electron Micr.* **4**, 127. [Pt. I is Nathan, 1971.]

Bonner, M. K. (1965). Technique for improving information quality in electron micrograph prints. *Appl. Opt.* **4**, 359.

Braggins, D. W., Gardner, G. M., and Gibbard, D. W. (1971). The applications of image analysis techniques to scanning electron microscopy. In: *Scanning Electron Microscopy* (Johari, O., and Corvin, I., eds.), pp. 393–400. IITRI, Chicago.

Brüders, R., Herrmann, K. H., Krahl, D., and Rust, H. P. (1976). Signal processing in a digital storage unit. *Proc. Sixth Eur. Cong. Electron Microscopy* (Jerusalem) **1**, 318.

Burge, R. E. (1976). Contrast and image formation of biological specimens. In: *Principles and Techniques of Electron Microscopy*, Vol. 6 (Hayat, M. A., ed.), pp. 85–116. Van Nostrand Reinhold Company, New York and London.

Burge, R. E., and Garrard, D. F. (1968). The resolution of photographic emulsions for electrons in the energy range 7–60 keV. *J. Phys. E: Sci. Instru.* **1**, 715.

Burge, R. E., and Scott, R. F. (1975). Binary filters for high resolution electron microscopy. *Optik* **43**, 53.

Burge, R. E., and Scott, R. F. (1976). Binary filters for high resolution electron microscopy. II. *Optik* **44**, 159.

Burge, R. E., Fiddy, M. A., Greenaway, A. H., and Ross, G. (1976). The phase problem. *Proc. Roy. Soc. London* **A350**, 191.

Burge, R. E., Garrard, D. F., and Browne, M. T. (1968). The response of photographic emulsions to electrons in the energy range 7–60 keV. *J. Phys. E: Sci. Instru.* **1**, 707.

Cole, M., and Bond, C. P. (1972). Recent advances in automatic analysis using a television system. *J. Microscopy* **96**, 89.

Coleman, C. I., and Boksenberg, A. (1976). Image intensifiers. *Contemp. Phys.* **17**, 209.

Cosslett, V. E., and Hoppe, W., eds. (1977). Unconventional electron microscopical methods for the investigation of molecular structures. *Adv. Struct. Res. Diffraction Methods* **7**.

Cosslett, V. E., Jones, G. L., and Camps, R. A. (1974). Image viewing and recording in high voltage electron microscopy. In: *High Voltage Electron Microscopy* (Swann, P. R., Humphreys, C. J., and Goringe, M. J. eds.), pp. 147–154. Academic Press, London and New York.

Cowley, J. M. (1976). The principles of high resolution electron microscopy. In: *Principles and Techniques of Electron Microscopy*, Vol. 6 (Hayat, M. A., ed.), pp. 40–84. Van Nostrand Reinhold Company, New York and London.

Crawley, J. A., and Gardner, G. M. (1972). The applications of image analysis techniques to transmission electron microscopy. *Proc. Fifth. Eur. Cong. Electron Microscopy* (Manchester), unnumbered supplementary sheet.

Crowther, R. A., and Amos, L. (1971a). Three-dimensional image reconstruction of some small spherical viruses. *Cold Spring Harbor Symp. Quant. Biol.* **36**, 489.

Crowther, R. A., and Amos, L. (1971b). Harmonic analysis of electron microscopic images with rotational symmetry. *J. Mol. Biol.* **60**, 123.

Crowther, R. A., and Klug, A. (1975). Structural analysis of macromolecular assemblies by image reconstruction from electron micrographs. *Ann. Rev. Biochem.* **44**, 161.

Crowther, R. A., Amos, L. A., and Finch, J. T. (1975). Three-dimensional image reconstructions of bacteriophages R17 and f2. *J. Mol. Biol.* **98**, 631.

Crowther, R. A., De Rosier, D. J., and Klug, A. (1970). The reconstruction of a three-dimensional structure from projections and its application to electron microscopy. *Proc. Roy. Soc. London* **A317**, 319.

Crowther, R. A., Geelen, J. L. M. C., and Mellema, J. E. (1974). A three-dimensional image reconstruction of cowpea mosaic virus. *Virology* **57**, 20.

Dainty, J. C., and Shaw, R. (1974). *Image Science: Principles, Analysis and Evaluation of Photographic-type Imaging Processes*. Academic Press, London and New York.

De Rosier, D. J. (1971). The reconstruction of three-dimensional images from electron micrographs. *Contemp. Phys.* **12**, 437.

De Rosier, D. J., and Klug, A. (1968). Reconstruction of three dimensional structures from electron micrographs. *Nature* **217**, 130.

Donelli, G., and Paoletti, L. (1977). Electron micrograph analysis by optical transforms. *Adv. Electron. Electron Phys.* **43**, 1.

Dorignac, D. (1974). Observation d'atomes isolés en microscopie électronique par transmission. *Proc. Eighth Int. Cong. Electron Microscopy* (Canberra) **1**, 270.

Dunn, R. F., O'Leary, D. P., and Kumley, W. E. (1975). Quantitative analysis of micrographs by computer graphics. *J. Microscopy* **105**, 205.

Eccles, M. J., McCarthy, B. D., Proffitt, D., and Rosen, D. (1976a). A programmable flying-spot microscope and picture preprocessor. *J. Microscopy* **106**, 33.

Eccles, M. J., McCarthy, B. D., Proffitt, D., and Rosen, D. (1976b). The spatial and grey-level resolution of flying-spot microscopes. *J. Microscopy* **106**, 43.

Erickson, H. P. (1973). The Fourier transform of an electron micrograph—first order and second order theory of image formation. *Adv. Opt. Electron Micr.* **5**, 163.

Ferwerda, H. A. (1976). The phase problem in electron microscopy. *Proc. Sixth Eur. Cong. Electron Microscopy* (Jerusalem) **1**, 1.

Finch, J. T., and Klug, A. (1971). Three-dimensional reconstruction of the stacked-disk aggregate of tobacco mosaic virus protein from electron micrographs. *Phil. Trans. Roy. Soc. London* **B261**, 211.

Finch, J. T., Crowther, R. A., Hendry, D. A., and Struthers, J. K. (1974). The structure of *Nudaurelia capensis* β virus: the first example of a capsid with icosahedral surface symmetry T = 4. *J. Gen. Virol.* **24**, 191.

Finlay, B., and Brown, I. A. (1971). On-line analysis of scanning electron microscopy images. *Beitr. Elektronenmikroskop. Direktabb. Oberfl.* **4/1**, 47.

Fiori, C. E., Yakowitz, H., and Newbury, D. E. (1974). Some techniques of signal processing in scanning electron microscopy. In: *Scanning Electron Microscopy* (Johari, O., and Corvin, I., eds.), pp. 167–174. IITRI, Chicago.

Frank, J. (1973). Computer processing of electron micrographs. In: *Advanced Techniques in Biological Electron Microscopy* (Koehler, J. K., ed.), pp. 215–274. Springer Verlag, Berlin and New York.

Frank, J. (1975). Averaging of low-exposure electron micrographs of non-periodic objects. *Ultramicroscopy* **1**, 159.

Frank, J., Salih, S. M., and Cosslett, V. E. (1974). Radiation damage assessment by digital correlation of images. *Proc. Eighth Int. Cong. Electron Microscopy* (Canberra) **2**, 678.

Fu, K. S., ed. (1976). *Digital Pattern Recognition*. Springer Verlag, Berlin and New York.

Gerchberg, R. W., and Saxton, W. O. (1972). A practical algorithm for the determination of phase from image and diffraction plane pictures. *Optik* **35**, 237.

Gerchberg, R. W., and Saxton, W. O. (1973). Wave phase from image and diffraction plane pictures. In: *Image Processing and Computer-aided Design in Electron Optics* (Hawkes, P. W., ed.), pp. 66–81. Academic Press, London and New York.

Gibbs, A. J., and Rowe, A. J. (1976). Optical analysis and reconstruction of images. In: *Principles and Techniques of Electron Microscopy*, Vol. 7 (Hayat, M. A., ed.), pp. 202–230. Van Nostrand Reinhold Company, New York and London.

Glaeser, R. M. (1971). Limitations to significant information in biological electron microscopy as a result of radiation damage. *J. Ultrastruct. Res.* **36**, 466.

Glaeser, R. M. (1975). Radiation damage and biological electron microscopy. In: *Physical Aspects of Electron Microscopy and Microbeam Analysis*. (Siegel, B. M., and Beaman, D. R., eds.), pp. 205–229. Wiley, New York and London.

Glaeser, R. M. (1976). Theoretical and practical considerations affecting high resolution HVEM of organic materials. In: *Microscopie électronique à haute tension, 1975* (Jouffrey, B., and Favard, P., eds.) pp. 165–170. Société Française de Microscopie Electronique, Paris.

Glaeser, R. M., Kuo, I., and Budinger, T. F. (1971). Method for processing of periodic images at reduced levels of electron irradiation. *Proc. 29th Ann. Meeting EMSA* (Boston), p. 466.

Goldfarb, W., and Siegel, B. M. (1973). A computer-controlled vidicon system for high resolution microscopy. *Proc. 31st Ann. Meeting EMSA* (New Orleans), p. 264.

Goldfarb, W., and Siegel, B. M. (1975). The properties and use of a computer interfaced video system for high resolution microscopy. *Proc. 33rd Ann. Meeting EMSA* (Las Vegas), p. 124.

Goldstein, I., and Yakowitz, H., eds. (1975). *Practical Scanning Electron Microscopy*. Plenum Press, New York and London.

Gordon, R., and Herman, G. T. (1974). Three-dimensional reconstruction from projections: a review of algorithms. *Int. Rev. Cytol.* **38**, 111.

Hahn, M., and Baumeister, W. (1974). High resolution negative staining of ferritin molecules on vermiculite single crystal supports. *Biochim. Biophys. Acta* **371**, 267.

Hanszen, K. J. (1971). The optical transfer theory of the electron microscope: fundamental principles and applications. *Adv. Opt. Electron Micr.* **4**, 1.

Hawkes, P. W. (1972). *Electron Optics and Electron Microscopy*. Taylor and Francis, London.

Hawkes, P. W. (1975). Computer processing of electron micrographs: a nonmathematical account. *Int. Rev. Cytol.* **42**, 103.

Hayat, M. A., ed. (1974–1978). *Principles and Techniques of Scanning Electron Microscopy*. Van Nostrand Reinhold Company, New York and London.

Hayat, M. A. (1978). *Introduction to Biological Scanning Electron Microscopy*. University Park Press, Baltimore, Maryland.

Hayes, T. L., and Pease, R. F. W. (1968). The scanning electron microscope: principles and applications in biology and medicine. *Adv. Biol. Med. Phys.* **12**, 85.

Hegerl, R., and Hoppe, W. (1976). Influence of electron noise on three-dimensional image reconstruction. *Z. Naturforsch* **31a**, 1717.

Henderson, R., and Unwin, P. N. T. (1975). Three-dimensional model of purple membrane obtained by electron microscopy. *Nature* **257**, 28.

Herrmann, K. H., and Krahl, D. (1973). Bildverstärker. In: *Methodensammlung der Elektronenmikroskopie* (Schimmel, G., and Vogell, W., eds.), § 1.2.2. Wissenschaftliche Verlagsgesellschaft, Stuttgart.

Herrmann, K. H., Krahl, D., and Rindfleisch, V. (1972). Use of TV image intensifiers in electron microscopy. *Siemens Forsch. und Entwickl. Ber.* **1**, 167.

Herrmann, K. H., Krahl, D., and Rust, H. P. (1976a). Digital TV storage unit for electron microscopy. *Proc. Sixth Eur. Cong. Electron Microscopy* (Jerusalem) **1**, 320.

Herrmann, K. H., Krahl, D., Rust, H. P., and Ulrichs, O. (1976b). Aufbau und Anwendung eines digitalen Fernsehbild-Halbleiterspeichers für die Elektronenmikroskopie. *Optik* **44**, 393.

Herzog, R. F., and Everhart, T. E. (1973). Scanning electron microscope–computer interactions. In: *Scanning Electron Microscopy: Systems and Applications 1973* (Nixon, W. C., ed.), pp. 2–5. Institute of Physics, London and Bristol.

Herzog, R. F., Lewis, B. L., and Everhart, T. E. (1974). Computer control and the scanning electron microscope. In: *Scanning Electron Microscopy* (Johari, O., and Corvin, I., eds.) pp. 175–182. IITRI, Chicago.

Holt, D. B., Muir, M. D., Grant, P. B., and Boswarva, I. M. (1974). *Quantitative Scanning Electron Microscopy*. Academic Press, London and New York.

Hoppe, W. (1972). Dreidimensional abbildende Elektronenmikroskope I. Prinzip der Geräte. *Z. Naturforsch.* **27a**, 919.

Hoppe, W. (1974). Towards three-dimensional "electron microscopy" at atomic resolution. *Naturwissenschaften* **61**, 239.

Hoppe, W. (1975). Principles of trace structure analysis in electron microscopy. *Z. Naturforsch.* **30a**, 1188.

Hoppe, W., and Grill, B. (1977). Prospects of three-dimensional high resolution electron microscopy of non-periodic specimens. *Ultramicroscopy* **2**, 153.

Hoppe, W., and Kostler, D. (1976). Experimental results in high resolution electron microscopy using the tilt image reconstruction method. *Proc. Sixth Eur. Cong. Electron Microscopy* (Jerusalem) **1**, 99.

Hoppe, W., Gassmann, J., Hunsmann, N., Schramm, H. J., and Sturm, M. (1974). Three-dimensional reconstruction of individual negatively stained yeast fatty-acid synthetase molecules from tilt series in the electron microscope. *Hoppe-Seylers Z. Physiol. Chem.* **355**, 1483.

Hoppe, W., Gassmann, J., Hunsmann, N., Schramm, H. J., and Sturm, M. (1975). Comments on the paper "Relevance of three-dimensional reconstructions of stain distributions for structural analysis of biomolecules." *Hoppe-Seylers Z. Physiol. Chem.* **356**, 1317.

Hoppe, W., Hunsmann, N., Schramm, H. J., Sturm, M., Grill, B., and Gassmann, J. (1976a). Three-dimensional electron microscopy of individual objects. *Proc. Sixth Eur. Cong. Electron Microscopy* (Jerusalem) **1**, 8.

Hoppe, W., Langer, R., Knesch, G., and Poppe, C. (1968). Protein-Kristallstrukturanalyse mit Elektronenstrahlen. *Naturwissenschaften* **55**, 333.

Hoppe, W., Möllenstedt, G., Perutz, M. F., and Ruska, E., eds (1970). Methoden zur Untersuchung der atomaren Struktur von biogenen Makromolekülen. *Ber. Bunsen-Ges. Phys. Chem.* **74**, Nr. 11, 1089–1224.

Hoppe, W., Schramm, H. J., Sturm, M., Hunsmann, N., and Gassmann, J. (1976b). Three-dimensional electron microscopy of individual biological objects. I. Methods. *Z. Naturforsch.* **31a**, 645.

Horne, R. W., and Markham, R. (1972). Application of optical diffraction and image reconstruction techniques to electron micrographs. In: *Practical Methods in Electron Microscopy*, Vol. 1 (Glauert, A. M., ed.), pp. 327-434. North Holland, Amsterdam and London.

Horner, M. F. (1976). Practical computer systems for processing electron microscope images. In: *Developments in Electron Microscopy and Analysis* (Venables, J. A., ed.), pp. 209–212. Academic Press, London and New York.

Huang, T. S., ed. (1975). *Picture Processing and Digital Filtering*. Springer Verlag, Berlin and New York.

Humphreys, C. (1976). High voltage electron microscopy. In: *Principles and Techniques of Electron Microscopy*, Vol. 6 (Hayat, M. A., ed.), pp. 1–39. Van Nostrand Reinhold Company, New York and London.

Huxley, H. E., and Klug, A., eds. (1971). A discussion on new developments in electron microscopy with special emphasis on their application in biology. *Phil. Trans. Roy. Soc. London* **B261**, No. 837, 1–230.

Isaacson, M. S. (1975). Inelastic scattering and beam damage of biological molecules. In: *Physical Aspects of Electron Microscopy and Microbeam Analysis* (Siegel, B. M., and Beaman, D. R., eds.), pp. 247–258, Wiley, New York and London.

Isaacson M. S. (1976). Specimen damage in the electron microscope. In: *Principles and Techniques of Electron Microscopy*, Vol. 7 (Hayat, M. A., ed.), pp. 1–70. Van Nostrand Reinhold Company, New York and London.

Jack, A., Harrison, S. C., and Crowther, R. A. (1975). Structure of tomato bushy stunt virus. II. Comparison of results obtained by electron microscopy and X-ray diffraction. *J. Mol. Biol.* **97**, 163.

Jakobs, R.-H., and Katterwe, H. (1976). Interpretation of photographically recorded electron diffraction patterns by means of electronic contrast compensation. *Microsc. Acta* **78**, 238.

Johansen, B. V. (1976). Bright field electron microscopv of biological specimens. VI. Signal-to-noise ratio in specimens prepared on amorphous carbon and graphite crystal supports. *Micron* **7**, 157.

Joy, D. C., and Verney, G. E. (1973). Digital image processing. In: *Scanning Electron Microscopy: Systems and Applications 1973* (Nixon, W. C., ed.), pp. 50–55. Institute of Physics, London and Bristol.

Klug, A., and De Rosier, D. J. (1966). Optical filtering of electron micrographs: reconstitution of one-sided images. *Nature* **212**, 29.

Kölbel, H. K. (1976). Kohle–Trägerfilme für die hochauflösende Elektronenmikroskopie– Verbesserung von Eigenschaften und Herstellungstechnik. *Mikroskopie* **32**, 1.

Kübler, O., Hahn, M., and Seredynski, J. (1976). Image processing as tool for radiation damage assessment and minimization. *Proc. 6th Eur. Cong. Electron Microscopy* (Jerusalem) **1**, 306.

Kuo, I. A. M., and Glaeser, R. M. (1975). Development of methodology for low-exposure, high-resolution electron microscopy of biological specimens. *Ultramicroscopy* **1**, 53.

Lange, R. H. (1976). Tilting experiments in the electron microscope. In: *Principles and Techniques of Electron Microscopy*, vol. 6 (Hayat, M. A., ed.), pp. 241–270. Van Nostrand Reinhold Company, New York and London.

Ledley, R. S. (1970). Use of computers in biomedical pattern recognition. *Adv. Comput.* **10**, 217.

Lenz, F. (1971). Transfer of image information in the electron microscope. In: *Electron Microscopy in Material Science* (Valdrè, U., ed.), pp. 540–569. Academic Press, New York and London.

Lewis, B. L., and Sakrison, D. J. (1975). Computer enhancement of scanning electron micrographs. *IEEE Trans.* **CAS-22**, 267.

Lohmann, A. W., and Paris, D. P. (1968). Computer generated spatial filters for coherent optical data processing. *Appl. Opt.* **7**, 651.

Lohmann, A. W., Paris, D. P., and Werlich, H. W. (1967). A computer generated spatial filter, applied to code translation. *Appl. Opt.* **6**, 1139.

MacDonald, N. C. (1968). Computer-controlled scanning electron microscopy. *Proc. 26th Ann. Meeting EMSA* (New Orleans), p. 362.

MacDonald, N. C., and Waldrop, J. R. (1971). Laboratory computer scanning electron microscope system. *Proc. 29th Ann. Meeting EMSA* (Boston), p. 86.

Markham, R., Frey, S., and Hills, G. J. (1963). Methods for the enhancement of image detail and accentuation of structure in electron microscopy. *Virology* **20**, 88.

Markham, R., Hitchborn, J. H., Hills, G. J., and Frey, S. (1964). The anatomy of tobacco mosaic virus. *Virology* **22**, 342.

Matson, W. L., McKinstry, H. A., Johnson, G. G., and White, E. W. (1970). Computer processing of SEM images by contour analyses. *Patt. Recog.* **2**, 303.

McMillan, R. E., Johnson, G. G., and White, E. W. (1969). Computer processing of binary maps of SEM images. In: *Scanning Electron Microscopy* (Johari, O., ed.), pp. 439–444. IITRI, Chicago.

Mellema, J. E. (1975). Model for the capsid structure of alfalfa mosaic virus. *J. Mol. Biol.* **94**, 643.

Mellema, J. E., and Amos, L. A. (1972). Three-dimensional image reconstruction of turnip yellow mosaic virus. *J. Mol. Biol.* **72**, 819.

Mellema, J. E., and Klug, A. (1972). Quaternary structure of gastropod haemocyanin. *Nature* **239**, 146.

Mellema, J. E., and Schepman, A. M. H. (1976). Digital image analysis in structural biology. *Proc. Sixth Eur. Cong. Electron Microscopy* (Jerusalem) **1**, 4.

Mendelsohn, M. L., Mayall, B. H., Prewitt, J. M. S., Bostrom, R. C., and Holcomb, W. G. (1968). Digital transformation and computer analysis of microscopic images. *Adv. Opt. Electron Micr.* **2**, 77.

Mihama, K., and Tanaka, M. (1976). Beryllium oxide specimen supporting films for high resolution electron microscopy and their application to the observation of fine gold particles. *J. Electron Microscopy* **25**, 65.

Misell, D. L. (1973a). A method for the solution of the phase problem in electron microscopy. *J. Phys. D.: Appl. Phys.* **6**, L6.

Misell, D. L. (1973b). An examination of an iterative method for the solution of the phase problem in optics and electron optics. I: Test calculations; II: Sources of error. *J. Phys. D.: Appl. Phys.* **6**, 2200, 2217.

Misell, D. L. (1976). On the validity of the weak-phase and other approximations in the analysis of electron microscope images. *J. Phys. D.: Appl. Phys.* **9**, 1849.

Müller, K. H. (1976). Phasenplatten für Elektronenmikroskope. *Optik* **45**, 73.

Nathan, R. (1971). Image processing for electron microscopy. I: Enhancement procedures. *Adv. Opt. Electron Micr.* **4**, 85. [Pt. II is Billingsley, 1971.]

Newbury, D. E. (1975). Techniques of signal processing in the scanning electron microscope. In: *Scanning Electron Microscopy* (Johari, O., and Corvin, I., eds.), pp. 727–736. IITRI, Chicago.

Newbury, D. E., and Joy, D. C. (1973). Characteristics of SEM signals and techniques for signal processing. In: *Scanning Electron Microscopy* (Johari, O., and Corvin, I., eds.), pp. 151–158. IITRI, Chicago.

Oatley, C. W. (1972). *The Scanning Electron Microscope.* University Press, Cambridge.

Oron, M., and Gilbert, D. (1976). Combined SEM-minicomputer system for digital image processing. In: *Scanning Electron Microscopy* (Johari, O., and Corvin, I., eds.), pp. 121–128. IITRI, Chicago.

Paden, R. S. Tillet, P. I., and Upton, J. M. (1973). New display techniques for the scanning electron microscope. In: *Scanning Electron Microscopy* (Johari, O., and Corvin, I., eds.), pp. 211–216. IITRI, Chicago.

Parsons, D. F. (1975). Radiation damage to biological materials. In: *Physical Aspects of Electron Microscopy and Microbeam Analysis* (Siegel, B. M., and Beaman, D. R., eds.), pp. 259–265. Wiley, New York and London.

Pressman, N. J. (1976). *Optical Texture Analysis for Automatic Cytology and Histology: a Markovian Approach.* Dissertation, University of California, available as UCRL-52155.

Pressman, N. J. (1977). Optical texture analysis for automatic cytology and histology. In: *Recent Developments in Biomedical Engineering* (Saha, S., ed.) Academic Press, New York and London.

Preston, K. (1976). Digital picture analysis in cytology. In: *Digital Picture Analysis* (Rosenfeld, A., ed.), pp. 209–294. Springer Verlag, Berlin and New York.

Prewitt, J. M. S. (1972). Parametric and non-parametric recognition by computer; an application to leukocyte image processing. *Adv. Comput.* **12**, 285.

Prosch, H. von (1974). Vergrössern photographischer Aufnahmen unter gleichzeitigem Kontrastausgleich. *Mikroskopie* **30**, 321.

Reimer, L. (1975). Review of the radiation damage problem of organic specimens in electron microscopy. In: *Physical Aspects of Electron Microscopy and Microbeam Analysis* (Siegel, B. M., and Beaman, D. R., eds.), pp. 231–245. Wiley, New York and London.

Reimer, L. and Pfefferkorn, G. (1977). *Raster-Elektronenmikroskopie* (2nd ed.). Springer Verlag, Berlin and New York.

Rose, H. (1974). Phase contrast in scanning transmission electron microscopy. *Optik* **39**, 416.

Rosenfeld, A., ed. (1976). *Digital Picture Analysis.* Springer Verlag, Berlin and New York.

Rosenfeld, A., and Kak, A. C. (1976). *Digital Picture Processing.* Academic Press, New York and London.

Rosenfeld, A., and Weszka, J. S. (1976). Picture recognition. In: *Digital Pattern Recognition* (Fu, K. S., ed.), pp. 135–166. Springer Verlag, Berlin and New York.

Sandler, S. S. (1975). *Picture Processing and Reconstruction.* Heath, Lexington (Mass.) and London.

Saxton, W. O., (1974). A new computer language for electron image processing. *Comp. Graph. Im. Proc.* **3**, 266.

Saxton, W. O., and Frank, J. (1977). Motif detection in quantum noise-limited electron micrographs by cross-correlation. *Ultramicroscopy* **2**, 219.

Scherzer, O. (1976). Reduction of radiation damage by image reconstruction. *Proc. Sixth Eur. Cong. Electron Microscopy* (Jerusalem) **1**, 91.

Schmiesser, H. (1974). Quantitative elektronische Bildanalyse–ein experimentelles Hilfsmittel zur Oberflächenanalyse. *Beitr. Elektronenmikroskop. Direktabb. Oberfl.* **7**, 205.

Siegel, B. M. (1976). The transmission electron microscope system: characteristics for high resolution and optimum signal to noise. *Proc. Sixth Eur. Cong. Electron Microscopy* (Jerusalem) **1**, 105.

Simon, R. (1969). Image processing for scanning electron microscopy. In: *Scanning Electron Microscopy* (Johari, O., ed.), pp. 445–457. IITRI, Chicago.

Sitte, P. (1976). Elektronenmikroskopie und Biologie–Schicksal einer Symbiose. *Mikroskopie* **32**, 145.

Smith, P. R., Aebi, U., Josephs, R., and Kessel, M. (1976). Studies of the structure of the T4 bacteriophage tail sheath. I: The recovery of three-dimensional structural information from the extended sheath. *J. Mol. Biol.* **106**, 243.

Steven, A. C., Aebi, U., and Showe, M. K., (1976a). Folding and capsomere morphology of the P23 surface shell of bacteriophage T4 polyheads from mutants in five different head genes. *J. Mol. Biol.* **102**, 373.

Steven, A. C., Couture, E., Aebi, U., and Showe, M. K. (1976b). Structure of T4 polyheads. II: A pathway of polyhead transformations for T4 capsid maturation. *J. Mol. Biol.* **106**, 187.

Stroke, G. W., Halioua, M., Thon, F., and Willasch, D. (1974). Image improvement in high-resolution electron microscopy using holographic image deconvolution. *Optik* **41**, 319.

Typke, D., Hoppe, W., Sessler, W., and Burger, M. (1976). Conception of a 3-D imaging electron microscope. *Proc. Sixth Eur. Cong. Electron Microscopy* (Jerusalem) **1**, 334.

Unwin, P. N. T. (1975). Beef liver catalase structure: interpretation of electron micrographs. *J. Mol. Biol.* **98**, 235.

Unwin, P. N. T., and Henderson, R. (1975). Molecular structure determination by electron microscopy of unstained crystalline specimens. *J. Mol. Biol.* **94**, 425.

Vainshtein, B. K. (1977). Electron microscopical analysis of the three-dimensional structure of assemblies of biological macromolecules. *Adv. Opt. Electron Micr.* **7**, in the press.

Wakabayashi, T., Huxley, H. E., Amos, L. A., and Klug, A. (1975). Three-dimensional image reconstruction of actin-tropomyosin complex and actin–tropomyosin-troponin T-troponin I complex. *J. Mol. Biol.* **93**, 477.

Weibel, E. R., and Bolender, R. P. (1973). Stereological techniques for electron microscopic morphometry. In: *Principles and Techniques of Electron Microscopy*, vol. 3 (Hayat, M. A., ed.), pp. 237–296. Van Nostrand Reinhold Company, New York and London.

Welton, T. A. (1971). Computational correction of aberrations in electron microscopy. *Proc. 29th Ann. Meeting EMSA* (Boston), p. 94.

White, E. W., Görz, H., and Johnson, G. G. (1971). CESEMI [Computer evaluation of SEM images] and its applications. *Beitr. Elektronenmikroskop. Direktabb. Oberfl.* **4/2**, 415.

White, E. W., Görz, H., Johnson, G. G. and Lebiedzik, J. (1972a). Characterization of aluminas by CESEMI [Computer evaluation of SEM images]. *Beitr. Elektronenmikroskop. Direktabb. Oberfl.* **5**, 609.

White, E. W., Görz, H., Johnson, G. G., and McMillan, R. E. (1970). Particle size distributions of particulate aluminas from computer-processed SEM images. In: *Scanning Electron Microscopy* (Johari, O., ed.), pp. 57–64. IITRI, Chicago.

White, E. W., Mayberry, K., and Johnson, G. G. (1972b). Computer analysis of multi-channel SEM and X-ray images from fine particles. *Patt. Recog.* **4**, 173.

White, E. W., McKinstry, H. A., and Johnson, G. G. (1968). Computer processing of SEM images. In: *Scanning Electron Microscopy* (Johari, O., ed.), pp. 95–103. IITRI, Chicago.

Willasch, D. (1975). High resolution electron microscopy with profiled phase plates. *Optik* **44**, 17.

Yew, N. C., and Pease, D. E. (1974). Signal storage and enhancement techniques for the SEM. In: *Scanning Electron Microscopy* (Johari, O., and Corvin, I., eds.), pp. 191–198. IITRI, Chicago.

Zeitler, E. (1968). Resolution in electron microscopy. *Adv. Electron. Electron Phys.* **25**, 277.

Zeitler, E. (1976). Scanning transmission electron microscopy. In: *Electron Microscopy in Materials Science* (Ruedl, E., and Valdrè, U., eds.), Part IV, pp. 1275–1302. Commission of the European Communities, Luxembourg.

Zeitler, E., and Hayes, J. R. (1965). Electronography. *Lab. Invest.* **14**, 1324. Also published in: *Quantitative Electron Microscopy* (Bahr, G. F., and Zeitler, E. H., eds.), pp. 586–595. Krieger, Huntington, N. Y.

AUTHOR INDEX

SUBJECT INDEX

Acetone, 26
Actin-tropomyosin, 286
Adrenocortical tissue, 254
Aluminum, 47, 190, 191
Aluminum-beryllium alloy, 3, 50-52, 82
Aluminum films, 62
Aluminum foil, 2, 192
Aluminum oxide, 3, 79, 82
Aluminum oxide films, 48, 49, 51, 52, 66
Aluminum replica, 160, 161
Amplitude contrast, 264
Amplitude image, 285
Amyl acetate, 12, 27
Angle of illumination, 205
Angular spread, 183
Apertures, 6, 13, 27
Aqueous suspension of replica fragments, 161
Arteries, 252
Artifacts, 47, 116, 119, 266, 280
Asbestos, 64
Astigmatism, 6, 168, 210, 211, 214, 216, 217, 281, 295
Axial illumination, 205

Bacitracin, 68
Bacteriophage, 166, 280, 286
Barrel distortion, 166-168
Benzene, 26, 27
Beryllium, 3, 47
Beryllium oxide, 267
Beryllium oxide films, 48, 50, 51
Bioden, 18
Bone, 253
Boron, 47
Boron film, 52
Bovine liver catalase crystals, 165
Bragg diffraction, 229
Brain, 117, 258, 259, 261

Camera lucida, 257
Capillaries, 252, 260
Carbon, 79, 182

Carbon evaporation, 38, 124
Carbon films, 3, 4, 6, 24, 34, 37, 40, 51, 52, 57, 58, 62, 77, 88, 133, 138, 190, 191, 193, 212, 214, 216, 217
Carbon film release, 44
Carbon microgrids, 10
Cartilage, 253
Cast films, 28, 33
Casting method, 28
Catalase, 163-165, 267, 288
Catalase unit cell, 165
Cellulose, 45
Central nervous system, 246, 257, 259
Charging, 4, 217
Chatter, 279
Chemical exfoliation, 60, 61
Chromaffin tissue, 254
Chromatic aberration, 84, 181-184, 186, 188-190, 193, 195-197, 233, 234, 236, 237, 242, 283
Chromatic coherence, 211
Cilia, 249
Ciliated epithelium, 257
Cleaning of apertures, 6
  molybdenum, 8
Cold cathode counter tubes, 171
Collagen, 253
Collodion, 21, 33
Collodion films, 10, 18
Composite films, 34
Conical illumination, 200, 224
Contamination, 4, 8, 10, 43, 44, 50, 52, 54, 59, 62, 66, 70, 79, 84, 93, 101, 142, 248, 283
Contrast, 73, 84, 181-197, 219, 223, 224, 231, 238, 264
Copper phthalocyanin, 164
Curvature of field, 169

Dark field, 71, 87, 88, 98, 183-224, 241, 242, 249
Defocus, 42, 81, 91-101, 187-210, 264, 265, 268, 281-284, 295
Detergent, 15, 66, 68